Bio-Inspired Computation in Telecommunications

Bio-Inspired Computation in Telecommunications

Edited by

Xin-She Yang

Su Fong Chien

Tiew On Ting

AMSTERDAM • BOSTON • HEIDELBERG • LONDON
NEW YORK • OXFORD • PARIS • SAN DIEGO
SAN FRANCISCO • SINGAPORE • SYDNEY • TOKYO

Morgan Kaufmann is an imprint of Elsevier

Executive Editor: Steven Elliot
Editorial Project Manager: Amy Invernizzi
Project Manager: Punithavathy Govindaradjane
Designer: Mark Rogers

Morgan Kaufmann is an imprint of Elsevier
225 Wyman Street, Waltham, MA 02451, USA

ISBN: 978-0-12-801538-4

British Library Cataloguing-in-Publication Data
A catalogue record for this book is available from the British Library

Library of Congress Cataloging-in-Publication Data
A catalogue record for this book is available from the Library of Congress

For information on all MK publications
visit our website at www.mkp.com

Working together
to grow libraries in
developing countries

www.elsevier.com • www.bookaid.org

Contents

Preface

Humankind has always been fascinated by new ways of communication. Modern telecommunications have a huge impact on the ways we live, work, and think. In fact, modern telecommunications involve a vast array of algorithms, approaches, and technologies. Among many objectives and requirements concerning telecommunications, a key objective is to transmit signals with the optimal quality, least energy consumption, maximum capacity, and highest speed. Such requirements pose a series of challenging optimization problems in telecommunications.

Many problems in telecommunications require sophisticated algorithms and approaches to solve them. In principle, such problems can be tackled by sophisticated optimization techniques; however, traditional approaches do not usually work well. In addition, the stringent time requirements and ever-increasing complexity in constraints and dimensionality have further complicated these challenging issues, which become even more relevant in the current networks and the future 5G networks. Such challenges necessitate new approaches and new methods such as bio-inspired computation.

Current trends in optimization tend to use bio-inspired algorithms and swarm intelligence. In fact, there have been significant developments in bio-inspired computation in recent years, and swarm-intelligence-based algorithms now form an increasingly important part of optimization methods and approaches. Such bio-inspired algorithms are flexible, versatile, and efficient. Consequently, these algorithms can be used to solve a wide range of problems in diverse applications.

The rapid advances in bio-inspired computation have resulted in a much richer literature in recent years. It is impossible to review and summarize a good fraction of the latest developments in both bio-inspired computation and telecommunications. Consequently, we must select from many topics, with the emphasis on state-of-the-art developments, so as to provide a timely snapshot of recent advances. Therefore, this book intends to provide a timely review on a selection of topics, including the analysis and overview of bio-inspired algorithms, bio-inspired approaches to telecommunications, firefly algorithm, intrusion detection systems, VoIP (Voice over Internet Protocol) quality prediction, IP-over-WDM (wavelength-division multiplexing) networks, radio resources management of 4G networks, robust transmission for heterogeneous networks with cognitive small cells, resource distribution for sustainable communication networks, multiobjective optimization in optical networks, cell coverage area optimization for green LTE (long-term evolution) cellular networks, optimal coverage optimization in wireless sensor networks, minimum interference channel assignment for multiradio wireless mesh networks, and others.

The diverse topics covered in this book can provide an ideal and timely source of literature and case studies for graduates, lecturers, engineers, and researchers in telecommunications, wireless communications, computer science, electrical and electronic engineering, computational intelligence, and neural computing. It is our hope that this book can inspire more research to improve existing methodologies and create innovative technologies in telecommunications via bio-inspired methodologies.

We would like to thank our editors, Steven Elliot and Kaitlin Herbert, and the staff at Elsevier for their help and professionalism. Last but not least, we thank our families for their support and encouragement.

Xin-She Yang, Su Fong Chien, Tiew On Ting
October 2014

List of Contributors

Taufik Abrão
Department of Computer Science, State University of Londrina (UEL), Londrina, Brazil

Mario H.A.C. Adaniya
Department of Computer Science, Centro Universitário Filadelfia (Unifil), Londrina, Brazil; Department of Computer Science, State University of Londrina (UEL), Londrina, Brazil

M. Jawad Alam
School of Electrical and Electronic Engineering, Nanyang Technological University, Singapore

Mohammed H. Alsharif
Department of Electrical, Electronics and Systems Engineering, Faculty of Engineering and Built Environment, Universiti Kebangsaan Malaysia, Bangi, Selangor, Malaysia

Benjamín Barán
Polytechnic, National University of Asuncion - UNA, Asuncion, Paraguay

Luiz F. Carvalho
Department of Computer Science, State University of Londrina (UEL), Londrina, Brazil

Gerardo Castañón
Department of Electrical and Computer Engineering, Tecnológico de Monterrey, Monterrey, Mexico

Cheong Loong Chan
Faculty of Engineering and Green Technology, Universiti Tunku Abdul Rahman, Kampar, Perak, Malaysia

Jonathan H. Chan
Data and Knowledge Engineering Laboratory (D-Lab), School of Information Technology, King Mongkut's University of Technology Thonburi, Bangkok, Thailand

Su Fong Chien
Strategic Advanced Research (StAR) Mathematical Modeling Lab, MIMOS Berhad, Kuala Lumpur, Malaysia

Carrson C. Fung
Department of Electronics Engineering, National Chiao Tung University, Hsinchu, Taiwan

Mahamod Ismail
Department of Electrical, Electronics and Systems Engineering, Faculty of Engineering and Built Environment, Universiti Kebangsaan Malaysia, Bangi, Selangor, Malaysia

Abbas Jamalipour
School of Electrical and Information Engineering, The University of Sydney, Sydney, NSW, Australia

Paul Jean E. Jeszensky
Polytechnic School, University of São Paulo, São Paulo, Brazil

Prasert Kanthamanon
IP Communications Laboratory (I-Lab), School of Information Technology, King Mongkut's University of Technology Thonburi, Bangkok, Thailand

Sheng Chyan Lee
Faculty of Engineering and Green Technology, Universiti Tunku Abdul Rahman, Kampar, Perak, Malaysia

Fernando Lezama
Department of Electrical and Computer Engineering, Tecnológico de Monterrey, Monterrey, Mexico

Maode Ma
School of Electrical and Electronic Engineering, Nanyang Technological University, Singapore

Kumudu S. Munasinghe
Faculty of Education, Science, Technology, and Mathematics, The University of Canberra, Canberra, Australia

Rosdiadee Nordin
Department of Electrical, Electronics and Systems Engineering, Faculty of Engineering and Built Environment, Universiti Kebangsaan Malaysia, Bangi, Selangor, Malaysia

Diego P. Pinto-Roa
Polytechnic, National University of Asuncion - UNA, Asuncion, Paraguay

Mario Lemes Proença
Department of Computer Science, State University of Londrina (UEL), Londrina, Brazil

Lucas D.H. Sampaio
Polytechnic School, University of São Paulo, São Paulo, Brazil

Ana Maria Sarmiento
Department of Industrial Engineering, Tecnológico de Monterrey, Monterrey, Mexico

Sevil Sen
Department of Computer Engineering, Hacettepe University, Ankara, Turkey

Su Wei Tan
Faculty of Engineering, Multimedia University, Cyberjaya, Selangor, Malaysia

Tiew On Ting
Department of Electrical and Engineering, Xi'an Jiaotong-Liverpool University, Suzhou, Jiangsu Province, China

Tuul Triyason
IP Communications Laboratory (I-Lab), School of Information Technology, King Mongkut's University of Technology Thonburi, Bangkok, Thailand

Sake Valaisathien
IP Communications Laboratory (I-Lab), School of Information Technology, King Mongkut's University of Technology Thonburi, Bangkok, Thailand

Vajirasak Vanijja
IP Communications Laboratory (I-Lab), School of Information Technology, King Mongkut's University of Technology Thonburi, Bangkok, Thailand

Xin-She Yang
School of Science and Technology, Middlesex University, London, UK

C.C. Zarakovitis
School of Computing and Communications, Lancaster University, Lancaster, UK

Bruno B. Zarpelão
Department of Computer Science, State University of Londrina (UEL), Londrina, Brazil

Zhi-Hui Zhan
Department of Computer Science, Sun Yat-sen University, Guangzhou, China; Key Laboratory of Machine Intelligence and Advanced Computing (Sun Yat-sen University), Ministry of Education, Guangzhou, China; Engineering Research Center of Supercomputing Engineering Software (Sun Yat-sen University), Ministry of Education, Guangzhou, China; Key Laboratory of Software Technology, Education Department of Guangdong Province, Guangzhou, China

Jun Zhang
Department of Computer Science, Sun Yat-sen University, Guangzhou, China; Key Laboratory of Machine Intelligence and Advanced Computing (Sun Yat-sen University), Ministry of Education, Guangzhou, China; Engineering Research Center of Supercomputing Engineering Software (Sun Yat-sen University), Ministry of Education, Guangzhou, China; Key Laboratory of Software Technology, Education Department of Guangdong Province, Guangzhou, China

Bio-Inspired Computation and Optimization: An Overview

1

Xin-She Yang[1], Su Fong Chien[2], and Tiew On Ting[3]

[1]*School of Science and Technology, Middlesex University, London, UK*
[2]*Strategic Advanced Research (StAR) Mathematical Modeling Lab, MIMOS Berhad, Kuala Lumpur, Malaysia*
[3]*Department of Electrical and Engineering, Xi' an Jiaotong-Liverpool University, Suzhou, Jiangsu Province, China*

CHAPTER CONTENTS

1.1 INTRODUCTION

One of the main aims of telecommunications is to transmit signals with the minimum noise, least energy consumption, maximum capacity, and optimal transmission quality. All these form a series of very challenging problems. In fact, optimization is almost everywhere, and in almost every application, especially engineering designs, we are always trying to optimize something—whether to minimize the cost and energy consumption, or to maximize the profit, output, performance, and efficiency. In reality, resources, time, and money are always limited; consequently, optimization is far more important in practice (Koziel and Yang, 2011; Yang, 2010b). Thus, the proper use of available resources of any sort requires a paradigm shift in scientific thinking and design innovation.

Obviously, real-world applications are usually subject to complex constraints, and many factors and parameters may affect how the system behaves. Design processes can be slow and expensive, and thus, any saving in terms of time and resources will make designs greener and more sustainable. In the context of telecommunications, there exist a myriad of optimization techniques applicable in this sector. Conventional methods include the Newton-Raphson method, linear programming, sequential quadratic programming, interior-point methods, Lagrangian duality, fractional programming, and many others. New methods tend to be evolutionary or bio-inspired. Good examples include evolutionary algorithms, artificial neural networks (ANNs), swarm intelligence (SI), cellular signaling pathways, and others. For example, genetic algorithms (GA) and SI have been used in many applications. We will include more examples in the next chapters.

This chapter is organized as follows: Section 1.2 provides a brief discussion of the formulation concerning the optimization problems in telecommunications. Section 1.3 discusses the key issues in optimization, followed by a detailed introduction of commonly used bio-inspired algorithms in Section 1.4. Section 1.5 discusses neural networks and Section 1.6 presents support vector machines (SVMs). Finally, Section 1.7 concludes with some discussion.

1.2 TELECOMMUNICATIONS AND OPTIMIZATION

In telecommunications, the objective in the optimization tasks can vary, depending on the requirements and communications architecture. For example, the quality of service in an orthogonal frequency division multiple access (OFDMA) system can concern the minimization of the energy consumption of the system or the maximum of the energy efficiency (Ting et al., 2014). For example, the quality of service in an

OFDMA system can still be preserved in order to minimize the energy consumption of the system or to maximize the energy efficiency (Ting et al., 2014),

$$E = \frac{\left(\sum_{n=1}^{N}\sum_{m=1}^{M}c_{m,n}x_{m,n}\right)}{\left(P_0 + \epsilon\sum_{n=1}^{N}\sum_{m=1}^{M}p_{m,n}x_{m,n}\right)}, \tag{1.1}$$

where N is the number of subcarriers and M is the number of users. P_0 is the power offset associated with the battery backup and signal processing. $p_{m,n}$ is the transmission power allocated for user m on subcarrier n with a subcarrier bandwidth B. In addition, $x_{m,n} \in \{0,1\}$, which means the optimization problem is a fractional, mixed-integer programming problem.

$c_{m,n}$ is the achievable transmission capacity, which can be estimated as

$$c_{m,n} = B\log_2(1 + p_{m,n}\text{CNR}_{m,n}),$$

where $\text{CNR}_{m,n}$ is the channel-to-noise ratio for each wireless channel. In addition, this is subject to the constraints

$$\sum_{n}^{N}c_{m,n}x_{m,n} \geq \overline{C}_m, \quad \sum_{n}^{N}\sum_{m}^{M}p_{m,n}x_{m,n} \leq P_T, \quad \sum_{m}^{M}x_{m,n} = 1, \tag{1.2}$$

where $\overline{C}_m$ is the minimum rate of the transmission speed that servers user m, and P_T is the total available transmission power.

In a nutshell, all optimization problems in telecommunications can be written in a more general form. For example, the most widely used formulation is to write a nonlinear optimization problem as

$$\text{minimize } f(x), \tag{1.3}$$

subject to the constraints

$$h_j(x) = 0, \quad (j = 1, 2, \dots, J),$$

$$g_k(x) \leq 0, \quad (k = 1, 2, \dots, K), \tag{1.4}$$

where all the functions are in general nonlinear. Here the design vector $x = (x_1, x_2, \dots, x_d)$ can be continuous, discrete, or mixed in a d-dimensional space (Yang, 2010b). It is worth pointing out that here we write the problem as a minimization problem, but it can also be written as a maximization problem by simply replacing $f(x)$ by $-f(x)$.

When all functions are nonlinear, we are dealing with nonlinear constrained problems. In some special cases when all functions are linear, the problem becomes linear, and we can use a widely known linear programming technique such as the simplex method. When some design variables can only take discrete values (often integers), while other variables are real continuous, the problem is of mixed type, which is often difficult to solve, especially for large-scale optimization problems.

A very special class of optimization is convex optimization, which has the guaranteed global optimality. Any optimal solution is also the global optimum, and most important, there are efficient algorithms of polynomial time to solve such problems

(Conn et al., 2009). Efficient algorithms such as the interior-point methods (Karmarkar, 1984) are widely used and have been implemented in many software packages.

1.3 KEY CHALLENGES IN OPTIMIZATION

Optimization, especially nonlinear optimization with complex constraints, can be very challenging to solve. There are many key issues that need to be addressed, and we highlight only four main issues here: heuristicity, the efficiency of an algorithm, choice of algorithms, and time constraints.

1.3.1 INFINITE MONKEY THEOREM AND HEURISTICITY

Heuristic algorithms are essentially methods by trial and error. Such heuristicity can be understood by first analyzing a well-known thought experiment called the infinite monkey theorem, which states that the probability of producing any given text will almost surely be one if an infinite number of monkeys randomly type for an infinitely long time (Gut, 2005; Marsaglia and Zaman, 1993). In other words, the infinite monkeys can be expected to reproduce the whole works of Shakespeare. For example, to reproduce the text "telecommunications" (18 characters) from a random typing sequence of n characters on a 101-key computer keyboard, the probability of a consecutive 18-character random string to be "swarm intelligence" is $p_s = (1/101)^{18} \approx 8.4 \times 10^{-37}$, which is extremely small. However, the importance here is this probability is not zero. Therefore, for an infinitely long sequence $n \to \infty$, the probability of reproducing the collected works of Shakespeare is one, though the formal rigorous mathematical analysis requires rigorous probability theory.

In many ways, the heuristic and metaheuristic algorithms have some similarities to the infinite monkey-typing approaches. Monkeys type randomly, and ultimately meaningful high-quality text may appear. Similarly, most stochastic algorithms use randomization to increase the search capability. If such algorithms are run for a sufficiently long time with multiple runs, it can be expected that the global optimality of a given problem can be reached or found. In theory, it may take infinitely long to guarantee such optimality, but in practice, it can take many thousands or even millions of iterations. If we consider the optimality as an important work of Shakespeare, the infinite monkeys should be able to reproduce or achieve it in an infinite amount of time.

However, there are some key differences between heuristic algorithms and the infinite monkey approach. First, monkeys randomly type without any memory or learning processing, and each key input is independent of another. Heuristic algorithms try to learn from history and the past moves so as to generate new, better moves or solutions. Second, random monkeys do not select what has been typed, while algorithms try to select the best solutions or the fittest solutions (Holland, 1975). Third, monkeys use purely stochastic components, while all heuristic algorithms use both deterministic and stochastic components. Finally, monkey typing at most is equivalent to a random search on a flat landscape, while heuristic algorithms are often cleverly constructed to use the landscape information in combination with history (memory) and selection. All these differences ensure that heuristic algorithms are far better than the random monkey-typing approach.

In addition, metaheuristics are usually considered as a higher level of heuristics, because metaheuristic algorithms are not simple trial-and-error approaches, and metaheuristics are designed to learn from past solutions, to be biased toward better moves, to select the best solutions, and to construct sophisticated search moves. Therefore, metaheuristics can be much better than heuristic algorithms and can definitely be far more efficient than random monkey-typing approaches.

1.3.2 EFFICIENCY OF AN ALGORITHM

The efficiency of an algorithm can depend on many factors, such as the intrinsic structure of the algorithm, the way it generates new solutions, and the setting of its algorithm-dependent parameters. The essence of an optimizer is a search or optimization algorithm implemented correctly so as to carry out the desired search (though not necessarily efficiently). It can be integrated and linked with other modeling components. There are many optimization algorithms in the literature and no single algorithm is suitable for all problems, as dictated by the No Free Lunch Theorems (Wolpert and Macready, 1997). In order to solve an optimization problem efficiently, an efficient optimization algorithm is needed.

Optimization algorithms can be classified in many ways, depending on the focus or the characteristics we are trying to compare. For example, from the mobility point of view, algorithms can be classified as local or global. Local search algorithms typically converge toward a local optimum, not necessarily (often not) the global optimum, and such algorithms are often deterministic and have no ability of escaping local optima. Simple hill-climbing is an example. On the other hand, we always try to find the global optimum for a given problem, and if this global optimality is robust, it is often the best, though it is not always possible to find such global optimality. For global optimization, local search algorithms are not suitable. We have to use a global search algorithm. Modern bio-inspired metaheuristic algorithms in most cases are intended for solving global optimization, though not always successfully or efficiently. A simple strategy such as hill-climbing with random restart may change a local search algorithm into a global search. In essence, randomization is an efficient component for global search algorithms. In this chapter, we will provide a brief review of most metaheuristic optimization algorithms.

From the optimization point of view, a crucial question is how to make the right choice of algorithms so as to find the optimal solutions quickly.

1.3.3 HOW TO CHOOSE ALGORITHMS

Obviously, the choice of the right optimizer or algorithm for a given problem is important. However, the choice of an algorithm for an optimization task will largely depend on the type of the problem, the nature of the algorithm, the desired quality of solutions, the available computing resources, time limit, availability of the algorithm implementation, and the expertise of the decision makers (Blum and Roli, 2003; Yang, 2010b, 2014).

The intrinsic nature of an algorithm may essentially determine if it is suitable for a particular type of problem. For example, gradient-based algorithms such as hill-climbing are not suitable for an optimization problem whose objective is discontinuous. On the other hand, the type of problem to be solved also determines the algorithms needed to obtain good quality solutions. If the objective function of an optimization problem at hand is highly nonlinear and multimodal, classic algorithms such as hill-climbing are usually not suitable because the results obtained by these local search methods tend to be dependent on the initial starting points. For nonlinear global optimization problems, SI-based algorithms such as firefly algorithm (FA) and particle swarm optimization (PSO) can be very effective (Yang, 2010a, 2014).

In addition, the desired solution quality and available computing resources may also affect the choice of the algorithms. In most applications, computing resources are limited and good solutions (not necessary the best) have to be obtained in a reasonable and practical time, which means that there is a certain compromise between the available resources and the required solution quality. For example, if the main aim is to find a feasible solution (not necessarily the best solution) in the least amount of time, greedy methods or gradient-based methods should be the first choice.

Moreover, the availability of the software packages and expertise of the designers are also key factors that can affect the algorithm choice. For example, Newton's method, hill-climbing, Nelder-Mead downhill simplex, trust-region methods (Conn et al., 2009), and interior-point methods are implemented in many software packages, which partly increases their popularity in applications. In practice, even with the best possible algorithms and well-crafted implementation, we may still not get the desired solutions. This is the nature of nonlinear global optimization, as most of these problems are NP-hard (nondeterministic polynomial-time hard), and no efficient solutions (in the polynomial sense) exist for a given problem.

Therefore, from a practical point of view, one of the main challenges in many applications is to find the right algorithm(s) most suitable for a given problem so as to obtain good solutions, hopefully also the global best solutions, in a reasonable timescale with a limited amount of resources. However, there is no simple answer to this problem. Consequently, the choice of the algorithms still largely depends on the expertise of the researchers involved and the available resources in a more or less heuristic way.

1.3.4 TIME CONSTRAINTS

Apart from the challenges mentioned above, one of the most pressing requirements in telecommunications is the speed for finding the solutions. As users may come and go in a dynamic manner, real-time dynamic allocation is needed in practice. Therefore, the time should be sufficiently short so that the solution methods can be useful in practice, and this time factor poses a key constraint to most algorithms. In a way, the choice of the algorithm all depends on the requirements. If the requirement is to solve the optimization problem in real time or *in situ*, then the number of steps to find the solution should be minimized. In this case, in addition to the critical design and choice of algorithms, the details of implementation and feasibility of realization in terms of both software and hardware can be also very important.

In the vast majority of optimization applications, the evaluations of the objective functions often form the most computationally expensive part. In simulations using the finite element methods and finite volume methods in engineering applications, the numerical solver typically takes a few hours up to a few weeks, depending on the problem size of interest. In this case, the use of the numerical evaluator or solver becomes crucial (Yang, 2008). On the other hand, in the combinatorial problems, the evaluation of a single design may not take long, but the number of possible combinations can be astronomical. In this case, the way to move from one solution to another solution becomes more critical, and thus optimizers can be more important than numerical solvers.

Therefore, any approach to save computational time either by reducing the number of evaluations or by increasing the simulator's efficiency will save time and money (Koziel and Yang, 2011).

1.4 BIO-INSPIRED OPTIMIZATION ALGORITHMS

Metaheuristic algorithms are often nature-inspired, and they are now among the most widely used algorithms for optimization. They have many advantages over conventional algorithms, as we can see from many case studies presented in later chapters in this book. There are a few recent books that are solely dedicated to metaheuristic algorithms (Dorigo and Stütle, 2004; Talbi, 2009; Yang, 2008, 2010a,b). Metaheuristic algorithms are very diverse, including GA, simulated annealing, differential evolution (DE), ant and bee algorithms, PSO, harmony search (HS), FA, cuckoo search (CS), and others. In general, they can be put into two categories: SI-based and non-SI-based. In the rest of this section, we will introduce some of these algorithms briefly.

1.4.1 SI-BASED ALGORITHMS

SI-based algorithms typically use multiple agents, and their characteristics have often been drawn from the inspiration of social insects, birds, fish, and other swarming behavior in the biological systems. As a multiagent system, they can possess certain characteristics of the collective intelligence.

1.4.1.1 Ant and bee algorithms

There is a class of algorithms based on the social behavior of ants. For example, ant colony optimization (Dorigo and Stütle, 2004) mimics the foraging behavior of social ants via the pheromone deposition, evaporation, and route marking. In fact, the pheromone concentrations along the paths in a transport problem can also be considered as the indicator of quality solutions. In addition, the movement of an ant is controlled by a pheromone that will evaporate over time. Without such time-dependent evaporation, ant algorithms will lead to premature convergence to the (often wrong) solutions. With proper pheromone evaporation, they usually behave very well. For example, exponential decay for pheromone exploration is often used, and the deposition is incremental for transverse routes such that

$$p^{t+1} = \delta + (1-\rho)p^t, \tag{1.5}$$

where ρ is the evaporation rate, while δ is the incremental deposition. Obviously, if the route is not visited during an iteration, then $\delta = 0$ should be used for nonvisited routes. In addition to the pheromone variation, the probability of choosing a route needs to be defined properly. Different variants and improvements of ant algorithms may largely differ in terms of ways of handling pheromone deposition, evaporation, and route-dependent probabilities.

On the other hand, bee algorithms are based on the foraging behavior of honey bees. Interesting characteristics such as the waggle dance and nectar maximization are often used to simulate the allocation of the foraging bees along flower patches and thus different search regions in the search space (Karaboga, 2005; Nakrani and Tovey, 2004). For a more comprehensive review, please refer to Yang (2010a) and Parpinelli and Lopes (2011).

1.4.1.2 Bat algorithm

The bat algorithm (BA) was developed by Xin-She Yang in 2010 (Yang, 2011), based on the echolocation behavior of microbats. Such echolocation is essentially frequency tuning. These bats emit a very loud sound pulse and listen for the echo that bounces back from the surrounding objects. In the BA, a bat flies randomly with a velocity v_i at position x_i with a fixed frequency range $[f_{min}, f_{max}]$, varying its emission rate $r \in (0, 1]$ and loudness A_0 to search for prey, depending on the proximity of its target. The main updating equations are

$$f_i = f_{min} + (f_{max} - f_{min})\varepsilon, \; v_i^{t+1} = v_i^t + (x_i^t - x^*)f_i, \; x_i^{t+1} = x_i^t + v_i^t, \tag{1.6}$$

where ε is a random number drawn from a uniform distribution, and x^* is the current best solution found so far during iterations. The loudness and pulse rate can vary with iteration t in the following way:

$$A_i^{t+1} = \alpha A_i^t, \; r_i^t = r_i^0[1 - \exp(-\beta t)]. \tag{1.7}$$

Here, α and γ are constants. In fact, α is similar to the cooling factor of a cooling schedule in the simulated annealing, to be discussed later. In the simplest case, we can use $\alpha = \beta$, and we have in fact used $\alpha = \beta = 0.9$ in most simulations. The BA has been extended to multiobjective bat algorithm by Yang (2011), and preliminary results suggest that it is very efficient (Yang and Gandomi, 2012). Yang and He provide a relatively comprehensive review of the BA and its variants (Yang and He, 2013).

1.4.1.3 Particle swarm optimization

PSO was developed by Kennedy and Eberhart in 1995 (Kennedy and Eberhart, 1995), based on the swarm behavior such as fish and bird schooling in nature. In PSO, each particle has a velocity v_i and position x_i, and their updates can be determined by the following formula:

$$v_i^{t+1} = v_i^t + \alpha \varepsilon_1 \left[g^* - x_i^t \right] + \beta \varepsilon_2 \left[x_i^* - x_i^t \right], \tag{1.8}$$

where g^* is the current best solution and x_i^* is the individual best solution for particle i. Here, ε_1 and ε_2 are two random variables drawn from the uniform distribution in $[0,1]$. In addition, α and β are the learning parameters. The position is updated as

$$x_i^{t+1} = x_i^t + v_i^{t+1}. \tag{1.9}$$

There are many variants that extend the standard PSO algorithm, and the most noticeable improvement is probably to use an inertia function. There is relatively vast literature about PSO (Kennedy et al., 2001; Yang, 2014).

1.4.1.4 Firefly algorithm

The FA was first developed by Xin-She Yang in 2007 (Yang, 2008, 2009), based on the flashing patterns and behavior of fireflies. In the FA, the movement of a firefly i that is attracted to another more attractive (brighter) firefly j is determined by the following nonlinear updating equation:

$$x_i^{t+1} = x_i^t + \beta_0 e^{-\gamma r_{ij}^2} \left(x_j^t - x_i^t \right) + \alpha \varepsilon_i^t, \tag{1.10}$$

where β_0 is the attractiveness at distance $r = 0$. The attractiveness of a firefly varies with distance from other fireflies. The variation of attractiveness β with the distance r can be defined by

$$\beta = \beta_0 e^{-\gamma r^2}. \tag{1.11}$$

A demo version of FA implementation, without Lévy flights, can be found at the Mathworks file exchange website.[1] The FA has attracted much attention and there exist some comprehensive reviews (Fister et al., 2013; Gandomi et al., 2011; Yang, 2014).

1.4.1.5 Cuckoo search

CS was developed in 2009 by Yang and Deb (2009). CS is based on the brood parasitism of some cuckoo species. In addition, this algorithm is enhanced by Lévy flights (Pavlyukevich, 2007), rather than by simple isotropic random walks. Recent studies show that CS is potentially far more efficient than PSO and GA (Yang and Deb, 2010).

---

[1]http://www.mathworks.com/matlabcentral/fileexchange/29693-firefly-algorithm

This algorithm uses a balanced combination of the local random walk and the global explorative random walk, controlled by a switching parameter p_a. The local random walk can be written as

$$x_i^{t+1} = x_i^t + \alpha s H(p_a - \varepsilon)\left(x_j^t - x_k^t\right),\qquad(1.12)$$

where x_j^t and x_k^t are two different solutions selected randomly by random permutation.

Here, $H(u)$ is the Heaviside function, and ϵ is a random number drawn from a uniform distribution, while the step size s is drawn from a Lévy distribution. On the other hand, the global random walk is carried out by using Lévy flights

$$x_i^{t+1} = x_i^t + \alpha L(s, \lambda),\qquad(1.13)$$

where

$$L(s, \lambda) = \frac{\lambda \Gamma(\lambda)\sin(\pi\lambda/2)}{\pi}\frac{1}{s^{1+\lambda}},\quad (s \gg s_0 > 0).\qquad(1.14)$$

Here, $\alpha > 0$ is a scaling factor controlling the scale of the step sizes, and s_0 is a small fixed step size.

For an algorithm to be efficient, a substantial fraction of the new solutions should be generated by far field randomization, with locations far enough from the current best solution. This will ensure that the system will not be trapped in a local optimum (Yang and Deb, 2010). A Matlab implementation is given by the author, and can be downloaded.[2] CS is very efficient in solving engineering optimization problems (Yang, 2014).

1.4.2 NON-SI-BASED ALGORITHMS

Not all bio-inspired algorithms are based on SI. In fact, their sources of inspiration can be very diverse from physical processes to biological processes. In this case, it is better to extend the bio-inspired computation to nature-inspired computation.

1.4.2.1 Simulated annealing

Simulated annealing, developed by Kirkpatrick et al. in 1983, was among the first metaheuristic algorithms (Kirkpatrick et al., 1983). It was essentially an extension of the traditional Metropolis-Hastings algorithm but applied in a different context. In deterministic algorithms such as hill-climbing, the new moves are generated by using gradients, and thus are always accepted. In contrast, simulated annealing used a stochastic approach for generating new moves and deciding the acceptance.

---

[2]www.mathworks.com/matlabcentral/fileexchange/29809-cuckoo-search-cs-algorithm

Loosely speaking, the probability of accepting a new move is determined by the Boltzmann-type probability

$$p = \exp\left[-\frac{\Delta E}{k_B T}\right],$$ (1.15)

where k_B is the Boltzmann's constant, and T is the temperature for controlling the annealing process. The energy change ΔE should be related to the objective function $f(x)$. In this case, the above probability becomes

$$p(\Delta f, T) = e^{-\Delta f / T}.$$ (1.16)

Whether or not a change is accepted, a random number (r) is often used as a threshold. Thus, if $p > r$, the move is accepted. It is worth pointing out the special case when $T \to 0$ corresponds to the classical hill-climbing because only better solutions are accepted, and the system is essentially climbing up or descending along a hill.

Proper selection of the initial temperature T_0 is very important so as to make the algorithm well behaved. Another important issue is how to control the annealing or cooling process so that the system cools down gradually from a higher temperature to ultimately freeze to a global minimum state. There are many ways of controlling the cooling rate or the decrease of temperature. Geometric cooling schedules are often widely used, which essentially decrease the temperature by a cooling factor $0 < \alpha < 1$, so that T is replaced by αT or

$$T(t) = T_0 \alpha^t.$$ (1.17)

The advantage of this method is that $t \to \infty$, and thus there is no need to specify the maximum number of iterations if a tolerance or accuracy is prescribed.

1.4.2.2 Genetic algorithms

GA are a class of algorithms based on the abstraction of Darwin's evolution of biological systems, pioneered by J. Holland and his collaborators in the 1960s and 1970s (Holland, 1975). GA use the genetic operators: crossover, mutation, and selection. Each solution is encoded as a string (often binary or decimal), called a chromosome. The crossover of two parent strings produces offspring (new solutions) by swapping parts or genes of the chromosomes. Crossover has a higher probability, typically 0.8-0.95. On the other hand, mutation is carried out by flipping some digits of a string, which generates new solutions. This mutation probability is typically low, from 0.001 to 0.05. New solutions generated in each generation will be evaluated by their fitness that is linked to the objective function of the optimization problem. The new solutions are selected according to their fitness—selection of the fittest. Sometimes, in order to make sure that the best solutions remain in the population, the best solutions are passed on to the next generation without much change. This is called elitism.

GA have been applied to almost all areas of optimization, design, and applications. There are hundreds of good books and thousands of research articles. There are many variants and hybridization with other algorithms, and interested readers can refer to more advanced literature, such as Goldberg (1989).

1.4.2.3 Differential evolution

DE was developed by R. Storn and K. Price (Storn, 1996; Storn and Price, 1997). It is a vector-based evolutionary algorithm with some similarity to the conventional pattern search idea. In DE, each solution x_i at any generation t is represented as

$$x_i^t = \left(x_{1,i}^t, x_{2,i}^t, \ldots, x_{d,i}^t \right), \tag{1.18}$$

which can be considered as the chromosomes or genomes.

DE consists of three main steps: mutation, crossover, and selection. Mutation is carried out by the mutation scheme. The mutation scheme can be updated as

$$v_i^{t+1} = x_p^t + F\left(x_q^t - x_r^t \right), \tag{1.19}$$

where x_p^t, x_q^t, x_r^t are three distinct solution vectors, and $F \in (0,2]$ is a parameter, often referred to as the differential weight. This requires that the minimum number of population size is $n \geq 4$. The crossover is controlled by a crossover probability C_r, and actual crossover can be carried out in two ways: binomial and exponential. Selection is essentially the same as that used in GA—it is to select the fittest, and for the minimization problem, the minimum objective value. Therefore, we have

$$x_i^{t+1} = \begin{cases} u_i^{t+1} & \text{if } f\left(u_i^{t+1} \right) \leq f\left(x_i^t \right), \\ x_i^t & \text{otherwise.} \end{cases} \tag{1.20}$$

There are many variants of DE, and some review studies can be found in the literature (Price et al., 2005).

1.4.2.4 Harmony search

Strictly speaking, HS is not a bio-inspired algorithm and should not be included in this chapter. However, because it has some interesting properties, we will discuss it briefly. HS is a music-inspired algorithm, first developed by Geem et al. (2001). HS can be explained in more detail with the aid of the discussion of the improvisation process by a musician. When a musician is improvising, he or she has three possible choices: (1) play any famous piece of music (a series of pitches in harmony) exactly from his or her memory; (2) play something similar to a known piece (thus adjusting the pitch slightly); or (3) compose new or random notes. If we formalize these three options for optimization, we have three corresponding components: usage of harmony memory, pitch adjusting, and randomization.

The usage of harmony memory is important because it is similar to choosing the best fit individuals in the GA. The pitch adjustment can essentially be considered as a random walk:

$$x_{\text{new}} = x_{\text{old}} + b(2\varepsilon - 1), \tag{1.21}$$

where ε is a random number drawn from a uniform distribution $[0,1]$. Here, b is the bandwidth, which controls the local range of pitch adjustment. There are also relatively extensive studies concerning the HS algorithm (Geem et al., 2001).

1.4.3 OTHER ALGORITHMS

There are many other metaheuristic algorithms that may be equally popular and powerful, and these include Tabu search (Glover and Laguna, 1997), artificial immune system (Farmer et al., 1986), and others (Koziel and Yang, 2011; Yang, 2010a,b).

The efficiency of bio-inspired algorithms can be attributed to the fact that they imitate the best features in nature, especially the selection of the fittest in biological systems that have evolved by natural selection over millions of years. There is no doubt that bio-inspired computation will continue to become even more popular in the coming years.

1.5 ARTIFICIAL NEURAL NETWORKS

Apart from the SI-based algorithms, telecommunications also involve other popular methods such as ANNs. As we will see, ANNs are in essence optimization algorithms, working in different context and applications (Gurney, 1997; Yang, 2010a).

1.5.1 BASIC IDEA

The basic mathematical model of an artificial neuron was first proposed by W. McCulloch and W. Pitts in 1943, and this fundamental model is referred to as the McCulloch-Pitts model (Gurney, 1997). Other models and neural networks are based on it. An artificial neuron with n inputs or impulses and an output y_k will be activated if the signal strength reaches a certain threshold θ. Each input has a corresponding weight w_i. The output of this neuron is given by

$$y_l = \Phi\left(\sum_{i=1}^{n} w_i u_i\right), \quad \xi = \sum_{i=1}^{n} w_i u_i, \tag{1.22}$$

where the weighted sum ξ is the total signal strength, and Ω is the activation function, which can be taken as a step function. That is, we have

$$\Phi(\xi) = \begin{cases} 1 & \text{if } \xi \geq \theta, \\ 0 & \text{if } \xi < \theta. \end{cases} \tag{1.23}$$

We can see that the output is only activated to a nonzero value if the overall signal strength is greater than the threshold θ. This function has discontinuity, so it is easier to use a smooth sigmoid function

$$S(\xi) = \frac{1}{1 + e^{-\xi}}, \tag{1.24}$$

which approaches 1 as $U \to \infty$, and becomes 0 as $U \to -\infty$. Interestingly, this form will lead to a useful property in terms of the first derivative

$$S'(\xi) = S(\xi)[1 - S(\xi)]. \tag{1.25}$$

1.5.2 NEURAL NETWORKS

The power of the ANN becomes evident with connections and combinations of multiple neurons. The structure of the network can be complicated, and one of the most widely used is to arrange the neurons in a layered structure, with an input layer, an output layer, and one or more hidden layers. The connection strength between two neurons is represented by its corresponding weight. Some ANNs can perform complex tasks and can simulate complex mathematical models, even if there is no explicit functional form mathematically. Neural networks have developed over the last few decades and have been applied in almost all areas of science and engineering.

The construction of a neural network involves the estimation of the suitable weights of a network system with some training/known data sets. The task of the training is to find the suitable weights w_{ij} so that the neural networks not only can best fit the known data, but also can predict outputs for new inputs. A good ANN should be able to minimize both errors simultaneously—the fitting/learning errors and the prediction errors. The errors can be defined as the difference between the calculated (or predicated) output o_k and real output y_k for all output neurons in the least-square sense

$$E = \frac{1}{2}\sum_{k=1}^{n_o}(o_k - y_k)^2. \tag{1.26}$$

Here, the output o_k is a function of inputs/activations and weights. In order to minimize this error, we can use the standard minimization techniques to find the solutions of the weights.

In the standard ANN, the steepest descent method is often used. For any initial random weights, the weight increment for w_{hk} is

$$\Delta w_{hk} = -\eta\frac{\partial E}{\partial w_{hk}} = -\eta\frac{\partial E}{\partial o_k}\frac{\partial o_k}{\partial w_{hk}}. \tag{1.27}$$

Typically, we can choose the learning rate as $\eta = 1$. From

$$S_k = \sum_{h=1}^{m}w_{hk}o_h, \quad (k=1,2,\ldots,n_o), \tag{1.28}$$

and

$$o_k = f(S_k) = \frac{1}{1+e^{-S_k}}, \tag{1.29}$$

we have

$$f' = f(1-f), \tag{1.30}$$

$$\frac{\partial o_k}{\partial w_{hk}} = \frac{\partial o_k}{\partial S_k}\frac{\partial S_k}{\partial w_{hk}} = o_k(1-o_k)o_h, \tag{1.31}$$

and

$$\frac{\partial E}{\partial o_k} = (o_k - y_k). \tag{1.32}$$

Therefore, we have

$$\Delta w_{hk} = -\eta \delta_k o_h, \quad \delta_k = o_k(1 - o_k)(o_k - y_k). \tag{1.33}$$

1.5.3 BACK PROPAGATION ALGORITHM

There are many ways of calculating weights by supervised learning. One of the simplest and most widely used methods is to use the back propagation algorithm for training neural networks, often called back propagation neural networks. Here, the basic idea is to start from the output layer and propagate backward so as to estimate and update the weights (Gurney, 1997). From any initial random weighting matrix w_{ih} (for connecting the input nodes to the hidden layer) and w_{hk} (for connecting the hidden layer to the output nodes), we can calculate the outputs of the hidden layer

$$o_h = \frac{1}{1 + \exp\left[-\sum_{i=1}^{n_j} w_{ih} u_i\right]}, \quad (h = 1, 2, \ldots, m) \tag{1.34}$$

and the outputs for the output nodes

$$o_k = \frac{1}{1 + \exp\left[-\sum_{h=1}^{m} w_{hk} o_h\right]}, \quad (k = 1, 2, \ldots, n_o). \tag{1.35}$$

The errors for the output nodes are given by

$$\delta_k = o_k(1 - o_k)(y_k - o_k), \quad (k = 1, 2, \ldots, n_o), \tag{1.36}$$

where $y_k(k = 1, 2, \ldots, n_o)$ are the data (real outputs) for the inputs $u_i(i = 1, 2, \ldots, n_i)$. Similarly, the errors for the hidden nodes can be written as

$$\delta_h = o_h(1 - o_h)\sum_{k=1}^{n_o} w_{hk} \delta_k, \quad (h = 1, 2, \ldots, m). \tag{1.37}$$

The updating formulae for weights at iteration t are

$$w_{hk}^{t+1} = w_{hk}^{t} + \eta \delta_k o_h \tag{1.38}$$

and

$$w_{ih}^{t+1} = w_{ih}^{t} + \eta \delta_h u_i, \tag{1.39}$$

where the learning rate is $0 < \eta \leq 1$.

Here we can see that the weight increments are

$$\Delta w_{ih} = \eta \delta_h u_i, \tag{1.40}$$

with similar updating formulae. An improved version is to use the weight momentum α to increase the learning efficiency

$$\Delta w_{ih} = \eta \delta_h u_i + \alpha w_{ih}(\tau - 1),$$ (1.41)

where τ is an extra parameter. There are many good software packages for ANNs, and there are dozens of good books fully dedicated to the implementation (Gurney, 1997).

1.6 SUPPORT VECTOR MACHINE

SVMs are an important class of methods that can be very powerful in classifications, regression, machine learning, and other applications (Vapnik, 1995).

1.6.1 LINEAR SVM

The basic idea of classification is to try to separate different samples into different classes. For binary classifications, we can try to construct a hyperplane

$$wx + b = 0,$$ (1.42)

so that the samples can be divided into two distinct classes. Here, the normal vector w and b have the same size as x, and they can be determined using the data, though the method of determining them is not straightforward. This requires the existence of a hyperplane.

In essence, if we can construct such a hyperplane, we should construct two hyperplanes so that the two hyperplanes are as far away as possible and no samples are between the two planes. Mathematically, this is equivalent to two equations

$$wx + b = +1$$ (1.43)

and

$$wx + b = -1.$$ (1.44)

From these two equations, it is straightforward to verify that the normal (perpendicular) distance between these two hyperplanes is related to the norm $\|w\|$ via

$$d = \frac{2}{\|w\|}.$$ (1.45)

A main objective of constructing these two hyperplanes is to maximize the distance or the margin between the two planes. The maximization of d is equivalent to the minimization of $\|w\|$ or more conveniently $\|w\|^2$. From the optimization point of view, the maximization of margins can be written as

$$\text{minimize } \frac{1}{2}\|w\|^2 = \frac{1}{2}(w\,w).$$ (1.46)

If we can classify all the samples completely, for any sample (x_i, y_i) where $i = 1, 2, \ldots, n$, we have

$$wx_i + b \geq +1, \quad \text{if} (x_i, y_i) \in \text{one class}$$ (1.47)

and

$$wx_i + b \leq -1, \quad \text{if} (x_i, y_i) \in \text{the other class}. \tag{1.48}$$

As $y_i \in \{+1, -1\}$, the above two equations can be combined as

$$y_i(wx_i + b) \geq 1, \quad (i = 1, 2, \ldots, n). \tag{1.49}$$

However, it is not always possible to construct such a separating hyperplane in practice. A very useful approach is to use nonnegative slack variables

$$\eta_i \geq 0, \quad (i = 1, 2, \ldots, n),$$

so that

$$y_i(wx_i + b) \geq 1 - \eta_i, \quad (i = 1, 2, \ldots, n). \tag{1.50}$$

Obviously, this optimization requirement for the SVM becomes

$$\text{minimize } \Psi = \frac{1}{2}||w||^2 + \lambda \sum_{i=1}^{n} \eta_i, \tag{1.51}$$

subject to

$$y_i(wx_i + b) \geq 1 - \eta_i, \tag{1.52}$$

$$\eta_i \geq 0, \quad (i = 1, 2, \ldots, n), \tag{1.53}$$

where $\lambda > 0$ is a parameter to be chosen appropriately. Here, the term $\sum_{i=1}^{n} \eta_i$ is essentially a measure of the upper bound of the number of misclassifications on the training data.

By using Lagrange multipliers $\alpha_i \geq 0$, we can rewrite the above constrained optimization into an unconstrained version, and we have

$$L = \frac{1}{2}||w||^2 + \lambda \sum_{i=1}^{n} \eta_i - \sum_{i=1}^{n} \alpha_i [y_i(wx_i + b) - (1 - \eta_i)]. \tag{1.54}$$

Based on this, it is straightforward to write the Karush-Kuhn-Tucker (KKT) conditions

$$\frac{\partial L}{\partial w} = w - \sum_{i=1}^{n} \alpha_i y_i x_i = 0, \tag{1.55}$$

$$\frac{\partial L}{\partial b} = -\sum_{i=1}^{n} \alpha_i y_i = 0, \tag{1.56}$$

$$y_i(wx_i + b) - (1 - \eta_i) \geq 0, \tag{1.57}$$

$$\alpha_i [y_i(wx_i + b) - (1 - \eta_i)] = 0, \quad (i = 1, 2, \ldots, n), \tag{1.58}$$

$$\alpha_i \geq 0, \ \eta_i \geq 0, \quad (i = 1, 2, \ldots, n). \tag{1.59}$$

From the first KKT condition, we can get

$$w = \sum_{i=1}^{n} y_i \alpha_i x_i. \tag{1.60}$$

It is worth pointing out here that only the nonzero coefficients α_i contribute to the overall solution. This comes from the KKT condition (Equation (1.58)), which implies that when $\alpha_i \neq 0$, the inequality (Equation (1.57)) must be satisfied exactly, while $\alpha_0 = 0$ means the inequality is automatically met. In this latter case, $\eta_i = 0$. Therefore, only the corresponding training data (x_i, y_i) with $\alpha_i > 0$ can contribute to the solution, and thus such x_i form the support vectors (hence the name support vector machine). All the other data with $\alpha_i = 0$ become irrelevant. In essence, the above is equivalent to solving the following quadratic programming:

$$\text{maximize} \sum_{i=1}^{n} \alpha_i - \frac{1}{2} \sum_{i,j=1}^{n} \alpha_i \alpha_j y_i y_j (x_i x_j), \tag{1.61}$$

subject to

$$\sum_{i=1}^{n} \alpha_i y_i = 0, \ 0 \leq \alpha_i \leq \lambda, \ (i = 1, 2, \ldots, n). \tag{1.62}$$

From the coefficients α_i, we can write the final classification or decision function as

$$f(x) = \text{sgn} \left[\sum_{i=1}^{n} \alpha_i y_i (x x_i) + b \right], \tag{1.63}$$

where sgn is the classic sign function.

1.6.2 KERNEL TRICKS AND NONLINEAR SVM

In reality, most problems are nonlinear, and the above linear SVM cannot be used. Ideally, we should find some nonlinear transformation ϕ so that the data can be mapped onto a high-dimensional space where the classification becomes linear. The transformation should be chosen in a certain way so that their dot product leads to a kernel-style function $K(x, x_i) = \phi(x) \dot{\phi}(x_i)$, which enables us to write our decision function as

$$f(x) = \text{sgn} \left[\sum_{i=1}^{n} \alpha_i y_i K(x, x_i) + b \right]. \tag{1.64}$$

From the theory of eigenfunctions, we know that it is possible to expand functions in terms of eigenfunctions. In fact, we do not need to know such transformations; we can directly use kernel functions $K(x, x_i)$ to complete this task. This is the kernel function trick. Now, the main task is to chose a suitable kernel function for a given problem.

Though there are polynomial classifiers and others, the most widely used kernel is the Gaussian radial basis function,

$$K(x,x_i) = \exp\left[-||x-x_i||^2/(2\sigma^2)\right] = \exp\left[-\gamma||x-x_i||^2\right], \qquad (1.65)$$

for nonlinear classifiers. This kernel can easily be extended to any high dimensions. Here, σ^2 is the variance and $\gamma = 1/(2\sigma^2)$ is a constant.

Following a similar procedure as discussed earlier for linear SVM, we can obtain the coefficients α_i by solving the following optimization problem:

$$\text{maximize} \sum_{i=1}^{n} \alpha_i - \frac{1}{2}\alpha_i\alpha_j y_i y_j K(x_i, x_j). \qquad (1.66)$$

It is worth pointing out under Mercer's conditions for kernel functions, the matrix $A = y_i y_j K(x_i, x_j)$ is a symmetric positive definite matrix, which implies that the above maximization is a quadratic programming problem, and can thus be solved efficiently by many standard quadratic programming techniques. There are many software packages (either commercial or open source) that are easily available, so we will not provide any discussion of the implementation. In addition, some methods and their variants are still an area of active research. Interested readers can refer to more advanced literature.

1.7 CONCLUSIONS

The extensive literature of bio-inspiration computation and its application in telecommunications is expanding. We have carried out a brief review of some bioinspired optimization algorithms in this chapter, but we have not focused sufficiently on the implementations, especially in the context of telecommunications. However, this chapter can still pave the way for introducing these methods in more detail in the rest of the book. The next chapter will start to emphasize more the design problems in telecommunications and highlight some key challenges. Then, more self-contained chapters will focus on various bio-inspired methods and their diverse range of applications in telecommunications.

REFERENCES

Blum, C., Roli, A., 2003. Metaheuristics in combinatorial optimization: overview and conceptual comparison. ACM Comput. Surv. 35, 268–308.

Conn, A.R., Schneinberg, K., Vicente, L.N., 2009. Introduction to derivative-free optimization. In: MPS-SIAM Series on Optimization, SIAM, Philadelphia, USA.

Dorigo, M., Stütle, T., 2004. Ant Colony Optimization. MIT Press, Cambridge, MA.

Farmer, J.D., Packard, N., Perelson, A., 1986. The immune system, adaptation and machine learning. Physica D 2, 187–204.

Fister, I., Fister Jr., I., Yang, X.S., Brest, J., 2013. A comprehensive review of firefly algorithms. Swarm Evol. Comput. 13 (1), 34–46.

Gandomi, A.H., Yang, X.S., Alavi, A.H., 2011. Mixed variable structural optimization using firefly algorithm. Comput. Struct. 89 (23/24), 2325–2336.

Geem, Z.W., Kim, J.H., Loganathan, G.V., 2001. A new heuristic optimization: harmony search. Simulation 76 (1), 60–68.

Glover, F., Laguna, M., 1997. Tabu Search. Kluwer Academic Publishers, Boston, USA.

Goldberg, D.E., 1989. Genetic Algorithms in Search, Optimization and Machine Learning. Addison Wesley, Reading, MA.

Gurney, K., 1997. An Introduction to Neural Networks. Routledge, London.

Gut, A., 2005. Probability: a graduate course. In: Springer Texts in Statistics, Springer, Berlin.

Holland, J., 1975. Adaptation in Natural and Artificial Systems. University of Michigan Press, Ann Anbor.

Karaboga, D., 2005. An idea based on honey bee swarm for numerical optimization. Technical report TR06, Erciyes University, Turkey.

Karmarkar, N., 1984. A new polynomial-time algorithm for linear programming. Combinatorica 4 (4), 373–395.

Kennedy, J., Eberhart, R.C., 1995. Particle swarm optimization. In: Proceedings of IEEE International Conference on Neural Networks, Piscataway, NJ, pp. 1942–1948.

Kennedy, J., Eberhart, R.C., Shi, Y., 2001. Swarm Intelligence. Morgan Kaufmann Publishers, San Francisco.

Kirkpatrick, S., Gelatt, C.D., Vecchi, M.P., 1983. Optimization by simulated annealing. Science 220 (4598), 671–680.

Koziel, S., Yang, X.S., 2011. Computational Optimization, Methods and Algorithms. Springer, Germany.

Marsaglia, G., Zaman, A., 1993. Monkey tests for random number generators. Comput. Math. Appl. 26 (9), 1–10.

Nakrani, S., Tovey, C., 2004. On honey bees and dynamic server allocation in internet hosting centers. Adapt. Behav. 12 (3-4), 223–240.

Parpinelli, R.S., Lopes, H.S., 2011. New inspirations in swarm intelligence: a survey. Int. J. Bio-Inspir. Com. 3, 1–16.

Pavlyukevich, I., 2007. Lévy flights, non-local search and simulated annealing. J. Comput. Phys. 226, 1830–1844.

Price, K., Storn, R., Lampinen, J., 2005. Differential Evolution: A Practical Approach to Global Optimization. Springer, Berlin.

Storn, R., 1996. On the usage of differential evolution for function optimization. In: Biennial Conference of the North American Fuzzy Information Processing Society (NAFIPS), pp. 519–523.

Storn, R., Price, K., 1997. Differential evolution – a simple and efficient heuristic for global optimization over continuous spaces. J. Glob. Optim. 11 (2), 341–359.

Talbi, E.G., 2009. Metaheuristics: From Design to Implementation. John Wiley & Sons, Hoboken, NJ.

Ting, T.O., Chien, S.F., Yang, X.S., Lee, S.H., 2014. Analysis of quality-of-service aware orthogonal frequency division multiple access system considering energy efficiency. IET Commun. 8 (11), 1947–1954.

Vapnik, V., 1995. The Nature of Statistical Learning Theory. Springer-Verlag, New York.

Wolpert, D.H., Macready, W.G., 1997. No free lunch theorems for optimization. IEEE Trans. Evol. Comput. 1 (1), 67–82.

Yang, X.S., 2008. Nature-Inspired Metaheuristic Algorithms, first ed. Luniver Press, UK.

Yang, X.S., 2009. Firefly algorithms for multimodal optimization. In: Watanabe, O., Zeugmann, T. (Eds.), 5th Symposium on Stochastic Algorithms, Foundation and Applications (SAGA 2009). In: LNCS, 5792. Springer, Heidelberg, pp. 169–178.

Yang, X.S., 2010a. Nature-Inspired Metaheuristic Algorithms, second ed. Luniver Press, UK.

Yang, X.S., 2010b. Engineering Optimization: An Introduction with Metaheuristic Applications. John Wiley & Sons, Hoboken, NJ.

Yang, X.S., 2011. Bat algorithm for multi-objective optimisation. Int. J. Bio-Inspir. Com. 3 (5), 267–274.

Yang, X.S., 2014. Nature-Inspired Optimization Algorithms. Elsevier, London.

Yang, X.S., Deb, S., 2009. Cuckoo search via Lévy flights. In: Proceedings of World Congress on Nature & Biologically Inspired Computing (NaBic 2009). IEEE Publications, USA, pp. 210–214.

Yang, X.S., Deb, S., 2010. Engineering optimization by cuckoo search. Int. J. Math. Model. Num. Opt. 1 (4), 330–343.

Yang, X.S., Gandomi, A.H., 2012. Bat algorithm: a novel approach for global engineering optimization. Eng. Comput. 29 (5), 1–18.

Yang, X.S., He, S., 2013. Bat algorithm: literature review and applications. Int. J. Bio-Inspir. Com. 5 (3), 141–149.

Bio-Inspired Approaches in Telecommunications

2

Su Fong Chien[1], C.C. Zarakovitis[2], Tiew On Ting[3], and Xin-She Yang[4]

[1]*Strategic Advanced Research (StAR) Mathematical Modeling Lab, MIMOS Berhad,*
Kuala Lumpur, Malaysia
[2]*School of Computing and Communications, Lancaster University, Lancaster, UK*
[3]*Department of Electrical and Engineering, Xi'an Jiaotong-Liverpool University,*
Suzhou, Jiangsu Province, China
[4]*School of Science and Technology, Middlesex University, London, UK*

CHAPTER CONTENTS

2.1 INTRODUCTION

Bio-inspired algorithms have become popular optimization tools to tackle complex design problems. With the steady advancement of computing facilities, both scientists and engineers have started to utilize bio-inspired algorithms due to their advantages over conventional methods (Yang and Koziel, 2011; Yang, 2014). A major characteristic of bio-inspired algorithms is that they are flexible and straightforward to implement, yet efficient to solve tough problems in applications such as engineering and telecommunications. In fact, bio-inspired computation in telecommunications has a rather rich history.

Probably the first application of a multiobjective bio-inspired algorithm was attempted by J.D. Schaffer in the mid-1980s (Schaffer, 1984). A considerable extension in this area is now known as *multiobjective evolutionary algorithm*. The activities have been reflected by the ever-increasing number of technical papers published in conferences, journals, and books. An important advantage of bio-inspired algorithms on the solutions of multiobjective optimization problems is that they are able to produce feasible solutions in which Pareto optimal sets are attainable in a single run of the algorithms (Coello, 1999). Compared with the conventional methods, bio-inspired methods may take longer to run but they can indeed produce satisfactory solutions (or even global optimal solutions). In addition, bio-inspired algorithms are less sensitive to the shape or continuity of the Pareto front, avoiding some disadvantages of conventional mathematical programming (Coello et al., 2002). The importance of bio-inspired algorithms in communications and networking can be seen from the extensive literature survey done by Kampstra et al. (2006), where more than 350 references were listed on the applications of bio-inspired techniques for solving telecommunication design problems.

Design problems in telecommunications tend to be complex and large-scale, and thus computationally demanding. Such issues become even more challenging due to the increasing demands of bandwidth as well as disruption-free services in wireless communications, which is in addition to quality-of-service (QoS) constraints for subscribers. The complexity of such problems means that conventional methods are not able to meet these challenges. In recent years, bio-inspired methods have become powerful alternative techniques to deal with design problems in telecommunications. In fact, both types of techniques complement each other in terms of simplicity, efficiency, and transparency in communication for the end users. However, architectural redesign efforts are very time-consuming and thus can be very costly, and consequently, there is an ever-increasing demand for efficient techniques that can support proper design requirements.

In essence, common problems that need to be addressed in telecommunications areas include node location problems, network topology design problems, routing and path restoration problems, efficient admission control mechanisms, channel and/or wavelength assignments and resource allocation problems, and so on. Due to the complicated nature of communications infrastructures, such design problems become even more complex. Hence, all these necessitate a multiobjective approach subject to noise, the dynamic behavior of parameters, and large solution spaces. As a result, conventional methods are not capable of solving such problems effectively (Routen, 1994).

The main aim of this chapter is to review the types of design problems in telecommunications and their solution strategies. Thus, Section 2.2 outlines the seven types of design problems, and Section 2.3 discusses green communications. Section 2.4 briefly introduces orthogonal frequency division multiplexing (OFDM), and Section 2.5 presents a case study of orthogonal frequency division multiple access (OFDMA) with the consideration of energy efficiency (EE). Finally, Section 2.6 draws some conclusions.

2.2 DESIGN PROBLEMS IN TELECOMMUNICATIONS

There is a wide range of design problems in telecommunications. A survey by P. Kampstra et al. has classified such design problems into seven major categories (Kampstra et al., 2006). The first category regards node location problems. One of the problems concerns the placement of concentrators in a local access network, where the genetic algorithm (GA) was applied in this case (Calégari et al., 1997). As discussed by Calégari et al. (1997), the key issue for a radio network is the proper positioning of locations of antennas and receivers, whereas the problem of the proper selection of base stations (BSs) was studied by Krzanowski and Raper (1999). In fact, most problems were attempted by GAs with satisfactory results.

The second category refers to topology design in computer networking: GAs where the key tool compared to others. Topologies for computer networks focused on reliability problems (Kumar et al., 1995). In their study, they used a variant of the GA, together with some problem-specific repair and crossover functions. Survivable military communication networks were also investigated, considering the damaging impact from the network with some satellite links (Sobotka, 1992). Nevertheless, network reliability is not the only important factor; backbone topologies must also take into account the economic costs (Deeter and Smith, 1998; Konak and Smith, 2004). To meet the ever-increasing demand for bandwidth and speed, different types of telecommunication technologies such as *asynchronous transfer mode* (ATM) networks had become an alternative for future networks. To a certain extent, ATM network topology designs have been rigorously studied (Tang et al., 1998; Thompson and Bilbro, 2000).

Moreover, as backbone technologies evolve, the subscriptions of video-on-demand services become prevalent. A proper network design with storage nodes for videos has become one of the important topics (Tanaka and Berlage, 1996). Another application area that is worth studying is the way of assigning terminals to concentrators. GAs were used to assign terminals to concentrators, powered by permutation encoding. It was found that GAs can outperform the greedy algorithm (Abuali et al., 1994). In the era of multimedia traffic, technology that has to support the huge bandwidth of at least several of tens of terabit per second becomes a must for fulfilling the needs of the market. Optical communications and networking are deemed a feasible technology for the requirements mentioned previously. For a network layer, concerns such as topology design, path protection, wavelength assignment problems, and wavelength converter placements are still hot topics to be investigated (Sinclair, 1995; Zhang et al., 2007; Banerjee et al., 2004; Teo et al., 2005). Sinclair (1995), one of the pioneers who designed optical network topologies, used GAs to optimize the links in the network topology considering time and costs.

The third category in P. Kampstra et al.'s survey regards tree construction problems for the network. More precisely, Chardaire et al. (1995) utilized the GA to determine Steiner points from a given set of nodes. A well-known tree-related problem is the minimum spanning tree problem, which received enormous attention from the research society. For instance, Hsinghua et al. (2001) investigated the effect of several properties of a GA to solve the constrained minimal-spanning tree problem.

The fourth category concerns the packet routing. In a network, routing plays a significant role that determines the performance of networks. Routing can have either static or dynamic characteristics. Routing designs are usually based on physical distances, path lengths in hop count, bandwidth, or traffic flow. For example, Mann et al. (1995) tackled the fixed routing to minimize costs, while balancing the traffic load. They found that the GA was more robust than simulated annealing. Another interesting issue is the point-to-multipoint routing. Hoelting et al. (1996) studied a tour for fault detecting in a point-to-point telecommunication network. They claimed that their GA approach outperforms the deterministic method. Hoelting et al. (1996) also studied broadcasting messages from the station to all nodes with the least possible broadcasting time. Their computed results showed that the GA with a permutation-based encoding achieved better results. In essence, the shortest path routing protocol is a routing protocol that connects between Internet Service Providers. Such routing protocol requires weights to be set for the connections. Buriol et al. (2005) tested a hybrid GA for the purpose mentioned above. They obtained better results than other algorithms in terms of efficiency and robustness. Besides point-to-point routing, multicast routing and QoS routing have also gained much interest for more than a decade. Wang et al. (2001) tried to solve the multicast routing problem under delay and bandwidth constraints.

The fifth category deals with the dimensioning issue, which determines the minimum system capacity needed in order that the incoming traffic meet the specified QoS. GAs were used to solve this problem almost two decades ago. The link capacity in the dimensioning problem becomes an important factor to be dealt with (Davis and Coombs, 1989; Mostafa and Eid, 2000). Furthermore, the dimensioning issue has also attracted much attention in optical communication and networking. For example, Chong and Kwong (2003) produced some encouraging results by applying GAs to allocate spare capacities. On the other hand, Mutafungwa (2002) managed to design link redundancy enhancements for optical cross-connected nodes. To ensure good QoS to subscribers, it should have speedy path restoration in case of link failure. For the problem in the restoration category, both genetic and hybrid GAs were applied. For instance, Kozdrowski et al. (1997) applied a hybrid GA to allocate link capacities and traffic streams in the network when link failures occur.

The sixth category addresses the call admission control or call management, which is the management of the calls to prevent oversubscription of the networks. Sherif et al. (2000) studied the call admission in ATM wireless networks, where the QoS of various multimedia services was considered, whereas Karabudak et al. (2004) studied the call admission for next generation wireless networks that involved multimedia QoS "calls."

The last category concerns frequency assignment and wavelength allocation problems. In wireless communications, the combination of frequency assignments (subcarriers/subchannel assignment) and power allocation can lead to a *nondeterministic polynomial-time hard*, mixed-binary-integer programming problem. Without strategic subchannel assignments, the total sum data rate in the system may be significantly affected. Similarly, due to the limited number of available wavelengths in optical communications, smart algorithms are preferred to ensure the smooth packet's transmission to its destination. From the survey by Kampstra et al. (2006), many frequency assignment problems in wireless communications reveal interference issues among subchannels (Ngo and Li, 1998; Crisan and Muhlenbein, 1998). For optical communications, Tan and Sinclair (1995) used the GA to search for routes between node pairs in a network. The number of wavelengths in their study was an integer, and the gene did represent how the traffic was routed. The work of Tan and Sinclair (1995) was extended by Ali et al. (2000), who took into account the power efficiency. In general, one can add more criteria in the problem introduced by Tan and Sinclair (1995) to formulate more general models: such criteria can be the average flow, the average delay, and the expected blocking. Another problem that has been considered in the early twenty-first century is traffic grooming. This problem inspired many researchers to search for a better way to combine multiple low-bandwidth traffic streams into one waveband, aiming to minimize the required number of waveband stoppers.

Some issues introduced by Kampstra et al. (2006) still remain open. For instance, the wireless communications literature focuses on hot topics such as network reliability (Khanbary and Vidyarthi, 2009), adaptive routing for wireless backhaul network (Anastasopoulos et al., 2008), and antenna issues (Suneel Varma et al., 2013). Similarly, in optical communications, key issues are the reliability issues addressed by Balazs et al. (2013), the routing and wavelength assignment problem by Rubio-Largo et al. (2012), and the traffic grooming by Rubio-Largo et al. (2013).

2.3 GREEN COMMUNICATIONS

OFDMA wireless networks provide ubiquitous access in wide coverage with a connection speed of up to 1 Gbps, providing the required QoS to network users. OFDMA networks are packet-based and employ IP to transport various types of traffic such as voice, video, and data. Due to their simplicity, OFDMA networks are cost-effective; most of the network providers are willing to upgrade from the existing technology to better technology, for example, from a 3G to 4G wireless system. In this information era, human-to-human or human-to-machine or machine-to-machine communications heavily rely on wireless communications. Such huge demands inevitably can cause environmental issues on a global scale, making energy consumption a key issue. Therefore, there is a strong need to shift to green communications.

2.3.1 ENERGY CONSUMPTION IN WIRELESS COMMUNICATIONS

According to the well-known Gartner report (Gartner, 2007), the *information communication technology* (ICT) market contributes 2% of global greenhouse gas (CO_2) emissions. A typical mobile phone network in the United Kingdom consumes about 40-50 MW. A service provider such as Vodafone uses more than 1 million gallons of diesel per day to power up its networks. This implies that wireless communications can create a significant proportion of the total energy consumption by the ICT infrastructure. In order to save costs, two major issues have to be considered in green communications:

1. To reduce the energy consumption so as to cut down the *operating expenditure* costs.
2. To create a more user-friendly environment by reducing carbon emissions. Therefore, it is necessary to develop innovative algorithms so as to reduce the total energy required for the operation of wireless access networks.

In addition, it is also important to address the relevant portions of wireless architectures that can affect the power consumption significantly. Based on Figure 2.1, the power consumed by the retail group, data centers, core transmission, mobile switches, and BS are about 2%, 8%, 15%, 20%, and 55%, respectively (Suarez et al., 2012).

This indicates that either a BS or an access point could be an important element of future research.

The escalation of energy usage in wireless networks causes serious greenhouse gas emissions that can be deemed a major threat to environmental protection and sustainable development. To tackle this issue, one effective way is to implement more energy-efficient wireless networks. In fact, before the *green radio* (GR) program, there were efforts to improve energy saving in wireless networks, such as

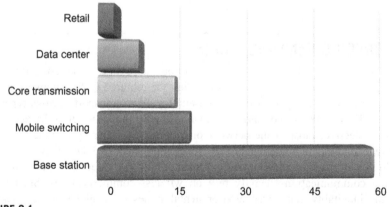

FIGURE 2.1

Percentage of power consumption in the cellular network infrastructure.

developing ultra-efficient power amplifiers, minimizing feeder losses, and implementing passive cooling. However, these efforts were not good enough to achieve the energy-saving target within 5-10 years. On the other hand, the GR program with innovative solutions, which is based on top-down architecture as well as the joint design across all system levels and protocol stacks, may certainly bring more fruitful results in the near future. Several international research projects dedicated to energy-efficient wireless communications include the GR (Grant, 2010), Energy Aware Radio and Network Technologies (EARTH, 2010), Optimizing Power Efficiency in Mobile Radio Networks (OPERANet, 2010), energy-efficient wireless networking (Ewin, 2008), and others. Feng et al. (2013) have tabulated the detailed solutions of these projects. After further analysis, we categorized these projects under five strategies in terms of EE, listed in Table 2.1.

2.3.2 METRICS FOR ENERGY EFFICIENCY

To the best of our knowledge, there has been no standard metric to determine the EE up to now. For example, Suarez et al. (2012) classified the metrics for EE into three levels: component level, access node level, and network level. The detailed descriptions for the metrics are shown in Figure 2.2. At the component level, by definition, the EE metric is the ratio of the attained utility to the consumed power, whereas the energy consumption metric corresponds to the power consumed per unit of the attainable utility. Another important element is the power amplifier efficiency metric, where the measurement is the ratio of the *peak average* output power to the supplied power. One possible metric for the power amplifier is the *peak-to-average power ratio* (PAPR). The reduction of PAPR produces better amplifier efficiency. Measurement units that have been used to measure the performance of computing processing associated to energy consumption are *millions of instructions per second per watt* and *millions of floating-point operations per second per watt*. Moreover, we may also include the energy consumption gain, which is the ratio of the consumed energy of a baseline device to the consumed energy of a given device under test.

At the access node level, new metrics have been introduced, especially for BSs. For instance, the *energy consumption rating* is the energy used for transmitting an amount of information measured in joules per bit. Some researchers would prefer the EE measure in bits per joule, which is the system throughput per unit-energy consumption. It is worth pointing out that the most common metrics used today actually target the attained utility of the different resources on the condition that there exist some trade-offs, such as the spectral efficiency (b/s/Hz) and the power efficiency (b/s/Hz/W). The metric aimed to cover all aspects in a more general form is the radio efficiency ((b·m)/s/Hz/W) that measures the data rate transmitted and the transmission distance attainable given the respective figures of the bandwidth and the supplied power resources. In the case of the access level, normally, the metrics evaluate the global attained service provision given a consumed power, introduced by *European Tele-Communications Standards Institute* (ETSI) for GSM networks; it is either in terms

Table 2.1 Research projects for energy-efficient wireless communications (Feng et al., 2013)

Energy-Efficient Strategies	Projects			
	Green Radio	EARTH	OPERANet	eWin
(i) Energy Metrics and Model This covers the works on energy metrics to quantify consumption in the communication network. The relevant communication models take into account the calculation for energy consumption.	Yes	Yes	No	No
(ii) Energy-Efficient Hardware This involves the integration of relevant hardware modules, the use of advanced power electronics methods, and efficient *Digital Signal Processing* (DSP) techniques to achieve low power consumption.	Yes	No	Yes	No
(iii) Energy-Efficient Architectures This concerns the heterogeneous network deployment (microcell, picocell, femtocell). The architectures also include multihop routing, network coding and relay, and cooperative communications. This architecture bounds the power requirements imposed by QoS and efficient backhaul.	Yes	Yes	No	Yes
(iv) Energy-Efficient Resource Management This aims to achieve the auto(re)-configuration of control software and network resources, in response to changes in infrastructure and demand. It also involves the radio resource management for cooperative and competitive heterogeneous environments, and the application of dynamic spectrum management.	Yes	Yes	Yes	Yes
(v) Radio Technologies and Components The advanced transmission techniques such as MIMO, OFDM, and adaptive antennas are capable of low energy consumption. There exists a mechanism for power control on the transceiver and relevant components.	No	Yes	Yes	No

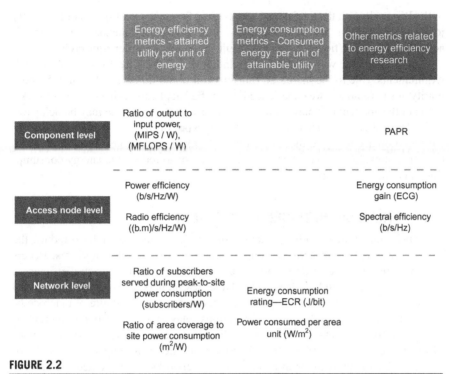

FIGURE 2.2

Classifications of the metrics used in the energy efficiency for wireless networks.

of the ratio of subscribers served or by means of the covered area to the area power consumption.

2.3.3 RADIO RESOURCE MANAGEMENT

Radio resource management (RRM) plays a significant role in controlling power consumption. It has been found that energy savings can be achieved during the low-traffic-load period. For example, Oh and Krishnamachari (2010) showed that the daily traffic loads at BSs change widely over time and space. Hence, when the traffic load is low, it may waste a lot of energy. Service providers can also implement a new feature in their software, dynamic power save, which can save up to 27% of the power consumption for the BSs. In the OPERANet project, it proposed solutions for energy saving via cell-size breathing and sleep modes based on traffic loads. Furthermore, Marsan et al. (2009) showed that switching off the active cells under circumstances during low traffic periods can enjoy 25-30% energy savings. Because the switch-off technique was efficient, many extension studies have been done by considering the blocking probability requirement, a minimum mode holding time, the effect of the mean and variance of the traffic load, and so on. Chen et al. (2010) have shown some potential approaches to reduce the energy consumption of the BSs scale with the traffic load across time, frequency, and spatial domains. By jointly reconfiguring the bandwidth and the number of antennas and channels according to the traffic load, it could result in the maximum energy savings.

Another possible way to save energy in RRM is to promote diversity of the QoS requirements. Due to the increased transmission in multimedia traffic and the popularity of smartphones, we experienced diversified applications in cellular networks. Some of the applications may require real-time services and some may be delay tolerant. Thus, it is necessary to differentiate the types of traffic and adapt the energy consumption according to the traffic type. Meshkati et al. (2009) and Kolios et al. (2009) exploited the service latency of applications to reduce the energy consumption in cellular networks.

2.3.4 STRATEGIC NETWORK DEPLOYMENT

Obviously, the strategic network deployment will certainly be able to reduce the energy consumption in cellular communications. Most of the time, researchers mainly focus on the network performance, such as the coverage, spectral efficiency, and capacity (Hanly and Mathar, 2002). Other researchers focus on the optimal cell size (Chen et al., 2010), emerging heterogeneous networks (mixture of macrocells, microcells, picocells, and femtocells), various relay and cooperative communications, and so on. Richter et al. have achieved a series of significant outcomes (Richter and Fettweis, 2009, 2010; Fehske et al., 2009; Richter et al., 2009) for energy-efficient applications in heterogeneous networks. They studied the optimal layout of microcells overlaying conventional macrocells. In particular, many researchers focus on the simple energy consumption models of different BSs, considering fewer scenarios as the case studies. In their research, the transmit power was separated into both dependent parts (amplifier, feeder, transmission-related cooling devices) and independent parts (circuit power for signal processing, battery backup, site cooling consumption, etc.). The energy consumption at the BS was modeled as the sum of these two types of power. Note that the energy consumption of the dependent part scales linearly with the average radiated power. Besides, the effect of inter-site distance and the average number of microsites per macrocell on area power consumption was also considered. In their later works, they investigated the potential energy reduction by changing the numbers of microsites and macrocell size to obtain the given spectral efficiency targets under full-load conditions. Their results showed that the deployment of microsites was much preferred because of a significant amount of reduction in the area power consumption in the network, while maintaining the required area throughput target. Furthermore, they made a comparison of the area power consumption with area spectral efficiency of homogeneous macrosites, homogeneous microsites, and heterogeneous networks. Their findings indicated the improvement of EE, while deploying additional microsites is necessary under the condition of higher area throughput targets with higher user densities. Besides this, they concluded that deploying picocells and femtocells is a good strategy to provide cost-effective services.

Besides, it may be advantageous to combine wireless and optical communication technologies in order to reduce the overall power consumption (Zhang et al., 2010). This technology can be seen as the *distributed antenna system* defined in the *long-term evolution advanced* (LTE-A) system. This idea allocates two main units for the BS: *baseband unit* (BBU) and *remote radio unit* (RRU). The RRUs are geographically separated points in the *coordinated multipoint* system that connects to the BBU through an optical fiber. The purpose for this technology is to decentralize the deployment of antennas to expand the cell coverage. Thus, this increases the capacity for the system, due to the fact that the distance between the user and antennas has been shortened, thereby reducing the power usage between the user and antennas. As the distance between a user and the antenna varies in a wide range, this can cause variations of the transmission power. Therefore, an efficient algorithm for the deployment of RRUs is necessary in order to achieve better EE.

On the other hand, the green cellular is another new architecture, proposed by Ezri and Shilo (2009), targeting the minimal emission from mobile stations without any additional radiation sources. This new architecture is equipped with "dubbed green antennas" at each transceiver in the BS. Mobile users close to green antennas can transmit at a lower transmit power, and thus reduce power consumption with fewer interference problems. Besides these advantages, there is no additional radiation produced by the green antennas, as the traffic relay only concerns the uplink.

One of the current hot topics concerning EE is the architecture regarding the relay and cooperative communications. The architecture with more relay nodes can save energy as it reduces path losses due to a shorter transmission range. Consequently, this architecture generates less interference because of the low transmission power (Bae and Stark, 2009; Miao et al., 2009). Unlike the conventional relay systems, each cooperative node in cooperative communications is capable of generating an information source and relaying information as well. This architecture exploits the channel diversity for potential energy savings. Cui et al. (2004) showed that, over some distance ranges, the cooperative *multiple-input multiple-output* (MIMO) transmission and reception could also promote energy savings. Moreover, they also showed that the constellation size for different transmission distances could improve the EE of cooperative communication, which was better than direct communications. However, the disadvantage of both relay and cooperative communications is that selection for optimal partners can be extremely challenging. Another prominent issue is the resource allocation to achieve the minimal energy consumption.

2.4 ORTHOGONAL FREQUENCY DIVISION MULTIPLEXING
2.4.1 OFDM SYSTEMS

The demand for broadband wireless communications is growing rapidly in the era of multimedia traffic and big data. Such trends will continue in the near future. The common technology for current wireless communications for high-rate multimedia transmission utilizes the multicarrier air interface based on OFDM (Morelli et al., 2007; Cvijetic 2012). The key idea behind OFDM is that it is able to obtain a set of frequency-flat subchannels via the conversion of frequency selective channels, where the converted spectra are partially overlapped. The system then divides the input high-rate data stream into a number of substreams for transmission in parallel over orthogonal subcarriers. Thus, OFDM is prone to multipath distortions compared to conventional single-carrier systems because a bank of one-tap multipliers can accomplish the channel equalization. Besides this, OFDM has higher flexibility in choosing modulation parameters such as the constellation size and coding scheme over each subcarrier. The OFDMA is an extension of OFDM, which allows multiple users to share an OFDM symbol.

The OFDMA system is beneficial due to its flexibility of utilizing radio resources. In such systems, each subcarrier not only serves a relevant user, but also is able to adjust the power on each subcarrier. As the probability for all users to experience a deep fading for particular subcarriers is low, OFDMA can be favored by the channel diversity because each subcarrier serves different users with the highest signal-to-noise ratio. Such a feature can greatly increase the system capacity. OFDMA is a dominant multiple-access scheme for next-generation wireless networks, adopted by two promising standards: the LTE-A and the 802.16 m. As mentioned previously, the inherent property of multiuser diversity in OFDMA systems not only enhances the network capacity, but it is also capable of reducing the energy consumption. When a channel with a high gain is detected, the channel may then be assigned to a corresponding user, reducing the transmit power. Consequently, there is ongoing research on power optimization for the optimal subcarrier, bit, and power allocations (Wong et al., 2004; Lo et al., 2014).

Despite OFDMA's appealing features, its design still has some technical challenges. One open issue is the stringent requirement on frequency and timing synchronization. Both OFDM and OFDMA are very sensitive to timing errors and carrier frequency offsets between the incoming signal and the local references used for signal demodulation. However, when the compensation of the frequency offset is not accurate enough, the orthogonal symbols among subcarriers will be damaged, and therefore, *interchannel interference* and *multiple-access interference* will appear. Furthermore, if there are timing errors, OFDMA will produce the *interblock interference*, and thus, prompt actions are necessary to avoid significant error-rate degradations. The solution for this defection is to use a sufficiently long guard interval between adjacent OFDM blocks that can intrinsically protect timing errors by sacrificing a certain amount of reduction in data throughput due to the extra overhead.

2.4.2 THREE-STEP PROCEDURE FOR TIMING AND FREQUENCY SYNCHRONIZATION

In OFDMA systems, the synchronization of time and frequency are of equal importance. There are three stages in this synchronization process, as described below.

1. Synchronization process

    The synchronization process is initiated during the downlink transmission period, with each *mobile terminal* utilizing a pilot signal transmitted by the BS to undergo frequency and timing estimation. This process reduces synchronization errors within a tolerable range. The predicted parameters on each user can further detect the downlink data stream and synchronization references for the uplink transmission. Nevertheless, both Doppler shifts and propagation delays that cause the residual synchronization errors can affect the uplink signals to the BS.

2. Frequency and timing estimation

    In this step, frequency and timing estimation is performed in the uplink. This is a challenging task because the uplink waveform consists of a mixture of signals sent by various users, each affected by exclusive synchronization errors. Accordingly, such signal recovery process in the uplink is a multiparameter estimation problem. Hence, the separation of users should be performed prior to the synchronization procedure. This can be achieved with an efficient subcarrier assignment algorithm.

3. Timing and frequency correction

    After accomplishing the uplink estimation for timing and frequency offsets, restoration for the orthogonality among subcarriers should be employed in a similar manner at the BS.

After the above procedure, the BS will send the estimated offsets back to the corresponding user, and, with this information, the user is able to adjust its transmitted signal. In such a dynamic environment, it is necessary for the BS to provide the latest estimated synchronization parameters periodically, which consequently may produce excessive overhead and outdated information due to feedback delays. Considering such circumstances, current OFDMA systems prefer to use advanced signal processing techniques, which are capable of compensating synchronization errors directly at the BS. Hence, this may solve the problem of returning timing and frequency estimates back to the subscriber terminals.

2.5 OFDMA MODEL CONSIDERING ENERGY EFFICIENCY AND QUALITY-OF-SERVICE

In this section, we present a mathematical model of OFDMA, considering EE and QoS parameters as a case study.

2.5.1 MATHEMATICAL FORMULATION

The QoS in the system guarantees the minimum required transmission rate for each subscriber. This formulation involves one BS and M users that share N subcarriers, each one of bandwidth B. The basic assumptions are given in Ting et al. (2013), Chien et al. (2013), and Zarakovitis and Ni (2013), which include the frequency-selective Rayleigh fading with a perfect-channel state information and flat-channel shading. In OFDMA, the transmission capacity is given by

$$c_{m,n} = B \cdot \log_2 \left(1 + p_{m,n} \times \frac{|h_{m,n}|^2}{\sigma^2} \right). \tag{2.1}$$

Variable σ^2 in Equation (2.1) is the additive white Gaussian noise on the relevant channel, and $h_{m,n}$ is the complex fading suffered by a user m on a subcarrier with $m = 1, \ldots, M$ and $n = 1, \ldots, N$. The variable $p_{m,n}$ represents the transmission power for user m on subcarrier n. The requirement of QoS is regarded as the satisfaction of the required minimum transmission rate. Thus, the QoS requirement for m users can be expressed as

$$\sum_{n=1}^{N} c_{m,n} \cdot x_{m,n} \geq \overline{C}_m, \quad \forall m, \tag{2.2}$$

where $\overline{C}_m$ represents the minimum transmission speed that must be satisfied to serve user m. The binary $x_{m,n} \in \{0, 1\}$ is an indicator of the allocation index, for example, $x_{m,n} = 1$ means that subcarrier n has been allocated to user m otherwise $x_{m,n} = 0$.

Our previous works aim to improve the energy emitted in the downlink of a BS as it consumes the major part of energy consumption of the system, with results shown in Figure 2.1. Therefore, it is important to optimize the energy emitted from the BS. The objective here is formulated as the maximization of EE. The power consumption at the BS is a linear model, given by

$$P_{BS} = \varepsilon_0 \cdot P_{tx} + P_0 \tag{2.3}$$

In Equation (2.3), coefficient ε_0 is the scale factor of the radiated power, which is closely related to feeder losses and amplifier inefficiency. P_0 is the power offset, which represents battery backup, signal processing, and so on, and it is not associated with the radiated power.

In our formulation, we use the standard metric for the EE, i.e., "bit/J." By the integration of both QoS and EE, we formulate a maximization problem as follows:

$$EE = \frac{\sum_{n=1}^{N} \sum_{m=1}^{M} c_{m,n} \cdot x_{m,n}}{\varepsilon_0 \sum_{n=1}^{N} \sum_{m=1}^{M} p_{m,n} \cdot x_{m,n} + P_0}, \tag{2.4}$$

subject to

$$\sum_{n=1}^{N} c_{m,n} \cdot x_{m,n} \geq \overline{C}_m, \quad \forall m, \tag{2.5}$$

$$\sum_{n=1}^{N} \sum_{m=1}^{M} p_{m,n} \cdot x_{m,n} \leq P_T, \tag{2.6}$$

$$\sum_{m=1}^{M} x_{m,n} = 1, \quad \forall n, \tag{2.7}$$

$$x_{m,n} \in \{0,1\}, \quad p_{m,n} \geq 0, \quad \forall n, \forall m. \tag{2.8}$$

In the problem (2.4)–(2.8), constraint (2.6) bounds the radiated power from BS not to be larger than the total transmission power P_T. In addition, constraints (2.7) and (2.8) follow the principal law of OFDMA, enabling only one user on a subcarrier to avoid *cochannel interference*.

2.5.2 RESULTS

We have used similar settings as in our earlier studies (Ting et al., 2013; Chien et al., 2013) in a small-scale OFDMA system. To search for the power allocation values, we utilize the GA and present a characteristic simulation result. This optimization process by GA is quite efficient in a manner that only a few iterations are required to bring the solution close to the optimal value. As shown in Figure 2.3, the EE is the negative of the fitness value, measured in a unit of bit/J. From this figure, after just five iterations, the solution is close enough to the optimal EE, which is 2240.7862 bit/J. Such heuristic approaches are straightforward for obtaining feasible solutions, compared to the tedious analytical solution (Ting et al., 2014).

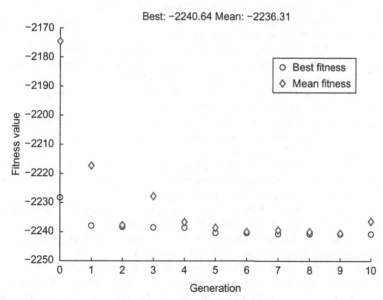

FIGURE 2.3

Convergence characteristics on the OFDMA model via a GA approach.

2.6 CONCLUSIONS

In this chapter, we have reviewed the relative extensive literature concerning the applications that involved the use of bio-inspired algorithms in the field of telecommunications, with a focus on GAs. However, it is worth pointing out that GAs are not the only class of bio-inspired algorithms used in telecommunications. In fact, the emphasis of this book is to provide a timely review of other bio-inspired computational methods in telecommunications.

Many areas in the telecommunication sectors are covered and discussed in general whereby the bio-inspired algorithms are potentially useful and powerful. Lastly, we hope that this content provides a good starting point for those venturing into the telecommunication fields, with emphasis on the energy-efficiency consideration. Our results show that near-optimal solutions can be easily obtained by bio-inspired algorithms. With some improvements, better results and even globally optimal solutions can be obtained with higher speed and higher accuracy. All these challenging topics will be investigated further in the rest of this book.

REFERENCES

Abuali, F.N., Schoenefeld, D.A., Wainwright, R.L., 1994. Terminal assignment in a communications network using genetic algorithms. In: Proceedings of the 22nd Annual ACM Computer Science Conference (CSC'94), Phoenix, AZ, USA, pp. 74–81.

Ali, M., Ramamurthy, B., Deogun, J.S., 2000. Routing and wavelength assignment with power considerations in optical networks. Comput. Netw. 32, 539–555.

Anastasopoulos, M.P., Arapoglou, P.-D.M., Kannan, R., Cottis, P.G., 2008. Adaptive routing strategies in IEEE 802.16 multi-hop wireless backhaul networks based on evolutionary game theory. IEEE J. Sel. Areas Commun. 26 (7), 1218–1225.

Bae, C., Stark, W., 2009. End-to-end energy–bandwidth tradeoff in multihop wireless networks. IEEE Trans. Inf. Theory 55 (9), 4051–4066.

Balazs, K., Soproni, P.B., Koczy, L.T., 2013. Improving system reliability in optical networks by failure localization using evolutionary optimization. In: 2013 IEEE International Systems Conference (SysCon), pp. 394–399.

Banerjee, D., Mehta, V., Pandey, S., 2004. A genetic algorithm approach for solving the routing and wavelength assignment problem in WDM networks. In: International Conference in Networks, ICN'04, Pointe-à-Pitre, Guadeloupe.

Buriol, L.S., Resende, M.G.C., Ribeiro, C.C., Thorup, M., 2005. A hybrid genetic algorithm for the weight setting problem in OSPF/IS-IS routing. Networks 46 (1), 36–56.

Calégari, P., Guidec, F., Kuonen, F., Kobler, D., 1997. Parallel island-based genetic algorithm for radio network design. J. Parallel Distrib. Comput. 47 (1), 86–90.

Chardaire, P., Kapsalis, A., Mann, J.W., Rayward-Smith, V.J., Smith, G.D., 1995. Applications of genetic algorithms in telecommunications. In: Proceedings of the 2nd International Workshop on Applications of Neural Networks to Telecommunications (IWANNT'95), Stockholm, Sweden, pp. 290–299.

Chen, T., Zhang, H., Zhao, Z., Chen, X., 2010. Towards green wireless access networks. In: 2010 5th IEEE International ICST Conference on Communications and Networking in China (CHINACOM), pp. 1–6.

Chien, S.F., Ting, T.O., Yang, X.S., Ting, A.K.N., Holtby, D.W., 2013. Optimizing energy efficiency in multi-user OFDMA systems with genetic algorithm. In: 2013 International Conference on Advances in Computing, Communications and Informatics (ICACCI), pp. 1330–1334.

Chong, H.W., Kwong, S., 2003. Optimization of spare capacity in survivable WDM networks. In: GECCO, pp. 2396–2397.

Coello, C.A.C., 1999. A comprehensive survey of evolutionary based multiobjective optimization techniques. Knowl. Inf. Syst. 1 (3), 269–308.

Coello, C.A.C., Van Veldhuizen, D.A., Lamont, G.B., 2002. Evolutionary Algorithms for Solving Multi-Objective Problems. Kluwer Academic Publishers, New York, pp. 89–102. 3 Chapra, Computer Methods, Spring.

Crisan, C., Muhlenbein, H., 1998. The breeder genetic algorithm for frequency assignment. In: Proceedings of the 5th International Conference on Parallel Problem Solving from Nature (PPSN V), Amsterdam, The Netherlands, pp. 897–906.

Cui, S., Goldsmith, A., Bahai, A., 2004. Energy-efficiency of MIMO and cooperative MIMO techniques in sensor networks. IEEE J. Sel. Areas Commun. 22 (6), 1089–1098.

Cvijetic, N., 2012. OFDM for next-generation optical access networks. J. Lightwave Technol. 30 (4), 384–398.

Davis, L., Coombs, S., 1989. Optimizing network link sizes with genetic algorithms. In: Elzas, M.S., Oren, T.I., Zeigler, B.P. (Eds.), Modelling and Simulation Methodology. Elsevier Science Publishers B.V., North Holland, ISBN: 0444880445, pp. 317–331 (Chapter IV.3).

Deeter, D.L., Smith, A.E., 1998. Economic design of reliable networks. IIE Trans. 30 (12), 1161–1174.

EARTH, 2010. Most promising tracks of green network technologies. https://bscw.ictearth.eu/pub/bscw.cgi/d31509/EARTHWP3D3.1.pdf. Accessed on June 2013.

Ewin, 2008. Energy-efficient wireless networking. http://www.wireless.kth.se/research/projects/19-ewin. Accessed on June 2013.

Ezri, D., Shilo, S., 2009. Green cellular-optimizing the cellular network for minimal emission from mobile stations. In: IEEE International Conference on Microwaves, Communications, Antennas and Electronics Systems, COMCAS 2009, pp. 1–5.

Fehske, A., Richter, F., Fettweis, G., 2009. Energy efficiency improvements through micro sites in cellular mobile radio networks. In: 2009 IEEE GLOBECOM Workshops, pp. 1–5.

Feng, D.Q., Jiang, C.Z., Lim, G.B., Cimini, L.J., Feng, G., Geoffrey Li, Y., 2013. A survey of energy-efficient wireless communications. IEEE Commun. Surv. Tutor. 15 (1), 167–178.

Gartner, 2007. Green IT the new industry shockwave. In: Symposium/ITXPO Conference.

Grant, P., 2010. MCVE Core 5 Programme, Green radio—the case for more efficient cellular base stations. In: Globecom'10.

Hanly, S., Mathar, R., 2002. On the optimal base-station density for CDMA cellular networks. IEEE Trans. Commun. 50 (8), 1274–1281.

Hoelting, C.J., Schoenefeld, D.A., Wainwright, R.L., 1996. Finding investigator tours in telecommunication networks using genetic algorithms. In: Proceedings of the ACM Symposium on Applied Computing (SAC'96), Philadelphia, PA, USA, pp. 82–87.

Hsinghua, C., Premkumar, G., Chu, C., 2001. Genetic algorithms for communications network design—an empirical study of the factors that influence performance. IEEE Trans. Evol. Comput. 5 (3), 236–249.

Kampstra, P., Van der Mei, R.D., Eiben, A.E., 2006. Evolutionary computing in telecommunication network design: a survey. http://www.few.vu.nl/~mei/publications.php. Accessed on June 2013.

Karabudak, D., Hung, C., Bing, B., 2004. A call admission control scheme using genetic algorithms. In: Proceedings of the ACM Symposium on Applied Computing (SAC'04), pp. 1151–1158.

Khanbary, L.M.O., Vidyarthi, D.P., 2009. Reliability-based channel allocation using genetic algorithm in mobile computing. IEEE Trans. Veh. Technol. 58 (8), 4248–4256.

Kolios, P., Friderikos, V., Papadaki, K., 2009. Ultra low energy storecarry and forward relaying within the cell. In: IEEE 70th Vehicular Technology Conference Fall (VTC 2009-Fall), pp. 1–5.

Konak, A., Smith, A.E., 2004. Capacitated network design considering survivability: an evolutionary approach. Eng. Optim. 36 (2), 189–205.

Kozdrowski, S., Pioro, M., Arabas, J., Szczesniak, M., 1997. Robust design of multicommodity integral flow networks. In: Proceedings of the 7th International Conference on Genetic Algorithms (ICGA'97). Michigan State University, East Lansing, MI, pp. 607–614.

Krzanowski, R.M., Raper, J., 1999. Hybrid genetic algorithm for transmitter location in wireless networks. Comput. Environ. Urban. Syst. 23, 359–382.

Kumar, A., Pathak, R.M., Gupta, Y.P., Parsaei, H.R., 1995. A genetic algorithm for distributed system topology design. Comput. Ind. Eng. 28 (3), 659–670.

Lo, K.K., Chien, S.F., Chieng, D.H.T., Lim, H.S., Kwong, K.H., 2014. Proportional resource allocation for OFDMA system. In: IEEE TENSYMP 2014, Region 10 Symposium, Kuala Lumpur, Malaysia.

Mann, J.W., Rayward-Smith, V.J., Smith, G.D., 1995. Telecommunications traffic routing: a case study in the use of genetic algorithms. In: Proceedings of the Applied Decision Technologies (ADT'95), London, pp. 315–325.

Marsan, M., Chiaraviglio, L., Ciullo, D., Meo, M., 2009. Optimal energy savings in cellular access networks. In: IEEE International Conference on Communications Workshops, pp. 1–5.

Meshkati, F., Poor, H., Schwartz, S., Balan, R., 2009. Energy-efficient resource allocation in wireless networks with quality-of-service constraints. IEEE Trans. Commun. 57 (11), 3406–3414.

Miao, G., Himayat, N., Li, Y., Swami, A., 2009. Cross-layer optimization for energy-efficient wireless communications: a survey. Wirel. Commun. Mob. Comput. 9 (4), 529–542.

Morelli, M., Kuo, C.-C.J., Pun, M.O., 2007. Synchronization techniques for orthogonal frequency division multiple access (OFDMA): a tutorial review. Proc. IEEE 95 (7), 1394–1427.

Mostafa, M.E., Eid, S.M.A., 2000. A genetic algorithm for joint optimization of capacity and flow assignment in packet switched networks. In: Proceedings of the 17th National Radio Science Conference, pp. C5-1–C5-6.

Mutafungwa, E., 2002. GORA: an algorithm for designing optical cross-connect nodes with improved dependability. Comput. Commun. 25 (16), 1454–1464.

Ngo, C.Y., Li, V.O.K., 1998. Fixed channel assignment in cellular radio networks using a modified genetic algorithm. IEEE Trans. Veh. Technol. 47 (1), 163–171.

Oh, E., Krishnamachari, B., 2010. Energy savings through dynamic base station switching in cellular wireless access networks. In: GLOBECOM 2010, IEEE Global Telecommunications Conference, pp. 1–5.

Optimising Power Efficiency in mobile RAdio Networks (OPERANet), 2010. OPERA-Net PROJECT STAND # 42. In: 2010 NEM Summit Towards Future Media Internet, Barcelona, Spain. http://opera-net.org/Documents/5026v1Opera-Nete-NEM%20Event%20Barcelona%202010Demos%20Presentation290710.pdf. Accessed on June 2013.

Richter, F., Fettweis, G., 2009. Energy efficiency aspects of base station deployment strategies for cellular networks. In: 2009 IEEE 70th Vehicular Technology Conference Fall (VTC 2009 Fall), pp. 1 5.

Richter, F., Fettweis, G., 2010. Cellular mobile network densification utilizing micro base stations. In: 2010 IEEE International Conference on Communications (ICC), pp. 1–6.

Richter, F., Fehske, A., Marsch, P., Fettweis, G., 2009. Traffic demand and energy efficiency in heterogeneous cellular mobile radio networks. In: 2010 IEEE 71st Vehicular Technology Conference (VTC 2010-Spring), pp. 1–6.

Routen, T., 1994. Genetic algorithm and neural network approaches to local access network design. In: Proceedings of the 2nd International Symposium on Modelling, Analysis & Simulation of Computer & Telecommunication Systems (MASCOTS'94), Durham, NC, USA, pp. 239–243.

Rubio-Largo, A., Vega-Rodriguez, M.A., Gomez-Pulido, J.A., Sanchez-Pérez, J.M., 2012. A comparative study on multiobjective swarm intelligence for the routing and wavelength assignment problem. IEEE Trans. Syst. Man Cybern. C Appl. Rev. 42 (6), 1644–1655.

Rubio-Largo, A., Vega-Rodriguez, M.A., Gomez-Pulido, J.A., Sanchez-Perez, J.M., 2013. Multiobjective metaheuristics for traffic grooming in optical networks. IEEE Trans. Evol. Comput. 17 (4), 457–473.

Schaffer, J.D., 1984. Multiple Objective Optimization with Vector Evaluated Genetic Algorithms (Ph.D. thesis). Vanderbilt University, Nashville, TN.

Sherif, M.R., Habib, I.W., Nagshineh, M., Kermani, P., 2000. Adaptive allocation of resources and call admission control for wireless ATM using genetic algorithms. IEEE J. Sel. Areas Commun. 18 (2), 268–282.

Sinclair, M.C., 1995. Minimum cost topology optimisation of the COST 239 European optical network. In: Proceedings of the 2nd International Conference on Artificial Neural Networks and Genetic Algorithms (ICANNGA'95), Alés, France, pp. 26–29.

Sobotka, M., 1992. Network design: an application of genetic algorithms. In: Proceedings of the 1992 Applied Defense Simulation Conference. ISBN: 1-56555-005-6, pp. 63–66.

Suarez, L., Nuaymi, L., Bonnin, J.M., 2012. An overview and classification of research approaches in green wireless networks. EURASIP J. Wirel. Commun. Netw. 2012, 142.

Suneel Varma, D., Subhashini, K.R., Lalitha, G., 2013. Design optimization analysis for multi constraint adaptive antenna array using harmony search and differential evolution techniques. In: 2013 IEEE Conference on Information & Communication Technologies (ICT), pp. 647–651.

Tan, L.G., Sinclair, M.C., 1995. Wavelength assignment between the central nodes of the COST 239 European optical network. In: Proceedings of the 11th UK Performance Engineering Workshop, Liverpool, UK, pp. 235–247.

Tanaka, Y., Berlage, O., 1996. Application of genetic algorithms to VOD network topology optimization. IEICE Trans. Commun. E79-B (8), 1046–1053.

Tang, K., Ko, K., Man, K.F., Kwong, S., 1998. Topology design and bandwidth allocation of embedded ATM networks using genetic algorithm. IEEE Commun. Lett. 2 (6), 171–173.

Teo, C.F., Foo, Y.C., Chien, S.F., Low, A.L.Y., Castañón, G., 2005. Wavelength converters placement in all optical networks using particle swarm optimization. Photon Netw. Commun. 10 (1), 23–37.

Thompson, D.R., Bilbro, G.L., 2000. Comparison of a genetic algorithm with a simulated annealing algorithm for the design of an ATM network. IEEE Commun. Lett. 4 (8), 267–269.

Ting, T.O., Chien, S.F., Yang, X.S., Ramli, N., 2013. Resource allocation schemes in energy efficient OFDMA system via genetic algorithm. In: 2013 19th Asia-Pacific Conference on Communications (APCC), pp. 723–727.

Ting, T.O., Chien, S.F., Yang, X.S., Lee, S., 2014. Analysis of QoS-aware OFDMA system considering energy efficiency. IET Commun. 8 (11), 1947–1954.

Wang, Z., Shi, B., Zhao, E., 2001. Bandwidth-delay-constrained least-cost multicast routing based on heuristic genetic algorithm. Comput. Commun. 24, 685–692.

Wong, I.C., Shen, Z.K., Evans, B.L., Andrews, J.G., 2004. A low complexity algorithm for proportional resource allocation in OFDMA systems. In: IEEE Workshop on Signal Processing Systems (SIPS), pp. 1–6.

Yang, X.S., 2014. Nature-Inspired Optimization Algorithms. Elsevier, Netherlands.

Yang, X.S., Koziel, S., 2011. Computational Optimization and Applications in Engineering and Industry. Springer, Heidelberg.

Zarakovitis, C.C., Ni, Q., 2013. Energy efficient designs for communication systems: resolutions on inverse resource allocation principles. IEEE Commun. Lett. 17 (12), 2264–2267.

Zhang, Q., Sun, J., Xiao, G., Tsang, E., 2007. Evolutionary algorithms refining a heuristic: hyper-heuristic for shared-path protections in WDM networks under SRLG constraints. IEEE Trans. Syst. Man Cybern. B 37 (1), 51–61.

Zhang, C., Zhang, T., Zeng, Z., Cuthbert, L., Xiao, L., 2010. Optimal locations of remote radio units in comp systems for energy efficiency. In: 2010 IEEE Vehicular Technology Conference Fall (VTC 2010-Fall), pp. 1–5.

Firefly Algorithm in Telecommunications

3

Mario H.A.C. Adaniya[1,2], Luiz F. Carvalho[2], Bruno B. Zarpelão[2],
Lucas D.H. Sampaio[3], Taufik Abrão[2], Paul Jean E. Jeszensky[3], and
Mario Lemes Proença, Jr.[2]

[1]Department of Computer Science, Centro Universitário Filadelfia (Unifil), Londrina, Brazil
[2]Department of Computer Science, State University of Londrina (UEL), Londrina, Brazil
[3]Polytechnic School, University of São Paulo, São Paulo, Brazil

CHAPTER CONTENTS

3.1 INTRODUCTION

Swarm intelligence and nature-inspired algorithms have become increasingly popular, and various types of animals and their social behavior inspire it. A classic algorithm in this category is ant colony optimization (ACO), presented by Dorigo in the late '90s, based on the behavior of the ant and the pheromone trail created by it. Basically, the ants randomly leave the nest in search of food, and in their path they leave a chemical component known as pheromone. As the ants use more a specific path than another one, they increase and strengthen the pheromone concentration. If the path is too long, the pheromone will slowly decrease the concentration until the path is no longer used and the shortest path from the nest to the food source is constructed (Dorigo et al., 1999). ACO is an algorithm included by the swarm intelligence technique.

The swarm intelligence field deals with artificial systems composed of many individuals that are coordinated in a decentralized manner and that have the property of self-organization. The field is inspired by collective behaviors that result from local interactions of individuals with each other and with their environment. Other examples of systems studied by swarm intelligence are schools of fish, flocks of birds, and herds of land animals, among others, which inspired a number of algorithms: ant algorithms, particle swarm optimization (PSO), bee algorithm, genetic algorithm (GA), and others.

The firefly algorithm (FA) is also a swarm-intelligence-based algorithm; Yang developed the algorithm based on the behavior of fireflies and the flashing light patterns. The algorithm essentially works based on two factors: the distance between the fireflies and the light intensity. As the distance between the fireflies increases, the light intensity decreases. The opposite is also true: as the distance decreases, more light intensity is felt. Therefore, the output of the algorithm can be the locations where the fireflies' population groups the positions or the values of the brightest firefly, depending on the modeling of the problem. Primarily, the modeling of the problem is linked with the problem of fireflies' attraction, which is intrinsically connected to the problem of light intensity. Thus, to implement FA in any area, these two issues need to be defined (Yang, 2008).

The use of a metaheuristic for a problem is acceptable considering the practical time in which it can find a feasible solution. It might not guarantee the best solution, but it can guarantee the nearly optimal solution, because, in some problems, it is impossible to search for every combination in a reasonable time. Two technical features comprise any metaheuristic algorithm: exploration and exploitation. Exploration is the action of exploring the global space, generating diverse solutions at global scale; and exploitation is the use of local and global information to construct the best solution. The balance between these two characteristics is important. If exploitation takes the lead, it can accelerate the process and respond with a local optimum. However, if exploration is the most used, the probability of finding the global optimum increases, but this slows down convergence. In a classical FA, the diversification of the solution is carried out by the randomization term. Usually, a random walk is applied, but more elaborated techniques such as Lévy flights can be adjusted. The global best is not used, because the fireflies exchange information among the adjacent fireflies, and this is realized by the attraction function related to the fireflies' brightness (Yang, 2012).

Since FA was presented by Yang, several works have explored the FA advantages in many areas. In Fister et al. (2013), the authors presented a survey of some areas and applications of FA, extracting three major areas: optimization, classification, and engineering applications. Optimization concentrates on combinatorial, constraint, continuous, and multiobjective problems. When the solution involves machine learning, data mining, and neural networks, it is a classification problem. Engineering applications, on the other hand, covers image processing, industrial optimization, wireless sensor networks, robotics, semantic web, chemistry, and others.

Regarding the optimization area, in continuous optimization, the classical benchmark's functions are used to measure FA efficiency, and usually it is compared to another metaheuristic such as PSO, ACO, or GA, among others. For further details, see Yang, 2008; Yang, 2009; Yang, 2011; and Yang and He, 2013. Another application of FA in continuous and multiobjective optimization is found in Gandomi et al. (2011). In this application, the authors use FA to solve mixed-variable structural optimization problems, where they achieve very efficient results, although oscillatory behavior was observed. The authors point out an important issue in conclusion: optimization algorithms should be tested independent of their numerical implementation, because of the random nature of metaheuristic algorithms. For example, if an algorithm is implemented using a vectorized implementation and another uses regular "FOR" loops, the first algorithm might run faster than the other.

In Sayadia et al. (2013), the authors proposed the discrete firefly algorithm (DFA) for solving discrete optimization problems. In the test problems presented, the DFA exhibited good results, implying that the DFA is potentially more powerful in solving nondeterministic polynomial-time hard (NP-hard) problems. Another work developed a discrete version of FA to solve NP-hard scheduling problems (Sayadia et al., 2010). The flow-shop scheduling problem (FSSP) is a complex combinatorial optimization problem, in which there is a set of n jobs $(1, \ldots, n)$ to be processed in a set of m machines $(1, \ldots, m)$ in the same order. The author compared the FA with ACO, and the preliminary results showed a better performance of FA.

Clustering is another area where FA can be employed. Clustering is a technique that has the characteristic of grouping data objects into clusters and finding hidden patterns that may exist in data sets. It is usually connected to classification problems. Through clustering techniques, it is possible to take an amount of data, find groups and classifications, and extract information to infer better conclusions about the problem. In Senthilnath et al. (2011), the authors present the FA compared with artificial bee colony (ABC) and PSO in 13 benchmark data sets from the University of California, Irvine (UCI) Machine Learning Repository (Bache and Lichman, 2013).

The performance is evaluated using the classification error percentage (CEP) and classification efficiency. The authors conclude that FA can be implemented to generate optimal cluster centers. Still in the classification, in Rajini and David (2012), the authors used FA to optimize the parameters encoded in the neural tree, with FA obtaining better results than PSO.

In the engineering application group, many areas used FA to solve problems in image processing, industrial optimization, and antenna design, among others. In Noor et al. (2011), a multilevel thresholding using Otsu's method based on FA is developed. The experimental results show that the Otsu-FA produced good separation of DNA bands and its background. In Hu (2012), the author uses FA in simulations to solve an energy-efficient mathematical model. It compares FA with other classical methods and demonstrates a better performance for FA, raising the consideration of putting it into practical use. In Dos Santos et al. (2013), the authors use FA to weight an ensemble of rainfall forecasts from daily precipitation simulations with the developments on the Brazilian Regional Atmospheric Modeling System (BRAMS).

This structure of this chapter is as follows: Section 3.2 discusses the foundations of the FA. Section 3.3 brings the FA applied in traffic characterization, where the FA is used to cluster. Section 3.4 provides examples of applications in wireless cooperative networks. The chapter concludes with a brief overview of the performance of FA, and recommendations for future works.

3.2 FIREFLY ALGORITHM

Designed by Yang in 2008 (Yang, 2008), the FA was developed based on the behavior of fireflies and the behavior of light emitted by them. Some firefly species use the flashes to attract partners for mating, as a security mechanism. Other species, for example, *Photuris*, use the flashes to attract future prey. The females *Photuris* mimic the pattern of flashes emitted by other species, which is unique and particular (Yang, 2009).

To develop the algorithm, Yang used the assumption of the following three rules:

- All fireflies are unisex and can attract and be attracted.
- Attractiveness is proportional to the brightness by moving the firefly fainter toward the brightest.
- The brightness is directly linked to the function of the problem treated.

Two important issues must be addressed: the variation of light intensity and the formulation of attractiveness. Through the problem modeling, it is possible to implement the FA in the manner best suited to the problems. The author suggests a simplifying assumption that the attractiveness of a firefly is determined by its brightness, which in turn is associated with the objective function encoded. The pseudocode presented by Yang (2008) is written in Algorithm 1.

ALGORITHM 1

Firefly Algorithm
Objective function $f(x)$, $x = (x_1, \ldots, x_d)^T$
Initialize a population of fireflies x_i ($i = 1, 2, \ldots, m$)
Define light absorption coefficient γ
while ($t <$ MaxGeneration)
  for $i = 1 : m$ all n fireflies
    for $j = 1 : i$ all m fireflies
       Light intensity I_i at x_i is determined by $f(x_i)$
       if ($I_j > I_i$)
          Move firefly i toward j in all d dimensions
       end if
       Attractiveness varies with distance r via $\exp[-\gamma r]$
       Evaluate new solutions and update light intensity
    end for j
  end for i
Rank the fireflies and find the current best
end while
Postprocess results and visualization

It is known that light intensity L observed from a distance r obeys $L \propto 1/r^2$, where if the distance of the observer increases, the intensity of light is smaller, and if the distance decreases, the intensity of light felt is increased. The air also assists in the process of absorbing the light, reducing the intensity according to the distance, making it weaker. These two factors combined make the fireflies visible at limited distances. Yang (2009) based it on the Euclidean distance between two fireflies m and n written by equation

$$r_{mn} = \|x_m - x_n\| = \sqrt{\sum_{k=1}^{d} (x_{m,k} - x_{n,k})^2}. \tag{3.1}$$

The movement of attraction run by firefly m against the brightest firefly n is described by the equation

$$x_m = x_m + \beta_0 e^{-\gamma r_{mn}^2}(x_n - x_m) + \alpha \left(\text{rand} - \frac{1}{2} \right), \tag{3.2}$$

where the first variable is the own x_m, the second part of the equation is responsible for the attraction and convergence of the algorithm, and the last term aims to assist in escaping the local optima process, adding the randomness process proportional to α. The **rand** is a random number generated from a uniform distribution, [0,1]. γ is the absorption coefficient of light. Yang adopts the values of $\beta_0 = 1$ and $\alpha \in [0,1]$. The particles in the FA have a setting visibility and versatile range of attractiveness, resulting in a high mobility for the search space, exploring better and with a capacity to escape local optima (Yang, 2008).

According to Yang (2008), the emission intensity of light from a firefly is proportional to the objective function, i.e., $I(x) \propto f(x)$, but the intensity with which light is perceived by the firefly decreases with the distance between the fireflies. Thus, the perceived intensity of a firefly is given by

$$I(r) = I_0 e^{\gamma r^2},\tag{3.3}$$

where I_0 is the intensity of light emitted; r is the Euclidean distance between i and j firefly, i being the bright and j the less bright; and γ the absorption coefficient. The attractiveness β of a firefly is defined by

$$\beta(r) = \beta_0 e^{-\gamma r^2}.\tag{3.4}$$

The agents in the FA have an adjustable visibility and are more versatile in attractiveness variations, which usually leads to higher mobility, and thus the search space is explored more efficiently, getting a better ability to escape the local minimum or maximum (Yang, 2008).

One special case is when $\gamma \to 0$, corresponding to the same efficiency as the accelerated PSO, a PSO variant, where the attractiveness is constant $\beta = \beta_0$, turning into a sky where the light intensity does not decrease and a flashing firefly can be seen anywhere in the region. On the other hand, when $\gamma \to \infty$, it leads to almost zero attractiveness for a given firefly performing a random search, because the fireflies can't see the others and will roam in a completely random way. The FA is in between these two extreme cases, and by adjusting the parameters γ and α, it is possible outstand the results of PSO and random search (Yang, 2008).

The ability to divide into subgroups is a great advantage for the FA, making each group swarm around the space, increasing the probability of finding the best global solution and the other optimal solutions simultaneously. As long as the population size is high enough, it is possible to deal with multimodal problems (Yang and He, 2013).

3.2.1 ALGORITHM COMPLEXITY

As presented in Algorithm 1, the FA has one loop for iteration t and two inner loops, when exploring the population m. So the complexity at the extreme case is $O(tm^2)$. The iteration t has a linear cost for the algorithm, thus it is not very computationally costly. The population m could turn out to be a problem in computational cost, because it is a quadratic polynomial, but usually the population m is small, for example, $m = 50$. In case of a high m, it is possible to use an inner loop, ordering fireflies by brightness or attractiveness function. Thus, it is possible to obtain an algorithm in $O(mt\log(m))$ (Yang and He, 2013).

3.2.2 VARIANTS OF FIREFLY ALGORITHM

Since Yang presented the FA, many versions of the FA have appeared in the literature. In Fister et al. (2013), the authors created a classification based on the following aspects:

- Representation of fireflies (binary, real-valued)
- Population scheme (swarm, multiswarm)
- Evaluation of the fitness function
- Determination of the best solution (nonelitism, elitism)
- Moving fireflies (uniform, Gaussian, Lévy flights, chaos distribution)

These aspects are influenced according to the problem and the area. Basically, the authors presented two groups: modified and hybrid. For more information, consult Fister et al. (2013), because the authors present a good review and discuss the modifications in detail.

3.3 TRAFFIC CHARACTERIZATION

Due to the speed and efficiency provided for the exchange of information, networks were highlighted in today's media communications, improving the interaction between people and providing the creation of essential services for daily use. However, progress in the development of network technologies implies the increasing complexity of their management. It increments the responsibility of the administrator to detect anomalies and problems that can cause any significant impact on the quality of network services or interruption of these services provided to users. To this end, the network management addresses various issues to ensure the reliability, integrity, and availability of communication. Concerning this situation, the International Standardization Organization (ISO) has developed an architecture for network management. According to Hunt (1997), this model contains the following features:

- Fault management: Necessary to detect the deterioration of system operating conditions, errors in applications, or failure of hardware components. This area may also include activities like isolation and fixing troubles.
- Accounting management: Desirable for create network usage records. It can be used to organize the use of resources for applications and appropriate operators.
- Configuration management: Tasks related to this area should maintain records of software and hardware. Furthermore, configuration management should provide information on maintenance and inclusion, and update relationships between devices during the network operation.
- Performance management: Essential for testing and monitoring, as well as providing statistics on network throughput, response time, and availability of equipment. This information is also intended to ensure that the network operates according to the quality of service signed by its users.
- Security management: This area of management ensures the integrity of traffic and the safe network operation. For this purpose, it includes activities such as constant monitoring of available resources and access restrictions to prevent misuse of resources by legitimate users and agents without authorization.

In addition to the areas previously presented, the network management divides its activities into two distinct categories (Thottan and Chuanyi, 2003): The first is traffic monitoring, which counts all network events by conducting constant traffic observation. The second category is related to the control, which aims to adjust the network parameters to ensure its best performance.

Comprising the five management areas proposed by the ISO, the manager has great control over the network, being provided with relevant information to assist in the constant monitoring of traffic behavior, facilitating the analysis and recognition of various events that harm the correct functioning of the network. However, due to the amount of information conferred from this process, it is infeasible to manually perform diagnostics and create reports on the state of the network, because this task requires increasingly skilled human resources and more effort to be completed.

The network monitoring can be performed in two ways (Evans and Filsfils, 2007). The passive approach ascertains the transmitted content on the network without interfering in data flow, and therefore does not affect its performance. This methodology analyzes a large amount of information, making it a challenge to restrict the volume of data that is needed for management. In active monitoring, packet probes are inserted on traffic in order to analyze the network functioning. The difficulty of this procedure is adjusting the volume of inserted packets for testing and metrics desired, without the impairment of network resources.

After choosing how the network will be monitored, the network administrator should consider the process of collecting traffic information. The sources of information used for traffic monitoring are essential for the success of network management, because its efficiency is dependent on the data contained in these sources. Therefore, the higher the precision of data used for network behavior modeling, the better the network management. Thottan and Chuanyi (2003) establish the various sources from which the network data can be obtained. They are

- Network probes: Specialized tools like ping and traceroute, able to obtain specific performance parameters of the network, for example, delays of connections and packet loss.
- Statistics based on flows using packet filtering: This approach performs a sampling of IP packet flows, capturing the header information. The data derived from this collection are used for management, as they provide an extensive set of attributes of traffic, including source and destination IP addresses, source and destination ports, and flag signaling, among others.
- Data from routing protocols: Approaches that follow this pattern may obtain information about network events through the protocols that act directly on the forwarding tables. Such information can be used for analyses based on network topology, because it provides constant monitoring of network links.
- Data from network management protocols: These protocols provide statistics on the network traffic. They use variables that correspond to the traffic counters and are collected directly from devices. Traffic statistics are passively extracted from monitored network elements. Although the information derived from

this source may not directly represent a metric of network performance, it is widely used for characterization of traffic and, consequently, to detect anomalies on network traffic behavior. Simple Network Management Protocol (SNMP) is the most notorious example of this category.

For effective traffic monitoring, the data source should provide important information for observing network activity and to aid in recognition of unusual traffic behavior. Thus, there is the predominant use of data from counters and packet flow analysis protocols.

3.3.1 NETWORK MANAGEMENT BASED ON FLOW ANALYSIS AND TRAFFIC CHARACTERIZATION

The development of techniques that contribute to the management of networks had been almost exclusively focused on the SNMP protocol for a long time. Since its introduction, this monitoring protocol has sought to provide the best possible visibility of network behavior for its managers. As a result, several improvements were included, making this protocol become a reference in the network management field. However, along with the concomitant improvement of SNMP, there have been more changes in behavior and network usage over the past decades. The advancement of real-time applications such as voice over IP (VoIP), video conferencing, virtual private networks (VPNs), and cloud computing, for example, reinforced the idea that a new tool was necessary to understand the data flow, and managing the networks would be required.

In this manner, in response to the monitoring of emerging technologies, a data source based on the analysis of packet flows was established. Cisco Systems introduced this approach via NetFlow proprietary protocol, which became a standard for adoption in various devices from multiple vendors. Another protocol used for the same purpose is sFlow, created by InMon. It can be defined as an acronym of Sampling Flows. Its name distinguishes the most important difference from NetFlow: packet sampling. sFlow provide a simple flow export protocol, capable of operating in networks that may transfer rates of 10 to 100 Gbps. Finally, the Internet Protocol Flow Information Export (IPFIX), defined in RFC 3917 (Quittek et al.), aims to monitor the applications considered of significant importance today and for the future of IP networks. It was developed by the Internet Engineering Task Force (IETF), based on the NetFlow version 9 protocol.

These approaches group the packets that share the same characteristics, such as transport protocol, IP source, and destination addresses and ports (Claise). Each of these groups is called a "flow." Because they can be identified by common characteristics, flows can be easily related to an application, network device, or user.

Based on the NetFlow, sFlow, and IPFIX approaches, it is possible to observe that these export protocols present common elements used in the monitoring and collecting flow proceedings. All these abovementioned protocols perform the processes of observation, measurement, and collection of exported packets for generating and recording information of flows. These tasks are liable to be carried out due to the intrinsic element of traffic monitoring called "sensor." Sensors are designed to capture packets, identify which of them belongs to a certain flow through the connection tracking, and then submit the information to the collector. They are network equipment, switches, or routers generally; however, software can play this role because it simulates the functions of such devices.

Management based on flow analysis monitors traffic passively, so a large volume of information is generated. In this respect, it is crucial to select the traffic that should be checked, resulting in the location in the network where the sensor should operate to export its information. The simplest approach is to implement a sensor in a gateway, resulting in the analysis of all traffic generated by the network.

Flow analysis allows the administrator to know who, what, when, where, and how network traffic is being transmitted. When the behavior of the network is well known, management is improved and an audit trail can be established. This increased understanding of network usage reduces the risk in relation to disruption and allows full operation. Furthermore, improvements in the operation of networks generate lower costs, because they ensure better use of their infrastructure. Thereby, the act of characterizing the behavior of traffic and creating network usage profiles provides quantitative and qualitative traffic analysis, that is, what types of streams are transmitted. This analysis is important because it allows the verification of whether or not a particular usage policy is being performed. Moreover, it provides subsidies to enhance security and detect abuses, denial of service attacks, configuration issues, and events that may be affecting proper network operation.

The Traffic network is currently composed of cycles consisting of bursts that have particular characteristics of their use. These behaviors are directly affected by working hours and the workday period of people who use the network (Proença et al., 2005). The firefly harmonic clustering algorithm (FHCA) methodology recognizes these behaviors and their characteristics, creating a traffic digital signature called Digital Signature of Network Segment using Flow (DSNSF) analysis. Such a signature is responsible for harboring information about the traffic profile in a network segment.

The use of DSNSF along with network management provides significant information for the solution of attacks or failures. Looking for changes in the normal traffic profile, it is possible to identify anomalies such as a Distributed Denial of Service-(DDoS) attack, flash crowd, and equipment failure in a nearly real-time interval. Furthermore, using the historical traffic, behaviors of malicious events can be assimilated through the succession of changes that they cause to traffic attributes monitored. Moreover, DSNSF allows constant monitoring of the traffic due to its ability to provide information about the use of resources utilized by network services. This information assists the network administrator in quick decision making, promoting the network reliability and availability of the services offered by it (De Assis et al. 2013).

3.3.2 FIREFLY HARMONIC CLUSTERING ALGORITHM

In this section, we briefly describe the FHCA, which was introduced in Adaniya et al. (2011) and Adaniya et al. (2012) to volume anomaly detection. We make use of the FA to find the solution to Equation (3.5).

$$\text{KHM}(x, c) = \sum_{i=1}^{n} \frac{k}{\sum_{j=1}^{k} \frac{1}{\|x_i - c_j\|^p}} \qquad (3.5)$$

Equation (3.5) describes the K-harmonic means (KHM) proposed by Zhang (Zhang et al., 1999). The main idea is to calculate a weight function, $w(x_i)$, to recalculate the centers where each point presents a certain weight to the final result, and a membership function, $m(c_j|x_i)$, to determine the relationship strength of x_i in relation to center c_j. As to the FA, one important feature is that the emission intensity of light from a firefly is proportional to the objective function, i.e., $I(X) \propto \text{KHM}(X,C)$, but the intensity with which light is perceived by the firefly decreases with the distance between the fireflies.

The objective of the joint KHM and FA procedures is to minimize the problem of initialization observed in K-means (KM) and to escape local optimal solutions, resulting in a more efficient algorithm to cluster the network traffic samples, in order to efficiently detect volume anomalies from management information base (MIB) objects. We named the proposed algorithm FHCA. The pseudocode is presented in Algorithm 2.

ALGORITHM 2

Firefly Harmonic Clustering Algorithm
Input: Real traffic samples
Output DSNSF
Initialize a population of fireflies in random positions
Define light absorption coefficient γ
While ($i <$ Iteration)$\|$(error $<$ errorAcepted(KHM(x,c)))
Calculate the objective function according to equation (7)
    For $i = 1$ to M
        For $j = 1$ to M
            Light intensity I_i at x_i is determined by $f(x_i)$
            if($I_j > I_i$)
                Move firefly i toward j in all d dimensions
            end if
            Attractiveness varies with distance
            Evaluate new solutions and update light intensity
        end For j
    Compute the membership function (Equation (6))
    Compute the weight function (Equation (7))
    Recompute c_j location based on Equation (8)
    end For i
Rank the fireflies and find the current best
end while

The KHM calculate the membership function (3.6) describing the proportion of data point x_i that belongs to center c_j:

$$m(c_j, x_i) = \frac{\|x_i - c_j\|^{-p-2}}{\sum_{j=1}^{k} \|x_i - c_j\|^{-p-2}} \qquad (3.6)$$

Increasing the parameter p in the membership function gives more weight to the points close to the center.

The weight function (3.7) defines how much influence data point x_i has in recomputing the center parameters in the next iteration:

$$w(x_i) = \frac{\sum_{j=1}^{k} \|x_i - c_j\|^{-p-2}}{\left(\sum_{j=1}^{k} \|x_i - c_j\|^{-p}\right)^2}. \qquad (3.7)$$

After calculating the membership function (3.6) and weight function (3.7), the algorithm calculates the new center location described by

$$c_j = \frac{\sum_{i=1}^{n} m(c_j|x_i) w(x_i) x_i}{\sum_{i=1}^{n} m(c_j|x_i) w(x_i)}. \qquad (3.8)$$

The new center will be calculated depending on the x_i and its $m(c_j|x_i)$ and $w(x_i)$. If x_i is closer to c_1 and far from c_2, it will present a $m(c_1|x_1) > m(c_2|x_1)$, and calculating the new center c_1, x_1 will be more representative in the final answer for c_1 and less representative for c_2.

3.3.3 RESULTS

To demonstrate the versatility and power of the FHCA, we used IP flows collected from the State University of Londrina (Brazil) network to build the DSNSF. This network is composed of approximately 7000 interconnected devices. The data were collected specifically from the BD8801 Extreme switch located in the core of the network, which aggregates all the university traffic. Due to a large information volume, it is sampled periodically, at a rate of 1 every 256 packets. We used the Softflowd network analyzer application to export flows to the collector in the sFlow sampled format (Phaal et al.). The collector saves the exported flows in binary files every five minutes to be processed by netflow dump (NFDUMP) tools (Haag, 2004), so they can be used later for network management.

The FHCA approach was developed using the bl-7 methodology, introduced by Proença et al. (2006), in which a single signature is generated for each day based on the history of its previous weeks. For traffic characterization, flow records collected from the workdays of September, October, and November of 2012 were analyzed, and this data set was separated into two groups. The first four weeks were used by FHCA as historical information for signature creating, and the workdays of the last three weeks (from October 22 to November 9) were used for traffic characterization evaluation. The choice to analyze only the days from Monday to Friday was due to the high utilization of the network on weekdays.

To measure the results of the DSNSF generated by FHCA, we adopted the normalized mean square error (NMSE) (Poli and Cirillo, 1993). This measure assesses the absolute difference between what was predicted by DSNSF and what is actually checked by actual traffic. By submitting values from zero to infinity, it is possible to verify that results near zero indicate excellent characterization of traffic, while high values indicate differences between the analyzed time series, DSNSF, and traffic movement. Table 3.1 lists the results for parameters γ and α.

In Figures 3.1 and 3.2, the green area represents the Real Network Traffic, and the blue line represents the DSNSF generated by FHCA. Figure 3.1 is the bit/s attribute and Figure 3.2 is the packets/s. We can observe that the DSNSF follows the network traffic most of the time. Some curves are far apart, but that can be explained because of the trained data. In the weeks before, the data volume was bigger than on the analyzed day.

3.4 APPLICATIONS IN WIRELESS COOPERATIVE NETWORKS

The popularization of mobile devices over the last years has led to an exponential growth in data rate demands. While telecommunication companies try to quickly expand their infrastructure to sustain the new scenario, many research groups over the world search for new solutions to improve mobile network performance, aiming to surpass the current scarcity in high data rate connectivity.

Wireless cooperative networks are among many technologies that emerged in order to fulfill these new company and user requirements. These networks are often more energy efficient than noncooperative systems, because power transmission is usually not linearly proportional to the distance between transmitter and receiver. In cooperative systems, each mobile terminal (MT) communicates with its respective base station (BS) through a relay station (RS) that may or may not be fixed. Besides energy efficiency (EE), other characteristics may be improved when cooperative scenarios are considered, such as spectral efficiency (SE), system throughput, and average transmission power (ATP).

In order to further improve network performance, one may use the multicarrier direct sequence code division multiple access (MC-DS/CDMA) concept, combined with multicarrier techniques such as orthogonal frequency division multiplexing (OFDM). Such telecommunication systems are characterized by the division of the total available spectrum into parallel, uncorrelated, nonselective CDMA subchannels, which improves granularity and is responsible for EE and SE enhancements, as well as lowering ATP of all users.

Table 3.1 NMSE for γ and α

γ/α	0.1	0.2	0.3	0.4	0.5	0.6	0.7	0.8	0.9	1
0.1	−0.049951	−0.06422	−0.060599	0.024221	−0.049259	0.033243	−0.047401	−0.053061	−0.058005	−0.010508
0.2	−0.043553	4,02E−01	−0.055895	−0.055831	−0.064408	0.067646	0.0083596	−0.051075	0.060875	−0.031668
0.3	0.0033968	−0.049791	0.068385	−0.070639	−0.055541	−0.044137	−0.057763	−0.025838	0.096634	−0.021878
0.4	−0.052125	0.084218	0.042767	−0.023859	−0.032733	−0.041453	0.10259	−0.051751	−0.037255	−0.06894
0.5	−0.046851	0.018417	−0.054153	−0.042471	−0.052008	0.028284	−0.047379	−0.048312	−0.033935	−0.043238
0.6	−0.0079522	−0.051398	0.054074	0.11191	−0.028014	−0.045529	−0.038187	−0.056166	0.064252	0.073859
0.7	−0.052952	0.061395	−0.060511	0.010917	−0.033931	−0.017493	−0.030388	−0.063884	−0.017482	−0.051635
0.8	0.027346	−0.02482	0.1021	−0.05204	−0.059534	6,41E−02	−0.035252	0.0098384	−0.056196	−0.021615
0.9	0.078149	0.071539	−0.065944	−0.054288	0.088321	−0.022493	−0.053909	−0.041752	−0.035451	−0.059797
1	−0.031729	0.098344	−0.047243	−0.023439	−0.050635	−0.04335	−0.050585	−0.031538	−0.052756	−0.064354

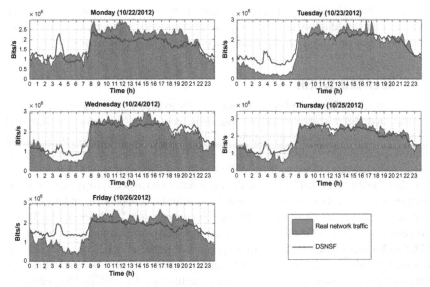

FIGURE 3.1

Real network traffic and DSNSF comparison for bits/s.

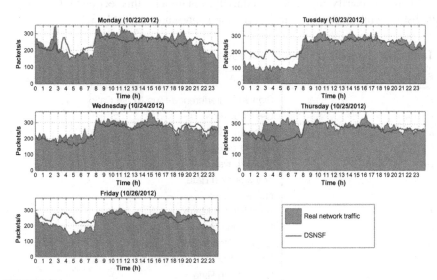

FIGURE 3.2

Real network traffic and DSNSF comparison for packets/s.

Because users may connect to the network to use different multimedia services such as voice calls, video calls, video streaming, data transfer, and so forth, the system must support a multirate scenario. MC-DS/CDMA networks may implement adaptive modulation or multiprocessing gain techniques in order to create such flexibility and support services from voice to high data rate ones.

The combination of all techniques presented above may prove able to solve the problems that would allow us to achieve the next network's generation, which aims to be *green* and profitable, and provide excellence in user experience.

3.4.1 RELATED WORK

Recently, many studies have been conducted to try to find a good resource allocation algorithm in different scenarios and systems. In this context, some works with analytical and heuristic solutions should be highlighted. In Abrão et al. (2011), the power control and throughput maximization problems in DS/CDMA networks were solved using the PSO algorithm, while in Sampaio et al. (2013), both problems were solved using the ACO algorithm. On the other hand, many analytical solutions have been proposed for the different resource allocation problems in interference-limited cooperative and noncooperative networks, such as Gross et al. (2011), Zaponne et al. (2011), Li et al. (2013), and Sampaio et al. (2014). In Gross et al. (2011), a distributed power control algorithm for the DS/CDMA system based on the Verhulst population growth mathematical model was presented. Besides, in Zaponne et al. (2011) and Sampaio et al. (2014), a game-theoretic approach for EE maximization was presented: the first one for DS/CDMA cooperative systems, and the second one for MC-DS/CDMA cooperative systems. Finally, Li et al. (2013) present an EE and SE trade-off analysis in interference limited networks.

In order to clarify and easily introduce the notation for this section of the chapter, the list of symbols used within this section is presented in Table 3.2.

Table 3.2 Symbols used in this section

Symbol	Meaning
δ	Signal-to-Interference-Plus-Noise Ratio
i	User Indexer
k	Subcarrier Indexer
j	Relay Indexer
p	Power Level
r	Transmission Rate
σ^2	Noise Power
$\boldsymbol{A}$	Amplification Matrix
F	Processing Gain
w	Channel Bandwidth
V	Bits Per Packet Transmitted
ℓ	Codification Rate
h	User-Relay Channel Power Gain
g	Relay-Base Station Channel Power Gain
ϱ	Power Amplifier Inefficiency
U	System Loading
N	Number of Subcarriers
R	Number of Relay Stations

3.4.2 SYSTEM MODEL AND PROBLEM STATEMENT

In MC-DS/CDMA networks, the total available spectrum, that is, the channel band-width B, is divided in N parallel uncorrelated flat CDMA subchannels (or subcarriers), each one with bandwidth w. Considering the uplink of such networks and a cooperative scheme, each user $i = 1, 2, \ldots, U$ in the system transmits to a relay that is responsible for forwarding the message to the respective BS. The equivalent baseband signal on each subchannel $k = 1, 2, \ldots, N$ received at the j-th RS may be mathematically described as

$$y_j(k) = \sum_{i=1}^{U} \sqrt{p_i(k)} h_{i,j}(k) b_i(k) s_i + \eta_j, \qquad (3.9)$$

where p is the transmission power; h is the complex slow and frequency nonselective channel gain between user and relay; b is the modulated symbol assumed constant during the symbol period; s is the spreading code with length $\mathcal{F}_i$ representing the processing gain; and η is the relay noise vector assumed to be additive white Gaussian noise (AWGN) with zero-mean and covariance matrix given by $\sigma^2 \mathbf{I}_U$.

Hence, the $U \times R$ uplink channel gain matrix considering path loss, shadowing, and fading effects at the k-th subcarrier is defined as

$$\mathbf{H}_k \triangleq \begin{bmatrix} h_{11}(k) & h_{12}(k) & \cdots & h_{1R}(k) \\ h_{21}(k) & h_{22}(k) & \cdots & h_{2R}(k) \\ \vdots & \vdots & \ddots & \vdots \\ h_{U1}(k) & h_{U2}(k) & \cdots & h_{UR}(k) \end{bmatrix}, \quad \text{with } k = 1, \ldots N, \qquad (3.10)$$

where h_{ij} is the channel gain between user i and relay j at the k-th subchannel. Similarly, the $R \times 1$ channel gains vector for the RS-BS hop may be mathematically defined as

$$\mathbf{g}_k \triangleq [g_1(k)\, g_2(k) \cdots g_R(k)]^\mathsf{T}, \quad \text{with } k = 1, \ldots, N, \qquad (3.11)$$

where g_i is the channel gain between the i-th RS and the BS.

In this work, we assume a simplified scenario with one fixed relay station (FRS). Figure 3.2 illustrates a macrocell environment and the positioning of the MTs, FRS, and BS. Note that because $R = 1$, $\mathbf{H}_k$ is a $U \times 1$ vector and $\mathbf{g}_k$ is a scalar. From now on, h_i will refer to the channel gain between user i and the FRS, while g^k will refer to the channel gain between FRS and BS at the k-th subcarrier.

According to the system configuration above and given the distance, natural obstacles between each MT and the BS, and the nonlinear growth of ATP with the distance, each MT must communicate with the BS through the FRS. Additionally, no interference from the MT's direct path to the BS is considered due to the hypothesis of great distance between the MT and BS.

In this scenario, the FRS is responsible for forwarding the MT message to the BS, which may be accomplished through three different protocols: decode-and-forward (DF), compress-and-forward (CF), and amplify-and-forward (AF). Because the AF protocol is the less complex protocol to implement and may operate with a constant transmission power (Zappone et al., 2011), in this work we have adopted AF protocol under small number of FRSs. Hence, the equivalent base-band signal received at the FRS is normalized by the square root of the average received power $\sqrt{P_N(k)} = \sqrt{\mathbb{E}\left[\|\mathbf{y}_k\|^2\right]}$. Considering noise and information symbols of each user to be uncorrelated, the normalized received power is equivalent to

$$P_N(k) = \sum_{i=1}^{U} p_i(k)|h_i(k)|^2 + \sigma^2 \mathcal{F}_i, \tag{3.12}$$

where i is the user indexer, k the subcarrier indexer, p is the transmission power level, $|h|^2$ is the channel power gain, σ^2 is the noise variance, and $\mathcal{F}$ is the processing gain.

After normalization, the received $\mathbf{y}(k)$ is amplified by the $U \times U$ amplification matrix $\mathbf{A}$ constrained by the total available transmission power p_R at the FRS, i.e., $\text{race}\left(\mathbf{A}\mathbf{A}^H\right) \leq p_R$, where $(\cdot)^H$ is the conjugated transposed matrix operator. Thus, the base-band signal received at the BS is

$$\mathbf{y}(k) = \frac{g^k}{\sqrt{P_N(k)}} \left(\sum_{i=1}^{U} \sqrt{p_i(k)} h_i(k) b_i(k) \mathbf{A}\mathbf{s}_i + \mathbf{A}\boldsymbol{\eta}_R \right) + \boldsymbol{\eta}_{BS}, \tag{3.13}$$

where $\boldsymbol{\eta}_R$ and $\boldsymbol{\eta}_{BS}$ are the FRS and BS thermal noise, respectively, assuming AWGN with covariance matrix $\sigma^2 \mathbf{I}_U$.

In multiple access systems limited by interference, a QoS measure of great importance is the signal-to-interference-plus-noise ratio (SINR), because all users transmit in the same spectrum at the same time. It causes what is known as multiple access interference (MAI). Considering the adoption of linear receivers, the postdetection SINR in the described scenario may be mathematically defined as

$$\delta_i(k) \triangleq \mathcal{F}_i \left(\frac{p_i(k)|h_i(k)|^2 |g(k)|^2 |\mathbf{d}_i^H \mathbf{A}\mathbf{s}_i|^2}{\mathcal{I}_i(k) + \mathcal{N}_i(k) + \sigma^2 |g(k)|^2 \|\mathbf{A}^H \mathbf{d}_i\|^2} \right), \tag{3.14}$$

where σ^2 is the thermal noise power and $\mathbf{d}$ is the linear filter at the receiver, such as single-user matched filter (MF) or multiuser decorrelator, minimum mean square error (MMSE) filter, and so on. According to the results in Souza et al. (2013), only the decorrelator filter will be considered here because among the linear multiuser detectors, it had the best results in terms of *performance-complexity* trade-off. The decorrelator filter is mathematically expressed as

$$\mathbf{D}_{DEC} = [\mathbf{d}_1, \ldots, \mathbf{d}_i, \ldots, \mathbf{d}_U] = \mathbf{S}(\mathbf{S}^T\mathbf{S})^{-1} = \mathbf{S}\mathbf{R}^{-1}, \tag{3.15}$$

where $\mathbf{d}_i$ is the linear filter for the i-th user, $\mathbf{S}$ is the spreading code matrix with each column representing a user spreading code, and $\mathbf{R}$ is the spreading sequence correlation matrix.

Additionally, the $\mathcal{I}_i(k)$ term in Equation (3.14) is the amplified MAI at the FRS that is forwarded to the BS and is mathematically defined as

$$\mathcal{I}_i(k) \triangleq |g(k)|^2 \sum_{j=1, j\neq i}^{U} p_j(k)|h_j(k)|^2 |\mathbf{d}_i^H \mathbf{A}\mathbf{s}_j|^2, \tag{3.16}$$

while the term $\mathcal{N}_i(k)$ in Equation (3.14) is the normalized thermal noise at the BS treated through the linear multiuser receiver, being defined as

$$\mathcal{N}_i(k) \triangleq \left(\sum_{i=1}^{U} p_i(k) |h_i(k)|^2 + \mathcal{F}_i \sigma^2 \right) \sigma^2 \|d_i\|^2. \tag{3.17}$$

Another important QoS measure that has to be considered is the bit error rate (BER) performance, which is directly related to the SINR. In multirate systems, users may deploy different modulation orders aiming to achieve different transmission rates under different channel propagation conditions. Indeed, higher order modulations enable the system to achieve higher data rates at the cost of an increase in BER if the SINR is constant. On the other hand, lower order modulations allow lower transfer rates with considerably better BER if SINR is a constant. Furthermore, considering M-QAM square constellation formats of order $M = M_i$, where i is the user indexer, and Gray coding, the BER for a given SINR $\delta_i(k)$ can be defined as (Simon and Alouini, 2004; Du and Swamy, 2010).

$$\text{BER}_i(k) = \frac{2(\sqrt{M_i} - 1)}{\sqrt{M_i} \log_2 M_i} \left(1 - \sqrt{\frac{3\delta_i(k) \log_2 M_i}{2(M-1) + 3\delta_i(k) \log_2 M_i}} \right). \tag{3.18}$$

3.4.2.1 Energy and spectral efficiencies

The SE of the MC-DS/CDMA system S_T may be computed as the sum of the SE of each user S_i throughout the N parallel subchannels. The SE of each user is defined by the number of bits per second transmitted in a single hertz of bandwidth. Hence, a theoretical bound for the system SE may be obtained considering the Shannon channel capacity equation

$$S_T = \sum_{i=1}^{U} S_i = \sum_{i=1}^{U} \sum_{k=1}^{N} \log_2[1 + \delta_i(k)] \left[\frac{\text{bits}}{\text{s Hz}} \right], \tag{3.19}$$

where i and k are the user and subcarrier indexers, respectively, and $\delta_i(k)$ is the postdetection SINR defined in Equation (3.14). Furthermore, still considering the Shannon capacity equation, the transmission rate of user i over the k-th subchannel is

$$r_i(k) = w \, \log_2[1 + \delta_i(k)] \left[\frac{\text{bits}}{\text{s}} \right]. \tag{3.20}$$

In MC-DS/CDMA systems with a single FRS, the practical EE of the system may be formulated as (Goodman and Mandayan, 2000)

$$\xi_T \triangleq \sum_{i=1}^{U} \xi_i \triangleq \sum_{i=1}^{U} \sum_{k=1}^{N} \frac{r_i(k) \ell_i(k) f(\delta_i(k))}{\varrho_i p_i(k) + \varrho_R p_R + p_C + p_{C_R}} \, (\text{bits/J}), \tag{3.21}$$

where $\ell_i(k) = L_i(k)/V_i(k) \leq 1$ is the codification rate, that is, the number of information bits $L_i(k)$ divided by the total number of transmitted bits $V_i(k)$. The power amplifier inefficiency for each user i is $\varrho_i \geq 1$, while the FRS power amplifier inefficiency is denoted by $\varrho_R \geq 1$. The powers p_R, p_C, and p_{C_R} are, respectively, the retransmission power deployed by the FRS, circuitry power at the MT transmitter, and the circuitry power at the FRS. Furthermore, the modulation efficiency function can be identified by the probability of error-free packet reception:

$$f(\delta_i(k)) = (1 - \text{BER}_i(k))^{V_i(k)}, \qquad (3.22)$$

where $\text{BER}_i(k)$ is defined in Equation (3.18). In this work, for simplicity of analysis, we have assumed that all users and subcarriers have the same packet size $V = 1$.

3.4.2.2 Problem statement

The EE maximization problem in MC-DS/CDMA cooperative networks consists in finding a power allocation policy for each user and each subcarrier that satisfies the minimum QoS requirements while maximizing the total EE as defined in Equation (3.21). Mathematically, we can formulate the following optimization problem:

$$\begin{aligned} \xi_T^* = \underset{\mathbf{P} \in \wp}{\text{maximize}} \ \ \xi_T \\ \text{s.t. c.1:} \quad \delta_i(k) \geq \delta_{i,\min}(k) \end{aligned}, \qquad (3.23)$$

where $\delta_{i,\min}(k)$ is the minimum SINR for each user and subcarrier associated with the minimum rate requirement, such that

$$\delta_{i,\min}(k) = \left[2^{\left(\frac{r_{i,\min}(k)}{w}\right)} - 1 \right]. \qquad (3.24)$$

Additionally, the transmission power of all users across the N subchannels must be bounded and nonnegative for any feasible power allocation policy, with $\mathbf{P}$ being the power allocation matrix defined as

$$\mathbf{P} \in \wp \triangleq \left\{ [p_i(k)]_{U \times N} \mid 0 \leq p_i(k) \leq p_{\max} \right\}, \qquad (3.25)$$

where $p_{\max}$ is the maximum transmission power per user subcarrier and $N \cdot p_{\max}$ is the maximum transmission power per MT transmitter.

In order to analytically solve the optimization problem in Equation (3.23), one may use the iterative Dinkelbach method (Dinkelbach, 1967) described below.

3.4.3 DINKELBACH METHOD

The Dinkelbach method is an iterative method to solve quasiconcave problems in a parameterized concave form (Dinkelbach, 1967). In order to apply the method to the optimization problem in Equation (3.23), it is reasonable to use a divide-and-conquer approach by solving N subproblems, which is equivalent to maximizing the EE at each subchannel $k = 1, \ldots, N$. Mathematically, solving (3.23) is equivalent to dealing with

$$\underset{0 \le p_i(k) \le p_{\max}}{\text{maximize}} \frac{\mathcal{C}(\boldsymbol{p})}{\mathcal{U}(\boldsymbol{p})} = \frac{\displaystyle\sum_{i=1}^{U} r_i(k)\ell_i(k)f(\delta_i(k))}{\varrho_R p_R + p_C + p_{C_R} + \displaystyle\sum_{i=1}^{U} \varrho_i p_i(k)}, \quad k = 1, \ldots, N. \tag{3.26}$$

$$\text{s.t. c.1:} \quad \delta_i(k) \ge \delta_{i,\min}(k), \ \forall i = 1, \ldots, U \text{ and } k = 1, \ldots, N$$

Note that each subproblem in (3.26) is a quasiconcave optimization problem (Sampaio et al., 2014). One may solve (3.26) through the associated parameterized concave program (Marques et al., 2013):

$$\underset{\boldsymbol{p} \in \mathcal{X}}{\text{maximize}} \ \mathcal{C}(\boldsymbol{p}) - \lambda \mathcal{U}(\boldsymbol{p}), \tag{3.27}$$

where $\mathcal{X} \triangleq \{\boldsymbol{p} \in \mathcal{X} | 0 \le p_i \le p_{\max} \ \forall i = 1, \ldots, U \}$. The objective function of the parameterized problem denoted by $\mathcal{Z}(\lambda) = \mathcal{C}(\boldsymbol{p}) - \lambda \mathcal{U}(\boldsymbol{p})$ is a convex, continuous, strictly decreasing function. The parameter λ is the EE at the last iteration of the Dinkelbach iterative process. Hence, without loss of generality, the maximum achievable EE of the parameterized problem is (Marques et al., 2013)

$$\lambda^* = \frac{\mathcal{C}(\boldsymbol{p}^*)}{\mathcal{U}(\boldsymbol{p}^*)} = \underset{\boldsymbol{p} \in \mathcal{X}}{\text{maximize}} \frac{\mathcal{C}(\boldsymbol{p})}{\mathcal{U}(\boldsymbol{p})}, \tag{3.28}$$

which is equivalent to finding $\mathcal{Z}(\lambda) = 0$. The Dinkelbach iterative method is basically the application of Newton's method to a nonlinear fractional program with super linear convergence rate (Schaible and Ibaraki, 1983). At each current iteration, EE value achieved with the power control policy found in the last iteration is deployed as the λ parameter. Therefore, on the inner loop of the method, another optimization problem arises, and can be formulated as

$$\mathcal{Z}(\lambda_n) = \underset{\boldsymbol{p} \in \mathcal{X}}{\max} \{\mathcal{C}(\boldsymbol{p}) - \lambda_n \mathcal{U}(\boldsymbol{p})\}, \tag{3.29}$$

where n is the iteration indexer. This updating process continues until some stopping criterion is reached: either the maximum number of iterations has been reached, or $\mathcal{Z}(\lambda)$ is smaller than a tolerance value ε. The algorithm for the Dinkelbach iterative method is described in Algorithm 1.

ALGORITHM 1

Dinkelbach's Method

Input: λ_0 satisfying $\mathcal{Z}(\lambda_0) \ge 0$; tolerance ϵ

Initialize: $n \leftarrow 0$

Repeat

  Solve problem (21), with $\lambda = \lambda_n$ to obtain $\boldsymbol{p}_n^*$

  $\lambda_{n+1} \leftarrow \dfrac{\mathcal{C}(\boldsymbol{p}_n^*)}{\mathcal{U}(\boldsymbol{p}_n^*)};$

  $n \leftarrow n + 1$

Until $|\mathcal{Z}(\lambda_n)| \le \epsilon$

Output: $\lambda_n; \boldsymbol{p}_n^*$

In order to expeditiously solve the inner-loop optimization problem in Algorithm 1, CVX tools were deployed (CVX Research, 2014).

3.4.4 FIREFLY ALGORITHM

In order to solve the original optimization problem presented in (3.23) using a heuristic optimization approach, such as the FA, a utility function should be defined. Because the problem (3.23) and the N subproblems presented in (3.26) are equivalent, the utility function may hold the same form as in (3.26), such that the achieved EE results from an analytical perspective and the heuristic approach are fairly compared. Therefore, the following utility function is proposed to solve the original EE maximization problem in (3.23):

$$J(P) = \sum_{k=1}^{N} \frac{\sum_{i=1}^{U} r_i(k)\ell_i(k)f(\delta_i(k))}{\varrho_R p_R + p_C + p_{C_R} + \sum_{i=1}^{U} \varrho_i p_i(k)},$$ (3.30)

where P is the power allocation matrix such that each row represents the transmission power for each user along all N subchannels. Hence, in order to guarantee a potential solution during each single algorithm iteration, the matrix P must satisfy the following requirements:

$$P_{i,k} = p_i(k) \leq p_{max} \text{ and } p_i(k) \geq 0.$$ (3.31)

The minimum SINR constraint in (3.23) is equivalent to a minimum transmission rate per user per subchannel. Note also that both expressions for the minimum transmission rate per user are equivalent:

$$r_{i,min} \leq w\sum_{k=1}^{N} \log_2(1+\delta_{i,min}(k)) \equiv \prod_{k=1}^{N}(1+\delta_{i,min}(k)) \geq 2^{(r_{i,min}/w)}.$$ (3.32)

It is noteworthy that each firefly for this problem will be identified as a power allocation matrix P and not an array, as commonly used in the literature and the term x_m on Equation (3.10). Also, for this problem, the Euclidean distance between two fireflies i and j is defined as

$$d_{ij} = \|P_i - P_j\|_F = \sqrt{\text{trace}\left[(P_i - P_j)^H - (P_i - P_j)\right]},$$ (3.33)

where $\|\cdot\|_F$ is the entrywise Frobenius norm.

Algorithm 2 describes the application of the FA on the MC-DS/CDMA cooperative network's EE maximization problem with power and rate constraints for each user.

ALGORITHM 2

Firefly Algorithm for EE Maximization in MC-DS/CDMA

Input: *Utility function J(P);*
         Firefly population $P_1, P_2, \ldots, P_m$;
         Max number of generations MaxGenerations,
         Step size α and the attractiveness β
         Absorption coefficient γ
Initialize: n ← 0
Repeat
    For *i = 1:m*
        For *j = 1:m*
            If*(J(P_j) > J(P_i))*
                Move the Firefly i towards j using eq. (2) and (25)
                If*(Conditions (23) and (24) are not satisfied)*
                    Replace P_i with a random generated P
                End If
            End if
        End For
    End For
    Rank the Fireflies and find the current Best P*
    n ← n + 1
Until *n ≥ MaxGenerations*
Output: ***P****

3.4.5 SIMULATIONS AND NUMERICAL RESULTS

In order to find the best FA parameters for the EE maximization problem in MC-DS/CDMA cooperative systems and to evaluate its performance when compared to the analytical solution given by the Dinkelbach method, simulations were conducted using the MatLab Platform. The simulation parameters are shown in Table 3.3. Note that all users were separated into three different QoS requirement classes, representing different multimedia services with different minimum acceptable transmission rates.

Three different figures of merit were considered to test the algorithm performance: the total EE; the ATP; and the normalized mean squared error (NMSE), which is mathematically defined as

$$\text{NMSE} = \frac{1}{\mathcal{T}} \sum_{n=1}^{\mathcal{T}} \frac{\|P_n - P^*\|^2}{\|P^*\|^2}, \tag{3.34}$$

where $\mathcal{T}$ is the number of trials in the Monte Carlo simulation.

Table 3.3 MC-DS/CDMA system simulation parameters

Parameters	Adopted Values
Fixed Relay MC-DS/CDMA System	
Noise Power	$\sigma^2 = -90$ (dBm)
MT Circuitry Power	$p_c = 0.1$ (W)
Relay Circuitry Power	$p_{C_R} = 0.5$ (W)
Relay Transmission Power	$p_R = 25$ (W)
Power Amplifier Inefficiency	$\varrho = \varrho_R = 2.5$
Codification Rate	$\ell = 3/4$
Processing Gain	$\mathcal{F} = 16$
Bits Per Packet	$V = 1$
Subcarriers	$N = 128$
Amplification Matrix	$\mathbf{A} = \mathbf{I}_F \cdot (p_R/\mathcal{F})$
Subchannel Bandwidth	$w = 78$ (KHz)
Channel Bandwidth	$N \cdot w = 10$ (Mhz)
Maximum Transmission Power	$p_{max} = 125$ (mW)
# Mobile Terminals	$U = 5$;
# Base Stations	$B = 1$
# Fixed Relay Stations	$R = 1$
Cooperative Protocol	AF
Mobile Terminal Distribution	Uniformly Distributed
Channel Gain	
Path Loss	$\propto$ distance^{-2}
Shadowing	Uncorrelated Log-Normal
Fading	Rayleigh
User Types	
User Rates	$r_{i,min} \in \{256, 512, 1024\}$ (kbps)
User Modulations	$M_i \in \{4, 16, 64\}$-QAM
# Users Per Class	3 Users 256 Kbps, 4-QAM
	1 User 512 Kbps, 16-QAM
	1 User 1024 Kbps, 64-QAM
Monte Carlo Simulation Parameters	
Trials	$T = 100$

Because the FA requires the adjustment of five input parameters, this parameter optimization process was divided into four steps. First, the population size m and number of generations were joint optimized, while the remaining parameters $\alpha = \beta = \gamma = 1$. Afterward, with the optimum values for m and MaxGenerations, each parameter was separately optimized in the following order: α, β, and γ.

The results for the best combinations of population size and generation number are presented in Figure 3.3. First, it is important to consider that there are $U \times N =$

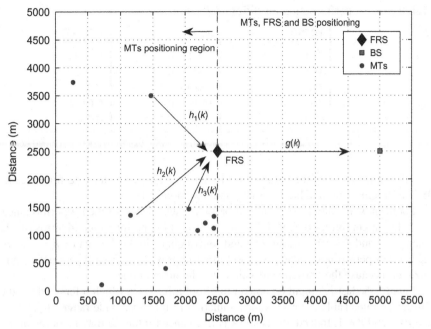

FIGURE 3.3

MT, FRS, and BS positioning in the considered MC-DS/CDMA network scenario.

640 decision variables, hence an NMSE around 10^{-1} and 10^{-2} is considered a good solution. Note that for the best parameters combination ($m = 50$ and MaxGenerations $= 500$), the NMSE is around 0.05 and it achieves almost 96% of the maximum EE, with a 9.5% greater ATP regarding the optimum solution obtained with disciplined optimization CVX tools. In the worst case ($m = 5$ and MaxGenerations $= 400$), the NMSE is almost 0.3, with 88% of the maximum EE achieved with CVX tools and 35% more ATP.

Thus, $m = 50$ and MaxGenerations $= 500$ were chosen as parameters for the next simulation results. The optimization of the α parameter was done considering $\beta = 1$ and $\gamma = 1$. The results are presented in Table 3.4 and the best value found was $\alpha = 10^{-3}$.

Table 3.4 FA performance for different α parameters

α	NMSE	$J(P)$	ATP (W)
1	0.05	6659×10^8	0.0194
10^{-1}	0.05	6678×10^8	0.0191
10^{-2}	0.03	6826×10^8	0.0187
10^{-3}	0.03	6829×10^8	0.0187
10^{-4}	0.09	6699×10^8	0.0197

Table 3.5 FA performance for different β parameters

β	NMSE	$J(P)$	ATP (W)
10	0.07	6601×10^8	0.0198
2	0.07	6607×10^8	0.0197
1	0.03	6829×10^8	0.0187
10^{-1}	0.03	6829×10^8	0.0187
10^{-2}	0.29	6205×10^8	0.0254

The optimization of the parameter α brings the best parameter's performance to 98% of the maximum achievable EE, and 5% more ATP.

For the next FA input parameter calibration, Table 3.5 shows the FA performance for different configurations of β. The difference in FA performance in terms of EE when $\beta = 1$ and $\beta = 10^{-1}$ is marginal and equivalent to 30 kb/J. Because $\beta = 1$ gives the best FA performance, there was no further improvement from β parameter optimization because the chosen start value was the optimal one.

Finally, the last FA input parameter to be optimized is γ, which refers to the light absorption rate. Table 3.6 shows the simulation results for a wide range of γ, while $\alpha = 10^{-3}$ and $\beta = 1$. Indeed, the best FA performance for the EE maximization problem is achieved with $\gamma = 1$, with a marginal performance difference between $\gamma = 1$ and $\gamma = 5$, in terms of system EE, that is, just 250 bits/J. With all the five FA input parameters calibrated, it is important to illustrate the FA convergence evolvement for an arbitrary trial. Figure 3.4 depicts the FA convergence evolvement compared to the maximal achievable EE.

Among the five FA input parameters, the population size and the number of generations are the only ones in which the performance is proportional to the chosen values; i.e., the bigger the population of fireflies and/or the number of generations, the better the results for the EE maximization problem at the cost of increasing the computational complexity. Hence, the other three parameters had their static optimal values hold for the EE optimization scenario evaluated in this section (Figure 3.5).

Table 3.6 FA performance for different values of the light absorption rate Parameter

γ	NMSE	$J(P)$	ATP (W)
10^{-1}	0.03	6827×10^8	0.0188
10^{-2}	0.03	6828×10^8	0.0188
1	0.03	6829×10^8	0.0187
5	0.03	6829×10^8	0.0187
10	0.03	6827×10^8	0.188

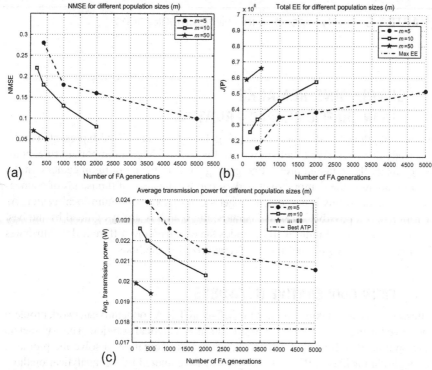

FIGURE 3.4

FA performance for different polation sizes (m) and values of maximum generations (MaxGenerations). Parameters $\alpha = 1$, $\beta = 1$, and $\gamma = 1$.

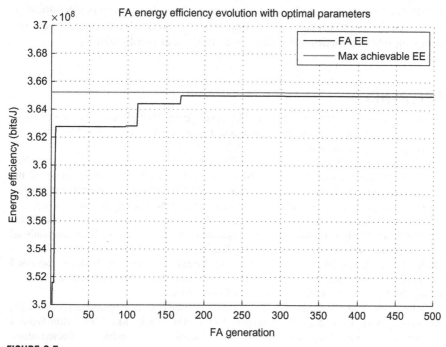

FIGURE 3.5

Utility function evolution with optimal parameters for an arbitrary trial.

3.5 CONCLUDING REMARKS

3.5.1 FA IN TRAFFIC CHARACTERIZATION

In this work, the DSNSF is generated using the FHCA. The DSNSF presented in the case study does not have excellent results, but FHCAs have an important rule, helping to cluster the data and preventing getting stuck in local optimum centers. In the task of managing the computer network, it can handle a set of thousands of connections per second; thus it is possible to get stuck in some optimum local center. As each firefly is a possible set of the centers, the search is always guided by the best answer found so far. Future work includes the expansion to different flow attributes and a comparison with other traffic characterization methods.

3.5.2 FA IN COOPERATIVE NETWORKS

In this work, the EE maximization in the MC-DS/CDMA cooperative network problem was solved using the FA and compared with the analytical solution. The parameters were optimized and the FA showed itself a robust alternative to solve the problem. The algorithm achieved 98% of the system EE computed by the analytical method, at the cost of 5% more ATP, which leads to an average NMSE of 3×10^{-2}. This may be considered a suitable result, given the fact that the problem in the proposed simulation scenario results in 640 decision variables.

Future work includes the expansion to different scenarios (multicellular), increasing the number of relays, and considering a joint relay selection and power allocation for EE maximization problems in multiple access networks. Furthermore, a comparison with other heuristics and analytical methods is also enlightening.

REFERENCES

Abrão, T., Sampaio, L.D.H., Proença Jr., M.L., et al., 2011. Multiple access network optimization aspects via swarm search algorithms. In: Nashat, M. (Ed.), Search Algorithms and Applications. Intech, Rijeka, pp. 261–298.

Adaniya, M.H.A.C., Lima, M.F., Sampaio, L.D.H., 2011. Anomaly detection using firefly harmonic clustering algorithm. In: DCNET 2011 and OPTICS 2011—Proceedings of the International Conference on Data Communication Networking and International Conference on Optical Communication Systems. SciTePress, Seville, pp. 63–68.

Adaniya, M.H.A.C., LIMA, M.F., Rodrigues, J.J.P.C., et al., 2012. Anomaly detection using DSNS and firefly harmonic clustering algorithm. In: Proceedings of IEEE International Conference on Communications. IEEE, Canada, pp. 1183–1187.

Bache, K., Lichman, M., 2013. UCI Machine Learning Repository. [online]. (accessed 14.03.13). Available from World Wide Web: http://archive.ics.uci.edu/ml&gt.

Claise, B., Cisco Systems NetFlow Services Export Version 9.

CVX Research, 2014. CVX: Matlab Software for Disciplined Convex Programming, version 2.1.

De Assis, M.V.O., Carvalho, L.F., Rodrigues, J.J.P.C., Proenca Jr., M.L., 2013. Holt-Winters statistical forecasting and ACO metaheuristic for traffic characterization.

In: Communications (ICC), 2013 IEEE International Conference on. IEEE, Budapest, pp. 2524–2528. http://ieeexplore.ieee.org/xpl/articleDetails.jsp?arnumber=6654913.

Dinkelbach, W., 1967. On nonlinear fractional programming. Manage. Sci. 12 (7), 492–498.

Dorigo, M., Caro, G., Gambardella, L.M., 1999. Ant algorithms for discrete optimization. Artif. Life 5 (2), 137–172.

Dos Santos, A.F., Freitas, S.R., De Mattos, J.G.Z., et al., 2013. Using the Firefly optimization method to weight an ensemble of rainfall forecasts from the Brazilian developments on the Regional Atmospheric Modeling System (BRAMS). Adv. Geosci. 35, 123–136.

Du, K., Swamy, N., 2010. Wireless Communication Systems: From RF Subsystems to 4G Enabling Technologies. Cambridge University Press, New York.

Evans, J., Filsfils, C., 2007. 5—SLA and Network Monitoring. Morgan Kaufmann, Oxford.

Fister, I., Iztok Fister, J.R., Yang, X.-S., Brest, J., 2013. A comprehensive review of firefly algorithms. Swarm Evol. Comput. 13, 34–46.

Gandomi, A.H., Yang, X.-S., Alavi, A.H., 2011. Mixed variable structural optimization using firefly algorithm. Comput. Struct. 89, 2325–2336.

Goodman, D.J., Mandayan, N.B., 2000. Power control for wireless data. IEEE Pers. Commun. Mag. 7 (4), 48–54.

Gross, T.J., Abrão, T., Jeszensky, P.J.E., 2011. Distributed power control algorithm for multiple access systems based on Verhulst model. AEU Int. J. Electron. Commun. 65 (4), 361–372.

Haag, P. 2004. NFDUMP–Flow Processing Tools. [online]. (accessed 14.06.13). Available from World Wide Web: http://nfdump.sourceforge.net/.

Hu, H., 2012. Fa-based optimal strategy of trains energy saving with energy materials. Adv. Mater. Res., 93–96.

Hunt, R., 1997. SNMP, SNMPv2 and CMIP—the technologies for multivendor network management. Comput. Commun. 20 (2), 73–88.

Li, Y., Sheng, M., Yang, C., Wang, X., 2013. Energy efficiency and spectral efficiency tradeoff in interference-limited wireless networks. IEEE Commun. Lett. 17 (10), 1924–1927.

Marques, M., Sampaio, L., Ciriaco, F., et al., 2013. Energy efficiency optimization in mpg ds/cdma. In: Simpósio Brasileiro de Telecomunicações (SBrT 13'). Sociedade Brasileira de Telecomunicações, Fortaleza, Brazil, pp. 1–5.

Noor, M.M., Hussain, Z., Ahmad, K.A., Ainihayati, A., 2011. Multilevel thresholding of gel electrophoresis images using firefly algorithm. In: IEEE International Conference on Control System, Computing and Engineering (ICCSCE). IEEE, pp. 18–21.

Phaal, P., Panchen, S., Mckee, N., InMon Corporation's sFlow: A Method for Monitoring Traffic in Switched and Routed Networks.

Poli, A.A., Cirillo, M.C., 1993. On the use of the normalized mean square error in evaluating dispersion model performance. Atmos. Environ., Part A 27 (15), 2427–2434.

Proença, M.L., Zarpelão, B.B., Mendes, L.S., 2005. Anomaly detection for network servers using digital signature of network segment. In: Advanced Industrial Conference on Telecommunications/Service Assurance with Partial and Intermittent Resources Conference/E-Learning on Telecommunications Workshop (AICT/SAPIR/ELETE'05), pp. 290–295.

Proença Jr., M.L., Coppelmans, C., Bottoli, M., de Souza Mendes, L., 2006. Baseline to help with network management. In: e-Business and Telecommunication Networks, pp. 58–166.

Quittek, J., Zseby, T., Claise, B., Zander, S., Requirements for IP Flow Information Export (IPFIX).

Rajini, A., David, V.K., 2012. A hybrid metaheuristic algorithm for classification using micro array data. Int. J. Sci. Eng. Res. 3 (2), 1–9.

Sampaio, L.D.H., Marques, M.P., Adanyia, M.H., et al., 2013. Ant colony optimization for resource allocation and anomaly detection in communication networks. In: Taufik, A. (Ed.), Search Algorithms for Engineering Optmization. Rijeka, Intech, pp. 1–34.

Sampaio, L., Souza, A., Abrão, T., Jeszensky, P.J.E., 2014. Game theoretic energy efficient design in MC-CDMA cooperative networks. IEEE Sens. J. 14 (9), 1–11.

Sayadia, M.K., Ramezaniana, R., Ghaffari-nasaba, N., 2010. A discrete firefly meta-heuristic with local search for makespan minimization in permutation flow shop scheduling problems. Int. J. Ind. Eng. Comput. 1, 1–10.

Sayadia, M.K., Hafezalkotobb, A., Seyed Gholamreza Jalali, N.A.I.N.I.A., 2013. Firefly-inspired algorithm for discrete optimization problems: an application to manufacturing cell formation. J. Manuf. Syst. 32, 78–84.

Schaible, S., Ibaraki, T., 1983. Fractional programming. Eur. J. Oper. Res. 12 (4), 325–338.

Senthilnath, J., Omkar, S.N., Mani, V., 2011. Clustering using firefly algorithm: performance study. Swarm Evol. Comput. 1 (1), 164–171.

Simon, M.K., Alouini, M.-S., 2004. Digital Communication Over Fading Channels. Wiley, Hoboken, NJ.

Souza, A.R.C., Abrão, T., Sampaio, L.H., et al., 2013. Energy and spectral efficiencies trade-off with filter optimisation in multiple access interference-aware networks. Trans. Emerg. Telecommun. Technol. http://arxiv.org/abs/1206.4176v1.

Thottan, M., Chuanyi, J., 2003. Anomaly detection in IP networks. IEEE Trans. Signal Process. 51 (8), 2191–2204.

Yang, X.-S., 2008. Nature-Inspired Metaheuristic Algorithms, second ed. Luniver Press, p. 160. ISBN10: 1905986289.

Yang, X.S., 2009. Firefly algorithms for multimodal optimization. In: Watanabe, O., Zeugmann, T. (Eds.), Stochastic Algorithms: Foundations and Applications. Lecture Notes in Computer Science, vol. 5792. Springer, Berlin, Heidelberg. pp. 169–178. ISBN 978-3-642-04943-9. http://dx.doi.org/10.1007/978-3-642-04944-6_14.

Yang, X.S., 2011. Firefly algorithm, stochastic test functions and design optimisation. Int. J. Bio-Inspired Comput. 2 (2), 78–84.

Yang, X-S., 2012. Free lunch or no free lunch: that is not just a question? Int. J. Artif. Intell. Tools. 21 (3), 1240010 (13 pages). http://dx.doi.org/10.1142/S0218213012400106.

Yang, X-S., He, X., 2013. Firefly algorithm: recent advances and applications. Int. J. Swarm Intell. 1 (1), 36–50.

Zaponne, A., Buzzi, S., Jorswieck, E., 2011. Energy-efficient power control and receiver design in relay-assisted ds/cdma wireless networks via game theory. IEEE Commun. Lett. 15 (7), 701–703.

Zhang B., Hsu M., Dayal U, October, 1999. K-Harmonic Means - A Data Clustering Algorithm, Software Technology Laboratory. HP Laboratories Palo Alto, HPL-1999-124.

A Survey of Intrusion Detection Systems Using Evolutionary Computation

4

Sevil Sen

Department of Computer Engineering, Hacettepe University,
Ankara, Turkey

CHAPTER CONTENTS

4.1 INTRODUCTION

Intrusion detection systems (IDSs), aptly called the "second line of defense," play a key role in providing comprehensive security. Because it is difficult to develop a complete solution for the prevention of attacks, especially on complex systems, and attackers are always trying to find new ways to bypass these prevention mechanisms, IDSs have become an inevitable component of security systems. IDSs come into the picture after an intrusion has occurred. The main roles of an IDS are to detect possible threats to the systems, and to give proper responses such as notifying security experts, terminating damaging network connections, and other similar means.

Intrusion detection has been a popular research topic in the security field since Denning proposed an intrusion detection model in 1987 (Denning, 1987). Many techniques have been introduced to detect intrusions effectively and efficiently, so the security goals of a system—confidentiality, integrity, and availability—can be satisfied. Researchers have been working on finding answers to the following questions: how to detect attacks effectively and efficiently, which responses to give against detected attacks, how to continuously adapt to new attack strategies, and so on. Major research areas on intrusion detections can be summarized as follows (Lundin and Jonsson, 2002): foundations, data collection, detection methods, response, IDS environment and architecture, IDS security, testing and evaluation, operational aspects, and social aspects. In this study, we examine the evolutionary computation-based approaches proposed for each research area on intrusion detection.

Evolutionary computation (EC) is a subfield of artificial intelligence, inspired from natural evolution. It has been successfully applied to many research areas, such as software testing, computer networks, medicine, and art. Intrusion detection is the most studied area in the security domain, and various intrusion detection techniques already exist in the literature. The following characteristics of EC attract researchers to investigate these techniques on intrusion detection: generating readable outputs by security experts; ease of representation; producing lightweight solutions; and creating a set of solutions providing different trade-offs between conflict objectives, such as detection rate vs. power usage. Furthermore, EC does not require assumptions about the solution space (Fogel, 2000). There are many promising applications of EC on intrusion detection. It is especially suitable for resource-constrained and highly dynamic environments, due to their need for solutions satisfying multiple objectives. In this study, the main proposed solutions in the literature are looked at in detail. For example, how candidate solutions are represented, how evolved solutions are evaluated, which data sets are used, and what advantages and disadvantages the proposed solutions have are all presented.

Although some areas such as detection methods have already been extensively studied, there are only a few studies on areas such as testing and evaluation, and response. This study covers all research areas of intrusion detection from the EC point of view. The suitability of proposed solutions is discussed for each problem. Furthermore, some future directions for researchers are given at the end of the study. To sum up, this research outlines the main issues of intrusion detection and the proposed solutions based on EC in the literature, and discusses the potential of EC for intrusion detection.

The chapter is organized as follows: the fundamentals of intrusion detection and the possible research areas on intrusion detection are presented in Section 4.2, and an introduction to EC is given in Section 4.3. Section 4.4 then outlines the evolutionary computation-based approaches proposed for intrusion detection, classified according to the research area to which they contribute. Finally, the conclusions of the study and the future directions for researchers are summarized in Section 4.5.

4.2 INTRUSION DETECTION SYSTEMS

An IDS is an indispensable part of network security. It is introduced as a system for detecting intrusions that attempt to compromise the main security goals, confidentiality, integrity, and availability of a resource. The development of an IDS is motivated by the following factors:

- Most existing systems have security flaws that render them susceptible to intrusions. Finding and fixing all these deficiencies is not feasible (Denning, 1987), and in particular, complex systems are prone to errors that could be exploited by malicious users.
- Prevention techniques are not sufficient. It is almost impossible to have an absolutely secure system (Denning, 1987). An IDS comes into the picture when an intrusion has occurred and cannot be prevented by existing security systems.
- Because insider threats are generally carried out by authorized users, even the most secure systems are susceptible to insiders. Furthermore, many organizations express that threats from inside can be much more harmful than outsider attacks (CERT, 2011).
- New intrusions continually emerge. Therefore, security solutions need to be improved or introduced to defend our systems against novel attacks. This is what makes intrusion detection such an active research area.

An IDS detects possible violations of a security policy by monitoring system activities and responding to these violations according to the policy. An IDS can be called host-based IDS (HIDS) or network-based IDS (NIDS), according to the system that it monitors. If an attack is detected when it enters the network, a response can be initiated to prevent or minimize damage to the system. Moreover, the prevention techniques can be improved with the feedback acquired from IDSs. Security solutions do not operate on their own, as they once did. Nowadays, prevention, detection, and response mechanisms generally communicate with each other in order to protect the system from complex attacks.

There are generally two metrics employed in order to evaluate IDSs: detection rate and false positive rate. Detection rate represents the ratio of malicious activities detected to all malicious activities. A missed intrusion could result in severe damage to the system. False positives indicate normal activities that are falsely detected as malicious by the IDS. A low false positive rate is just as important as a high detection rate. When an intrusion is detected, it usually raises an alarm to the system administrator. High false positives result in excessive burden to the administrator and, as a result, might not be analyzed by security experts in real time. Another metric, called intrusion capability metric (C_{ID}), was introduced in 2006 in order to evaluate IDSs (Gu et al., 2006). The authors define C_{ID} as the ratio of the mutual information between IDS input and output to the entropy of the input. It naturally includes both the detection rate and the false positive rate. Even though many approaches still use the conventional intrusion detection metrics (i.e., detection and false positive rates),

C_{ID} has important characteristics to compare IDSs, and it is expected to be more commonplace in the near future.

4.2.1 IDS COMPONENTS

The three main components of an IDS—data collection, detection, and response—are depicted in Figure 4.1. The data collection component is responsible for the collection and preprocessing of data tasks, such as transforming data to a common format, data storage, and sending data to the detection module (Lundin and Jonsson, 2002). Various data from different sources, such as system logs, network packets, and management information base data, can be collected and formatted to send to the intrusion detection module.

The detection module analyzes and processes the formatted data obtained from the data collection model in order to detect intrusion attempts, and forwards the events flagged as malicious to the response module. There are three intrusion detection techniques: anomaly-based, misuse-based, and specification-based. The anomaly-based intrusion detection technique defines the normal behaviors of the system, such as usage frequency of commands or system calls, resource usage for programs, and so on. The activities falling out of the normal behaviors of the system are labeled as intrusions. Various techniques have been applied for anomaly detection, such as classification-based (e.g., neural networks, naive Bayes, support vector machines (SVM)), clustering-based techniques. Because the normal behavior can change over time, one of the biggest challenges in this approach is to define the normal behavior of a system. It is particularly challenging in highly dynamic networks, such as mobile ad hoc networks (MANETs) and vehicular ad hoc networks (VANETs). Another disadvantage of this technique is the high number of false positives. How to update the system profile automatically is another challenge. Concept drift, the problem of distinguishing malicious behaviors from the natural change in user/system behaviors, is an issue in anomaly-based detection systems. The conventional approaches mainly overcome this issue through the updating of user/system profiles, which is particularly crucial for the ongoing detection of attackers.

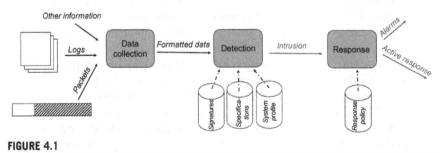

FIGURE 4.1

IDS components.

The updating system generally uses unlabeled data in retraining, due to the large amount of data. Therefore, the updating system has to trust the decisions that the anomaly-based detection system makes. For instance, if the detector misses an intrusive behavior, it will be added to the training data as benign datum. The ability to adapt to the concept drift depends on the accuracy of the detector. The authors showed that misclassified instances included in updating could considerably decrease the performance of anomaly-based detection approaches (Sen, 2014).

A misuse-based (or signature-based) IDS is based on defined signatures in order to detect known attacks. It is the most commercially employed approach due to its efficiency. Although it has a low false positive rate, the biggest disadvantage of this approach is that it cannot detect novel attacks and unknown variants of existing attacks. Many proposed approaches have low resilience against even the simplest obfuscation techniques. Another issue is to frequently update an attack signatures database. Because large numbers of attacks are introduced every day, the function of automatically generating new signatures is an essential characteristic of an IDS. Nowadays, both misuse-based and anomaly-based intrusion detection techniques are employed together. While the misuse-based systems are efficient in detecting known attacks, anomaly-based detection systems are employed to detect attacks missed from these systems.

The last intrusion detection technique is a specification-based method, in which attacks are detected as violations of well-defined specifications of a program/protocol. Since its introduction in 2001 (Uppuluri and Sekar, 2001), this technique has mainly been used for ad hoc networks. It detects both known and unknown attacks with a low false positive rate (Uppuluri and Sekar, 2001). Because the routing protocols proposed for ad hoc networks are vulnerable to attacks, due to their dynamic and collaborative nature, the specification-based intrusion detection is quite suitable for such networks. It is the most employed technique in ad hoc networks and is proposed as a way for different types of ad hoc routing protocols to be kept up-to-date. However, this technique cannot detect denial of service (DoS) attacks, because these types of attacks follow the system specifications. Generally, it cannot detect legitimate activities, even if they are unusual (Uppuluri and Sekar, 2001). Another disadvantage of this technique is the requirement to define specifications for each protocol used in the system. Therefore, it does not attract much interest in wired networks due to this time-consuming task requirement.

When an event is classified as malicious, it is sent to the response module. The module behaves according to the response policy defined. Intrusion detection responses are divided into two groups (Axelsson, 2000): active and passive responses. There are still many systems that give only passive responses: notifying the proper authority. On the other hand, an active response attempts to mitigate or prevent the damage of an intrusion by controlling either the attacked system or the attacking system (Axelsson, 2000). Blocking the IP address attacking the system, or terminating network connections for a while, are examples of commonly used active responses. These types of system are typically called Intrusion Prevention Systems.

4.2.2 RESEARCH AREAS AND CHALLENGES IN INTRUSION DETECTION

Intrusion detection has been an appealing research area since Denning first introduced a formal model for the problem (Denning, 1987). Intrusion detection is a challenging research area due to its very nature, and a great deal of research has emerged in this domain. Lundin et al. (Lundin and Jonsson, 2002) classify major research areas on intrusion detection as follows: foundations, data collection, detection methods, response, IDS environment and architecture, IDS security, testing and evaluation, operational aspects, and social aspects.

Foundations cover the research carried out on intrusions, intruders, and vulnerability. The main challenge here is to update IDSs against emerging new attacks every day. A good IDS must perform continuous adaptation to new attacks, changes in the system, and the like. *Data collection* deals with selecting data sources and features, how to collect data, logging, and formatting data. One of the main problems of IDSs is analyzing and processing highly imbalanced and large amounts of network data efficiently. Researchers mainly work on selecting appropriate features for intrusion detection and reducing redundant features. The majority of research has been carried out on *detection methods*. The main challenges of each detection technique are given in detail in the previous section. The difficulty of distinguishing normal data from abnormal data and developing systems that are robust against unknown attacks are among the most important challenges. Studies on *response* aim to answer the following questions: how to respond to detected intrusions (i.e., passively or actively, temporarily or permanently), and how to represent detected intrusions to the proper authority.

How to distribute IDS agents and facilitate interoperability between IDS agents are subresearch areas in *IDS environment and architecture*. It is a particularly active research area in networks with a lack of central points where we could monitor and analyze all network data. Three main intrusion detection architectures are proposed for such networks: stand-alone, distributed and collaborative, and hierarchical. There are few studies that define information exchanges between IDS agents. Mobile agents, which carry both data and software from one system to another system autonomously and continue their execution on the destination system, are another method of communication with many advantages, such as reducing the network load, and adapting dynamically (Lange and Oshima, 1999). *IDS security* is related to protecting IDS communication and the IDS itself from attacks. This "secure security" concept is especially important in critical domains such as health care and tactical systems. The studies on this immature research area have accelerated in recent years, and a survey on adversarial attacks against IDSs was recently proposed (Corona et al., 2013). *Testing and evaluation* takes into account how to evaluate IDSs. There are many comparisons available in the literature. The Knowledge Discovery in Databases (KDD) data set (Lippman et al., 2000) is considered benchmark data in these studies. *Operational aspects* cover technical issues such as maintenance, portability, and upgradeability of IDSs. *Social aspects* are related to ethical and legal issues of deploying IDSs (Lundin and Jonsson, 2002). Operational and social aspects are excluded due to their irrelevance in this study.

4.3 THE METHOD: EVOLUTIONARY COMPUTATION

EC is a computational intelligence technique inspired from natural evolution. An EC algorithm starts with creating a population consisting of individuals that represent solutions to the problem. The first population could be created randomly or fed into the algorithm. Individuals are evaluated with a fitness function, and the output of the function shows how well this individual solves or comes close to solving the problem. Then, some operators inspired from natural evolution, such as crossover, mutation, selection, and reproduction, are applied to individuals. Based on the fitness values of newly evolved individuals, a new population is generated. Because the population size has to be preserved as in nature, some individuals are eliminated. This process goes on until the termination criterion is met. Reaching the number of generations defined is the most used criterion to stop the algorithm. The best individual with the highest fitness value is selected as the solution. The general steps of an EC algorithm are shown below.

> *initilize population*
> *evaluate the fitness value of each individual*
> **while** *the optimal solution is not found and*
>   *the number of generations defined is not reached*
>   *select parents*
>   *apply genetic operators to the selected individuals*
>   *evaluate fitness values of new individuals*
>   *select individuals for the next generation*
> **end while**
> **return** *the best individual*

There are various EC techniques, such as Genetic Programming (GP), Genetic Algorithms (GAs), Grammatical Evolution (GE), Evolutionary Algorithms (EA), and the like. These techniques generally differ from each other based on how to represent the individuals. For example, while GP uses trees, GE uses Backus-Naur Form (BNF) grammar in order to define individuals.

One of the most popular EC techniques in the literature is GP. Since introduced by Koza (1992), it has been applied to many problems, and it has been shown that GP produces better solutions for complex problems than humans do. In GP, the crossover operator swaps subtrees of two individuals, and the mutation operator exchanges a subtree with another tree created. One of the problems in GP is bloating, which is the uncontrolled growth of the average size of trees in a population. It is generally controlled by limiting the depth of the individuals. However, bloating could show a positive effect on some problems. While GP uses trees, GAs represent each individual as an array of bits called chromosomes. In GA, the genetic operators are applied on subarrays of individuals selected.

GE is a technique, inspired largely by the biological process of generating a protein from the genetic material of an organism, which allows us to generate complete programs in an arbitrary language by evolving programs written in a BNF grammar (Ryan et al., 1998). However, the GE technique performs the evolution process on variable-length binary strings, not on the actual programs (O'Neill and Ryan, 2003). This transformation from variable-length binary strings to actual programs provides mapping from the genotype to the phenotype as in molecular biology. One of the benefits of GE is that this mapping simplifies the application of search to different programming languages and other structures. Another benefit of GE from the security point of view is the production of readable outputs in well-defined grammars.

Multiobjective evolutionary computation (MOEC) is used to create solutions satisfying more than one objective. It creates a set of solutions providing different trade-offs between objectives. Another way to achieve that is to create a weighted fitness function. Because the population-based nature of EA allows the generation of several elements of the Pareto optimal set in a single run, EA are highly motivated for solving problems with multiobjectives (Coello et al., 2007).

4.4 EVOLUTIONARY COMPUTATION APPLICATIONS ON INTRUSION DETECTION

The section outlines and discusses some representative examples of the EC applications to intrusion detection. Each solution is classified according to the subfield of intrusion detection to which it contributes. Please note that some solutions could be placed in more than one subfield.

4.4.1 FOUNDATIONS

This section covers the studies carried out on intrusions. Attackers have two prime motivations: to damage the targeted system and to avoid being identified. In order to achieve their goals, they continuously create new attacking strategies. On the other hand, IDSs generally are unable to detect these "new" attacks. Particularly, misuse-based detection systems are ineffective against new attacks or unknown variants of existing attacks. Therefore, researchers have been also working on developing new variants of existing attacks automatically, using EC techniques, and as a result, security solutions can be reassessed and strengthened. As far as we know, the first work on this area was proposed by Kayacik et al. (2005a). The authors aim to evolve successful stack overflow attacks by employing GE. The problem is represented as a simple C program that determines the size of the no-operation (NoOP) sled, the main evasion technique used by attackers, the offset, and the number of desired return addresses that state the address of the shellcode aimed to be executed by the attacker. Six criteria based on the success of the attack, the size of the NoOP sled, and the accuracy of the desired return address are incorporated into the fitness function. In some experiments, they also utilize fitness sharing in order to eliminate similar individuals in the population. Moreover, they carry out some experiments in order to discourage long NoOP sleds, which helps evasion of the evolved attacks. Snort (Roesch, 1999; Snort, 2014) is chosen as an exemplar IDS in order to evaluate their results. The results show that GE techniques are applied successfully in order to generate both successful and evasive stack overflow attacks. It is observed that the number of invalid programs among the evolved ones is quite small. However, it is stated that GE representation of malicious programs was not good enough to modify register references (Kayacik et al., 2005b). Therefore, the same authors employ Linear GP (LGP) to generate new buffer overflow attacks. At this time, the malicious programs are represented in assembly language. Instructions (opcodes and their operands) are represented by LGP. The positive effect of the bloating property of GP is observed in the results. Although bloating is generally minimized to increase readability of GP outputs in other domains, it is used here to hide malicious code parts in evolved programs. The authors also analyze the effects of different instruction sets (arithmetic, logic, etc.) in the representation on the results.

LGP is also used to automatically generate mimicry attacks, in which the exploits are represented as a sequence of system calls (Kayacik et al., 2009). The proposed method follows a black-box approach in which the attacker only has information about the output of the anomaly-based IDS. Four anomaly-based IDSs (Stide, Process Homeostasis, Process Homeostasis with Scheme Masks, and Markov Mode) are employed to obtain the anomaly rate used in the fitness function, together with the attack success ratio and attack length. The results show that mimicry attacks evolved with GP produce lower anomaly rates than the original attacks. Nonetheless, none of the attacks produced were completely undetected. An extended study compares the evolved attacks based on the black-box approach with the created attacks based on the white-box approach in which the attacker has the knowledge of the internal behavior of the IDS (Kayacik et al., 2011). They also include one additional objective in the fitness function: delay (a particular type of response against an identified attack). Even a detector that has a low anomaly rate could prevent an attack by deferring the attack with long delays. The evolved attacks have comparable results with the white-box approach, but the latter produces lower anomaly rates. Nonetheless, the black-box approach generates many attacks with different trade-offs in a Pareto front, while the white-box approach has one exploit per attack detector pair. Finally, an attacker might not have easy access to the detector where he or she could cost-effectively gain information about its internal behavior. These features make EC attractive for evolving evasive attacks and are summarized as ease of representation, multiobjective optimization (MOO), and natural obfuscation. Besides the studies given here, in the literature, there are applications of EC on generating variants of known malware (Noreen et al., 2009).

4.4.2 DATA COLLECTION

Studies applying EC techniques in this research domain mainly focus on feature selection and feature reduction. The choice of features is very important for any research domain. On the one hand, the features generally contain sufficient and expressive information that is enough to generate effective models; but on the other hand, too many features could in fact confuse the learning algorithm and degrade its performance. Therefore, superfluous features should be eliminated, while keeping necessary ones. This elimination speeds up the feature extraction by only spending time to get necessary features, and the training process by reducing the search space. From the intrusion detection point of view, features could also give some information about attacks and attackers' behaviors. If we reduce the number of features, we could better understand the attack motives and techniques.

We mainly divide feature selection approaches into two groups: filter and wrapper approaches. The wrapper approach selects features according to the performance of the learning algorithm. The filter approach only gives general insights into the features based on some measures, and it does not take into account the performance of the algorithms. The studies, based on feature selection for intrusion detection in the literature, mainly follow the wrapper approach.

As far as we know, the first wrapper approach using GAs for intrusion detection was proposed in Helmer et al. (1999), and extended by the same authors in Helmer et al. (2002). The authors used the RIPPER algorithm for learning, and showed that the performance of the algorithm is not affected even when the features used in training are reduced to half. In Hofmann et al. (2004), the number of features is decreased from 187 to 8 features by employing EA. In Kim et al. (2005), GA is employed to select both the optimal features and the optimal parameters for a kernel function of SVM for detecting DoS attacks on the KDD data set (Lippman et al., 2000). The results provided a better detection rate than the approaches that only adopted SVM in the IDS. A similar approach in order to improve the performance of SVM, by reducing the number of features, is also proposed for detection of a specific DoS attack against ad hoc networks (Sen and Dogmus, 2012). A recent GA application (Ahmad et al., 2014) works on principal components instead of working on features directly in order to both increase the performance of SVM and to use a lower number of features. Principal components are computed using Principal Component Analysis (PCA), a conventional technique for feature subset selection.

EC techniques for feature selection/reduction could also be employed with or after other techniques. In some approaches, it is combined with a filter approach, correlation-based feature selection (CFS) (Shazzad and Park, 2005; Nguyen et al., 2010), especially in the presence of a high number of features, as in the KDD data set. An approach to increase the performance of SVM is introduced (Shazzad and Park, 2005), in which the feature set is first evaluated by CFS. There are also GA-based wrapper approaches proposed for different classification techniques, such as decision trees (C4.5) (Nguyen et al., 2010; Stein et al., 2005), BayesNet (Nguyen et al., 2010), artificial neural networks (ANN) (Mukkamala et al., 2004), fuzzy data mining (Bridges and Vaughn, 2000), and the like.

The authors determine the weight of features for the k-nearest neighbor classifier in Middlemiss and Dick (2003). They run GA many times and get the average of the weighted feature sets. Unlike with feature selection, they include all features in the training because the final averaged feature set does not include any zero weights.

They analyze the top five ranked features for each attack class in the KDD data set. Moreover, they carry out additional experiments to remove the five features with the highest number of zero weights. However, they show that removing zero-weighted features decreases the performance of the classifier. This situation could be discovered by changing the fitness function, taking the number of features removed into account. Furthermore, we could apply EC-based approaches in order to eliminate costly features that require a considerable amount of time to compute, especially when monitoring a large amount of traffic. In order to analyze features, Mukkamala et al. (2004) removed one feature out in each run of LGP. The results help us see the effects of each feature on the results.

4.4.3 DETECTION TECHNIQUES AND RESPONSE

A great deal of research has emerged in the intrusion detection field; however, the majority of research has been carried out on detection techniques. Various techniques, such as statistical approaches, expert systems, SVM, ANN, EC, and the like, have been proposed for the problem. In recent years, computational intelligence methods have attracted considerable interest due to their characteristics suitable for intrusion detection, such as adaptation, fault tolerance, high computational speed, and error resilience in the face of noisy information (Wu and Banzhaf, 2010). In this section, we focus on EC techniques proposed in the literature. EC is one of the promising approaches on intrusion detection due to some advantages over other classical mechanisms: producing lightweight detection rules and the simple output that is understandable by security experts (Orfila et al., 2009).

4.4.3.1 Intrusion detection on conventional networks

The first GP application to intrusion detection was given by Crosbie and Spafford in 1995 (Crosbie and Spafford, 1995). The main idea in that research is to train autonomous agents based on the features related to network connections and the functions (arithmetic, logical, conditional) given to detect intrusive behaviors. An agent is evaluated with a fitness function that compares the output of the agent with the expected output. A weight parameter is also included in the fitness function to penalize the agents based on the difficulty of detecting an intrusion. Obvious intrusions that are misclassified during the evolution process are heavily penalized. This system is proposed to detect port flooding, port walking, probing, and password cracking attacks.

Applications of EC techniques to intrusion detection on conventional networks have usually employed either GP or a GA. GASSATA (Me, 1998) is one of the earliest works that employs GA to intrusion detection. It is a misuse-based detection system, using GA in order to detect 24 known attacks that are represented as sets of events (i.e., user commands). Because the system does not locate attacks precisely, a security expert is needed to analyze the audit trail and to pinpoint the attacks. As an improvement on GASSATA, a HIDS was introduced by Diaz-Gomez and Hougen (2005a). The authors propose an improved fitness function that is more general and independent of the audit trail. The results show that the model evolved with the improved fitness function produces no false positives and a low number of false negatives. The mathematical justification of the fitness function proposed is presented by Diaz-Gomez and Hougen (2005b).

One of the first proposals was the Network Exploration Detection Analyst Assistant (NEDAA), another GA-based approach (Sinclair et al., 1999). In NEDAA, automatically generated intrusion detection rules by GA and decision trees are fed into a deployed IDS. In the GA component, a rule (a chromosome) is represented by the source IP address and port, the destination IP and port, and the protocol. Each value in the chromosome could be a number specified in the range or a wild card. A simple chromosome representation for a rule (source ip: 193.140.216.*, source port: 1454, destination port: 53, protocol: 2-TCP) is shown in Figure 4.2. The fitness function is evaluated based on the performance of each rule on a preclassified data set (normal and anomalous connections). However, this GA application generates only one rule in each run. Because one single rule is not enough to identify different types of anomalous connections, the authors transfer the problem from finding global maxima to multiple local maxima of the fitness function by employing niching techniques. The rules generated by GA are compared with the rules generated by using decision trees in this study. While GA produces a set of rules, decision trees generate a single metarule with a number of different rules. This study investigates the use of GA on generating intrusion detection rules automatically; however, it does not present any experimental results. The authors discuss future plans on the alteration of the GA component in NEDAA, such as finding complex rules in order to detect advanced attacks disseminated in space and time, generating rule chaining, and generating dynamic rules in order to detect new attacks.

The first GE application to detect attacks is employed on the KDD data set (Wilson and Kaur, 2007). Abraham et al. (2007) employ two GP techniques, namely LGP and Multiexpression Programming (MEP), on the same data set. While MEP (a technique that allows us to encode multiple expressions) is more successful in the identification of some attack types, LGP is better in the detection of other types. Tahta et al. (2014) employ GP in order to differentiate malicious peers from benign ones in peer-to-peer (P2P) networks. They run P2P simulation for each individual to see how derived solutions are effective in preventing malicious peers from participating in the network.

The techniques based on artificial intelligence have the ability to solve complex problems cost-effectively. However, they face training a model on imbalanced and large data sets in intrusion detection. Some researchers investigate the suitability of EC to work on large data sets (Dam et al., 2005). Another study proposes to use

1	9	3	1	4	0	2	1	6	*	1	4	5	4	5	3	2

FIGURE 4.2

A simple chromosome representation of a rule.

RSS-DSS (random subset selection—dynamic subset selection) algorithms in order to train GP computationally efficient (Song et al., 2005). The solution is useful in terms of both fitting training data into the memory and processing large amounts of data. The RSS algorithm randomly selects a block of data from KDD, which includes approximately half a million patterns. Then a subset of the block is processed with the DSS algorithm, and this subset is given to the GP algorithm for evolution. The subset selection process is parameterized. The results make the GP algorithm a very practical solution for intrusion detection by showing that performing one run takes only 15 min on a PC. The authors also analyze different fitness functions based on the recognition that different types of attacks are not uniformly distributed in the data set. Using parallel GAs is another way of speeding up training time for complex problems with large data sets (Abadeh et al., 2007a).

A system that evolves attack signatures by using GP is proposed in Lu and Traore (2004). The results show that the derived rules are better at detecting both known and unknown attacks. Another signature-based intrusion detection was proposed recently (Gomez et al., 2013), generating attack signatures automatically and working in an integrated manner with Snort. Multiobjective EA are employed to obtain a set of solutions providing different trade-offs between false positives and false negatives. The performance of a weighted single fitness function is also shown in the results. Another recent approach uses Genetic Network Programming (GNP) in order to develop models both for misuse-based detection and anomaly-based detection (Mabu et al., 2011). GNP uses graphs in order to represent individuals in EC. EC could be integrated with other techniques. For example, the clustering GA introduced in Zhao et al. (2005) classifies activities into groups by employing clustering techniques in the first phase, then employs GA in order to distinguish normal activities from abnormal ones in the clusters. Another example is to represent individuals as fuzzy if-else rules, and then apply GA on these rules (Abadeh et al., 2007b). These are just some representative examples among the many EC applications to intrusion detection to be found in the literature.

4.4.3.2 Intrusion detection on wireless and resource-constrained networks
Although EC techniques have many potential application areas for wireless and ad hoc networks, such as data aggregation, forming clusters, and security, there are few studies on intrusion detection. Multiobjective optimization techniques are an especially apt candidate for problems on ad hoc networks due to their very characteristics, such as nodes with limited energy, dynamic topology, and the like (Sen and Clark, 2009a). Hence, possible trade-offs to make between functional (i.e., accuracy) and nonfunctional (i.e., power usage, bandwidth) properties of solutions could exist. Please also note that the solutions proposed for these highly constrained and dynamic networks should be lightweight.

The first work based on EC techniques proposed for intrusion detection in MANETs uses GE in order to detect attacks against MANETs such as dropping, ad hoc flooding, and route disruption attacks (Sen and Clark, 2009b). It is shown that the evolved programs detect both flooding and route disruption attacks successfully on simulated networks with varying traffic and mobility patterns. Furthermore, the grammars presented for each attack type are in a readable format so that security experts can easily understand the attack signatures. The authors extend their work by taking into account other objectives besides the intrusion detection capability of evolved programs. In Sen and Clark (2011), they propose a power-aware IDS in order to obtain a set of optimal solutions offering different trade-offs between detection ability and power usage by employing the Strength Pareto Evolutionary Algorithm 2 (SPEA2) (Zitzler et al., 2001). The programs are also compared with hand-coded detection programs that produce an almost perfect detection rate, with nonnegligible false positive rates for networks under high mobility (Sen and Clark, 2011). On the other hand, EC techniques not only produce acceptable false positive rates, but also consider nonfunctional properties of intrusion detection programs. The authors also explore a suitable intrusion detection architecture for MANETs. Two different architectures are considered: stand-alone, and distributed and cooperative architecture in the neighborhood. Other architecture proposals (Hassanzadeh and Stoleru, 2011b, 2013) employing MOEC techniques also exist in the literature. The details of the architectures are presented in Section 4.4.4.

EC is also an emerging research area for wireless sensor networks. The first study on this area (Johnson et al., 2005) investigates a suitable framework for performing GP on wireless sensor nodes. EC could be a good candidate for many nondeterministic polynomial-time hard (NP-hard) problems in the field. Some problems are given as follows (Nan and Li, 2008): node placement and layout optimization, energy efficient routing, clustering, scheduling, position estimation, analyzing the lifetime of sensor networks, and information fusion. In the literature, few approaches exist for intrusion detection (Khanna et al., 2007, 2009). These approaches, as covered in Section 4.4.4, mainly focus on optimally placing monitoring nodes.

A successful application of GP against a type of DoS attack (the deauthentication attack) at the link layer on Wi-Fi networks is given in Makanju et al. (2007). It is shown that GP is able to detect both the original attack and its modified versions with an almost 100% detection rate and a less than 1% false positive rate.

4.4.4 IDS ARCHITECTURE

There are few studies exploring the suitability of EC techniques for how to place IDSs on a network. On the contrary, MOEC is a favorable candidate for taking into account different objectives while deploying IDSs.

Chen et al. (2010) discover the applicability of GA with MOO techniques in order to place IDS sensors by satisfying conflict objectives. Various attack detection rates are traded off with false alarm rates and costs. Probing and information-gathering attack scenarios are evaluated based on the assumption that a placement for a type of attack could not be optimal for another type of attack. They try to satisfy the following four objectives: number of sensors, detection rate, false alarm rate, and monitoring costs calculated based on monitored traffic. The experimental results show that different trade-offs are possible for IDS sensor placement. While the nodes have equal characteristics such as detection capabilities, the level of risk against them in the experiments, exploring the placement of IDS sensors with different characteristics, could be investigated in the future.

One of the most eligible areas to apply MOO techniques for IDS placement is ad hoc networks, in which IDSs must be distributed and cooperative, due to their very nature. Different architectures are proposed in the literature: stand-alone, distributed, and hierarchical, as shown in Figure 4.3. While stand-alone architecture consists of only local monitoring (LM) nodes, nodes communicate with each other in order to obtain data and to distribute alarms in a distributed architecture. To achieve global detection of attacks, distributed and hierarchical architectures must attract the attention of researchers. In hierarchical architecture, while some nodes are only responsible for local detection, some particular nodes are more responsible than others by carrying out data aggregation and making decisions. The choice of these particular nodes is generally performed as random in order to provide security. Other criteria are residual energy, degree of connectivity, and the like. The first approach exploring an intrusion detection architecture suited to the distributed and resource-constrained environments is presented in Sen and Clark (2011). A distributed and collaborative architecture in a neighborhood where nodes communicate with one-hop-away nodes in order to carry out intrusion detection is analyzed and compared with stand-alone architecture. While the classification accuracy is aimed to be maximized, the number of nodes in cooperation is minimized in order to save both energy and bandwidth. While the classification accuracy is denoted with detection and false positive rate, the usage of bandwidth is denoted by the number of nodes in cooperation. The energy consumption is modeled mathematically (Feeney, 2001). The potential uses of EC techniques to discover complex properties of MANETs (such as limited power and bandwidth) are shown in the results.

Another important contribution of IDS deployment on MANETs by using GA is proposed in Hassanzadeh and Stoleru (2011b). It is proven that optimal monitoring node selection problem is NP-hard (Hassanzadeh et al., 2011a). Therefore, MOO techniques are proposed to assign intrusion detection nodes in a cluster tree organization, based on some criteria (energy consumption, event reporting delay, network coverage, and quality of data). Distinct roles are defined for intrusion detection, and each node is designed to perform only one role. It is shown that MOO techniques are necessary to optimize objectives simultaneously without negatively affecting others. Because the execution time of GA could be high, especially for large networks, an extra phase is introduced in their extended study (Hassanzadeh and Stoleru, 2013).

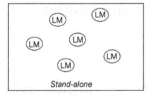 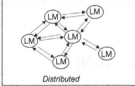 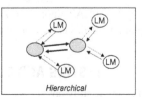

Stand-alone   *Distributed*   *Hierarchical*

FIGURE 4.3

IDS architectures.

In the two-phase hybrid algorithm for MOO, a set of suitable nodes as leaders (the nodes with the highest responsibilities) is determined in the first phase, and roles for n hop-away neighbors of each leader are assigned by using GA in the second phase. It is empirically shown that the hybrid algorithm outperforms only the GA solution, in terms of both execution time and the optimality of the solutions evolved in large networks. While the experiments are carried out with nodes performing misuse-based intrusion detection, this is applicable for any type of detection technique. Khanna et al. (2009) also take into account the security of monitoring nodes besides other objectives, node coverage, and residual battery power. A weighted fitness function is employed to assess the individuals generated by the GA algorithm. The detection of compromised nodes is fastened up with this approach.

4.4.5 IDS SECURITY

IDS itself could be the target of attacks. Corona et al. (2013) divide attack goals against IDSs into six categories: evasion, overstimulation, poisoning, DoS, reverse hijacking, and reverse engineering. The authors suggest focusing on techniques based on adversarial machine learning, because there is an expected increase in the machine learning techniques proposed for intrusion detection. The applications of EC mainly focus on evasion attacks (Kayacik et al., 2011), as covered in Section 4.4.1. Pastrana et al. employ GP in order to reverse engineer the IDS behavior (Pastrana et al., 2011). It is assumed that the attacker does not know the internal behavior of the targeted IDS. One of the main reasons for employing GP in this study is to generate easy-to-understand models. In order to achieve this objective, the depth of GP trees is restricted to five. The evolved IDS models (i.e., attack signatures) are used to generate evasion attacks against a C4.5-based IDS. The authors make legitimate modifications to evolved signatures in order to avoid being detected while actualizing the attack. The modifications are not made automatically, but it is shown that the GP evolves successful IDS models, similar to the IDS under study.

Recently, the adversarial capabilities against intrusion detection networks (IDNs) were presented in Pastrana (2014). IDNs consist of various monitoring and detection nodes distributed through the network in order to be able to detect distributed attacks by aggregating/correlating data obtained from various sources. The study is one of the most comprehensive analyses on IDNs to evaluate the risks against these networks. Some countermeasures are considered based on this analysis. The trade-offs between cost and risk for each solution are represented in a Pareto front by employing the SPEA2 algorithm (Zitzler et al., 2001). So, security experts can decide on the placement of countermeasures.

4.4.6 TESTING AND EVALUATION

Testing mainly covers issues related to setting benchmarks and creating data sets for evaluating IDSs objectively; however, it is out of the scope of this survey. In this section, we will outline the studies that compare different types of IDSs, particularly

EC proposals. One of these studies is proposed by Abraham et al. (2007), in which they compare GP performance on intrusion detection with decision tree and SVM approaches on the KDD data set (Lippman et al., 2000), which is the most widely used benchmark evaluation data. They said that GP outperforms both techniques. It is noted that SVM performs well on detection for all kinds of attacks in the data set. Mukkamala et al. (2004) compared LGP with SVM and resilient back-propagation neural networks (NN) on the KDD data set. LGP presents better detection accuracy than the other two techniques; however, it is a costly approach in terms of training time. Similar to the other comparisons (Abraham and Grosan, 2006; Pastrana et al., 2012), SVM is the best approach that is closest to GP. The authors also underline the possible effects of the parameters such as the program size, and the mutation and crossover rates of the results.

A comprehensive study on evaluating different classification algorithms for intrusion detection in MANETs is given in Pastrana et al. (2012). The four popular attacks on ad hoc networks—dropping, blackhole, flooding, and forging—are used in the experiments. The following six classifiers are compared: multilayer perceptron (MLP), a linear classifier, the Gaussian mixture model (GMM), the naive Bayes classifier, SVM model, and GP. The networks using an ad hoc on-demand distance vector (AODV) routing protocol (Perkins and Royer, 1999) are simulated; however, different mobility levels are not taken into account. In the results, two classifiers, namely SVM and GP, surpass other techniques. Based on the results, it is suggested to use GP in order to differentiate attacks from normal behaviors (two-class classification). In the multiclass case, SVM performs better than GP. Sen and Clark (2011) also compared their EC-based intrusion detection proposals for MANETs with manually crafted programs. A performance comparison of the techniques GP and GE on the same attacks can also be seen in the study. Makanju et al. (2007) compare the performance of GP on the detection of a DoS attack at the link layer on Wi-Fi networks with Snort-Wireless, a widely used misuse-based IDS. The results show that while the variants of the attacks cannot be detected by Snort, GP shows a high detection rate with a negligible false positive rate.

4.5 CONCLUSION AND FUTURE DIRECTIONS

In this study, we present a survey on EC techniques applied to intrusion detection. Each research area on intrusion detection is covered in detail. In this section, the highlights of the study are given.

Most studies in the literature are carried out for the development of effective and efficient intrusion detection methods. It is the same from the evolutionary computation-based approaches' point of view. These approaches attract significant interest from researchers by generating readable outputs. This is a critical characteristic because security experts mainly tend to use the systems they can understand. While GE techniques produce outputs in the specified grammar (Wilson and Kaur, 2007; Sen and Clark, 2009b), the outputs of GP could become more readable by limiting the depth of GP trees evolved (Orfila et al., 2009). The comparison studies show that EC techniques mainly perform equal to or greater than other machine learning techniques in this domain. The outputs of EC techniques are comparable with SVM, the technique showing the best performance in some studies.

Using fewer features is shown as another benefit of EC. We could also employ EC techniques in order to select the feature set that best represents the problem, or is the most cost-effective. Furthermore, MOEC techniques are among the best candidates for resource-constrained or dynamic environments in which different objectives such as less power usage and less communication between IDS agents need to be satisfied. Some studies also show that the outputs of EC techniques are lightweight.

There are few studies on IDS architecture. How to deploy intrusion detection nodes effectively and efficiently is an area worth investigating. Moreover, the proposed approaches should not introduce new vulnerabilities to the system; therefore, proposed architectures should also be evaluated in terms of security. Furthermore, the optimal solution of the problem could change over time. For example, the location of nodes should adapt to the system due to changes in mobility, traffic, and resources in some environments. The applicability of EC techniques could be investigated for such cases.

Testing is another area that needs to be explored. Even though studies that evolve new attacks or unknown variants of attacks exist in the literature, how to use a derived attack in order to develop our solutions has not been considered. It is believed that better solutions could be developed when evolved new attacks are taken into account. That's why a coevolutionary arms race mechanism could be employed to concurrently develop both intrusions and IDSs. Even though coevolutionary computation has a potential application area in security, there are few applications in this area (Ostaszewski et al., 2007).

Research on IDS security has accelerated in recent years. Attackers search for new ways to attack security solutions and to escape from such solutions. Especially, the security of machine learning-based IDSs should be investigated (Corona et al., 2013). As far as we know, there is no study exploring how EC-based solutions are resilient to attacks such as evasion and DoS attacks.

To sum up, IDS is a research area extensively explored. However, some subfields of intrusion detection such as IDS testing and IDS security still need to be the subject of further studies. With the introduction of new attack strategies, new types of networks, and the like, it is a field that has been evolving continuously. As shown in this study, EC techniques are suited to developing effective, efficient, and adaptive IDSs. The suitability of evolutionary dynamic optimization and coevolutionary computation techniques are promising areas that could be explored in future studies.

ACKNOWLEDGMENTS

This study is supported by the Scientific and Technological Research Council of Turkey (TUBITAK Project No: 112E354). We would like to thank TUBITAK for its support.

REFERENCES

Abadeh, M.S., Habibi, J., Barzegar, Z., Sergi, M., 2007a. A parallel genetic local search algorithm for intrusion detection in computer networks. Eng. Appl. Artif. Intell. 20, 1058–1069.

Abadeh, M.S., Habibi, J., Lucas, C., 2007b. Intrusion detection using a fuzzy genetics-based learning algorithm. J. Netw. Comput. Appl. 30, 414–428.

Abraham, A., Grosan, C., 2006. Evolving Intrusion Detection Systems. In: Nedjah, N., Abraham, A., de Macedo Mourelles, L. (Eds.), Genetic Systems Programming: Theory and Experiences, volume 13 of Studies in Computational Intelligence. Springer, Germany, pp. 57–79.

Abraham, A., Grosan, C., Martiv-Vide, C., 2007. Evolutionary design of intrusion detection programs. Int. J. Netw. Secur. 4, 328–339.

Ahmad, I., Hussain, M., Alhgamdi, A., Alelaiwi, A., 2014. Enhancing SVM performance in intrusion detection using optimal feature subset selection based on genetic principal components. Neural Comput. Appl. 24 (7–8), 1671–1682.

Axelsson, S., 2000. Intrusion detection systems: a survey and taxonomy. Technical report 99-15, Department of Computer Engineering, Chalmers University of Technology.

Bridges, S.M., Vaughn, R.B., 2000. Fuzzy data mining and genetic algorithms applied to intrusion detection. In: Proc. of the 23rd National Information Systems Security Conference, pp. 13–31.

CERT, 2011. Cybersecurity Watch Survey. http://www.cert.org/insider-threat/.

Chen, H., Clark, J.A., Shaikh, S.A., Chivers, H., Nobles, P., 2010. Optimising IDS sensor placement. In: Proc. of International Conference on Availability, Reliability, and Security.

Coello, C.A.C., Lamont, G.B., van Veldhuizen, D.A., 2007. Evolutionary algorithms for solving multi-objective problems. Springer, USA.

Corona, I., Giacinto, G., Roli, F., 2013. Adversarial attacks against intrusion detection systems: taxonomy, solutions and open issues. Inf. Sci. 239, 201–225.

Crosbie, M., Spafford, E.H., 1995. Applying genetic programming to intrusion detection. In: Working Notes for the AAAI Symposium on GP, pp. 1–8.

Dam, H.H., Shafi, K., Abbass, H.A., 2005. Can evolutionary computation handle large datasets? A study into network intrusion detection. In: Proc. of Australian Conference on Artificial Intelligence, LNAI 3809. Springer, pp. 1092–1095.

Denning, D., 1987. An intrusion detection model. IEEE Trans. Softw. Eng. 13 (2), 222–232.

Diaz-Gomez, P.A., Hougen, D., 2005a. Improved off-line intrusion detection using a genetic algorithms. In: Proc. of the Seventh International Conference on Enterprise Information Systems, pp. 66–73.

Diaz-Gomez, P.A., Hougen, D., 2005b. Analysis and mathematical justification of a fitness function used in an intrusion detection system. In: Proc. of the Genetic and Evolutionary Computation, pp. 1591–1592.

Feeney, L.M., 2001. An energy consumption model for performance analysis of routing protocols for mobile ad hoc networks. Mob. Netw. Appl. 6, 239–249.

Fogel, D.B., 2000. What is evolutionary computation? IEEE Spectr. 37, 28–32.

Gomez, J., Gil, C., Banos, R., Marquez, A.L., Montoya, F.G., Montoya, M.G., 2013. A Pareto-based multi-objective evolutionary algorithms for automatic rule generation in network intrusion detection systems. Soft. Comput. 17, 255–263.

Gu, G., Fogla, P., Dagon, D., Lee, W., Skoric, B., 2006. Measuring intrusion detection capability: an information-theoretic approach. In: Proc. of the 2006 ACM Symposium on Information, Computer and Communications Security, ACM, pp. 90–101.

Hassanzadeh, A., Stoleru, R., 2011b. Towards optimal monitoring in cooperative ids for resource constrained wireless networks. In: Proc. of the 20th International Conference on Computer Communications and Networks.

Hassanzadeh, A., Stoleru, R., 2013. On the optimality of cooperative intrusion detection for resource constrained wireless networks. Comput. Secur. 34, 16–35.

Hassanzadeh, A., Stoleru, R., Shihada, B., 2011a. Energy efficient monitoring for intrusion detection in battery-powered wireless mesh networks. In: Proc. of ADHOC-NOW, LNCS 6811, Springer, pp. 44–57.

Helmer, G., Wong, J., Honavar, V., Miller, L., 1999. Feature selection using a genetic algorithm for intrusion detection. In: Proc. of the Genetic and Evolutionary Computation Conference, vol. 2.

Helmer, G., Wong, S.K., Honavar, V., Miller, L., 2002. Automated discovery of concise predictive rules for intrusion detection. J. Syst. Softw. 6, 165–175.

Hofmann, A., Horeis, T., Sick, B., 2004. Feature selection for intrusion detection: an evolutionary wrapper approach. In: Proc. of IEEE International Joint Conference on Neural Networks, pp. 1563–1568.

Johnson, D.M., Teredesai, A.M., Saltarelli, R.T., 2005. Genetic programming in wireless sensor networks. In: Proc. of the European Conference on Genetic Programming, LNCS 3447, pp. 96–107.

Kayacik, H.G., Zincir-Heywood, A.N., Heywood, M.I., Burschka, S., 2009. Generating mimicry attacks using genetic programming: a benchmarking study. In: Proc. of the 2009 IEEE Symposium on Computational Intelligence in Cyber Security (CICS-2009).

Kayacik, H.G., Zincir-Heywood, A.N., Heywood, M.I., 2011. Evolutionary computation as an artificial attacker: generating evasion attacks for detector vulnerability testing. Evol. Intel. 4 (4), 243–266.

Kayacık, H.G., Heywood, M.I., Zincir-Heywood, A.N., 2005a. Evolving successful stack overflow attacks for vulnerability testing. In: Proc. of the IEEE 21st Annual Computer Security Applications Conference (ACSAC).

Kayacık, H.G., Heywood, M.I., Zincir-Heywood, A.N., 2005b. On evolving buffer overflow attacks using genetic programming. In: Proc. of the 11th Genetic and Evolutionary Computation Conference (GECCO-2006).

Khanna, R., Liu, H., Chen, H., 2007. Dynamic optimization of secure mobile sensor networks: a genetic algorithm. In: Proc. of IEEE International Conference on Communications.

Khanna, R., Liu, H., Chen, H., 2009. Reduced complexity intrusion detection in sensor networks using genetic algorithm. In: Proc. of IEEE International Conference on Communications.

Kim, D.S., Nguyen, H., Park, J.S., 2005. Genetic algorithm to improve SVM based network intrusion detection system. In: Proc. of the 19th International Conference on Advandced Information Networking and Applications.

Koza, J.R., 1992. Genetic Programming: On the Programming of Computers by Means of Natural Selection. MIT Press, Cambridge, MA.

Lange, D.B., Oshima, M., 1999. Seven good reasons for mobile agents. Commun. ACM 42 (3), 88–89.

Lippman, R.P., Haines, J.W., Fried, D.J., Korba, J., Das, K., 2000. Analysis and results of the 1999 darpa off-line intrusion detection evaluation. In: Proc. of the International Symposium on Recent Advances in Intrusion Detection, LNCS 2212, Springer, pp. 162–182.

Lu, W., Traore, I., 2004. Detecting new forms of network intrusion using genetic programming. Comput. Intell. 20 (3), 475–494.

Lundin, E., Jonsson, E., 2002. Survey of intrusion detection research. Technical report 02-04, Department of Computer Engineering, Chalmers University of Technology.

Mabu, S., Chen, C., Lu, N., Shimada, K., Hirasawa, K., 2011. An intrusion detection model based on fuzzy class-association-rule mining using genetic network programming. IEEE Trans. Syst. Man Cybern. C Appl. Rev. 41 (1), 130–139.

Makanju, A., LaRoche, P., Zincir-Heywood, A.N., 2007. A comparison between signature and GP-based IDSs for link layer attacks on WiFi networks. In: Proc. of the 2007 IEEE Symposium on Computational Intelligence in Security and Defense Applications.

Me, L., 1998. GASSATA: a genetic algorithm as an alternative tool for security audit trails analysis. In: Proc. of the International Symposium on Recent Advances in Intrusion Detection.

Middlemiss, M.J., Dick, G., 2003. Weighted feature extraction using a genetic algorithm for intrusion detection. In: Proc. of the 2003 Congress on Evolutionary Computation, pp. 1669–1675.

Mukkamala, S., Sung, A.H., Abraham, A., 2004. Modeling intrusion detection systems using linear genetic programming approach. In: Proc. of the 17th International Conference on Industrial and Engineering Applications of Artificial Intelligence and Expert Systems, LNCS 3029, Springer, pp. 633–642.

Nan, G., Li, M., 2008. Evolutionary based approaches in wireless sensor networks: a survey. In: Proc. of the 4th International Conference on Natural Computation.

Nguyen, H., Franke, K., Slobodan, Petrovic, 2010. Improving effectiveness of intrusion detection by correlation feature selection. In: Proc. of the International Conference on Availability, Reliability, and Security, pp. 17–24.

Noreen, S., Murtaza, S., Shafiq, M.Z., Farooq, M., 2009. Evolvable malware. In: Proc. of the 14th Genetic and Evolutionary Computation Conference (GECCO-2009).

O'Neill, M., Ryan, C., 2003. Grammatical Evolution: Evolutionary Automation Programming in an Arbitrary Language. Kluwer Academic Publishers, Dordrecht.

Orfila, A., Estevez-Tapiador, J.M., Ribagorda, A., 2009. Evolving high-speed, easy-to-understand network intrusion detection rules with genetic programming. In: Proc. of Evo-Workshops on Applications of Evolutionary Computations, LNCS Series, Vol. 5484, Springer, pp. 93–98.

Ostaszewski, M., Seredynski, F., Bouvry, P., 2007. Coevolutionary-based mechanisms for network anomaly detection. J. Math. Model. Algorithm. 6 (4), 411–431.

Pastrana, S., 2014. Attacks against intrusion detection networks: evasion, reverse engineering and optimal countermeasures. PhD Thesis, Computer Science and Engineering Department, Universidad Carlos III De Madrid.

Pastrana, S., Orfila, A., Ribagorda, A., 2011. Modeling NIDS evasion with genetic programming. In: Proc. of the 9th International Conference on Security and Management.

Pastrana, S., Mitrokotsa, A., Orfila, A., Peris-Lopez, P., 2012. Evaluation of classification algorithms for intrusion detection in MANETs. Knowl.-Based Syst. 36, 217–225.

Perkins, C.E., Royer, E.M., 1999. Ad-hoc on-demand distance vector routing. In: Proc. of the 2nd IEEE Workshop on Mobile Computing Systems and Applications, IEEE, pp. 90–100.

Roesch, M., 1999. Snort—lightweight intrusion detection for networks. In: Proc. of the 13th USENIX Conference on System Administration Conference, pp. 229–238.

Ryan, C., Colline, J.J., O'Neill, M., 1998. Grammatical evolution: evolving programs for an arbitrary language. In: Proc. of the First European Workshop on Genetic Programming, LNCS 1391, Springer, pp. 83–95.

Sen, S., 2014. Using instance weighted naive Bayes for adapting concept drift in masquerade detection. Int. J. Inf. Secur. 13, 583–590.

Sen, S., Clark, J.A., 2009a. Intrusion detection in mobile ad hoc networks. In: Misra, S., Woungang, I., Misra, S.C. (Eds.), Guide to Wireless Ad Hoc Networks. Springer, London, UK, pp. 427–454.

Sen, S., Clark, J.A., 2009b. A grammatical evolution approach to intrusion detection on mobile ad hoc networks. In: Proc. of the Second ACM Conference on Wireless Network Security, pp. 95–102.

Sen, S., Clark, J.A., 2011. Evolutionary computation techniques for intrusion detection in mobile ad hoc networks. Comput. Netw. 55 (15), 3441–3457.

Sen, S., Dogmus, Z., 2012. Feature selection for detection of ad hoc flooding attacks. Advances in Computing and Information Technology176, Springer, Germany, pp. 507–513.

Shazzad, K.M., Park, J.S., 2005. Optimization of intrusion detection through fast hybrid feature selection. In: Proc. of the Sixth International Conference on Parallel and Distributed Computing, Applications and Technologies.

Sinclair, C., Pierce, L., Matzner, S., 1999. An application of machine learning to network intrusion detection. In: Proc. of the 15th Annual Computer Security Applications Conference, pp. 371–377.

Snort, https://www.snort.org/ (accessed on July, 2014).

Song, D., Heywood, M.I., Zincir-Heywood, A.N., 2005. Training genetic programming on half a million patterns: an example from anomaly detection. IEEE Trans. Evol. Comput. 9 (3), 225–239.

Stein, G., Chen, B., Wu, A.S., Hua, K.A., 2005. Decision tree classifier for network intrusion detection with GA-based feature selection. In: Proc. of the 43rd ACM Annual Southeast Regional Conference, vol 2, pp. 136–141.

Tahta, U.E., Can, A.B., Sen, S., 2014. Evolving a trust model for peer-to-peer networks using genetic programming. In: Proc. of EvoApplications, LNCS 8602, Springer, pp. 3–14.

Uppuluri, P., Sekar, R., 2001. Experiences with specification-based intrusion detection. In: Proc. of the Recent Advances in Intrusion Detection, LNCS 2212, Springer, pp. 172–189.

Wilson, D., Kaur, D., 2007. Knowledge extraction from KDD'99 intrusion data using grammatical evolution. WSEAS Trans. Inf. Sci. Appl. 4, 237–244.

Wu, S.X., Banzhaf, W., 2010. The use of computational intelligence in intrusion detection systems: a review. Appl. Soft Comput. 10 (1), 1–35.

Zhao, J., Zhao, J., Li, J., 2005. Intrusion detection based on clustering genetic algorithm. In: Proc. of the Fourth International Conference on Machine Learning and Cybernetics, IEEE, pp. 3911–3914.

Zitzler, E., Laumans, M., Thiele, L., 2001. Spea2: improving the strength pareto evolutionary algorithm. Technical report 103, Swiss Federal Institute of Technology.

VoIP Quality Prediction Model by Bio-Inspired Methods

5

Tuul Triyason[1], Sake Valaisathien[1], Vajirasak Vanijja[1], Prasert Kanthamanon[1], and Jonathan H. Chan[2]

[1]*IP Communications Laboratory (I-Lab), School of Information Technology, King Mongkut's University of Technology Thonburi, Bangkok, Thailand*
[2]*Data and Knowledge Engineering Laboratory (D-Lab), School of Information Technology, King Mongkut's University of Technology Thonburi, Bangkok, Thailand*

CHAPTER CONTENTS

5.1 INTRODUCTION

The fundamentals of speech quality measurement have been researched and developed for many years to keep up with the rapid growth of telecommunication technology. Voice over Internet Protocol (VoIP) is a technology that carries voice in the form of packet data and sends it across the IP network. VoIP has become extremely popular because it is a low-cost service compared to the traditional telephone services, especially for making long-distance calls. However, VoIP has many limitations, such as difficulty guaranteeing quality of service (QoS) on the Internet, codec (compression/decompression algorithm) selection to minimize the bandwidth and maximize the voice quality, voice quality measurement, and so on. These issues are currently active research fields.

The IP network is a connectionless network that uses the best-effort paradigm to transmit data over a packet switched network (Chong and Matthews, 2004). Especially in the public Internet, it is almost impossible to control and stabilize the quality of the data transmission. Voice communications have been implemented on the circuit switch network for more than a century. The circuit switched network is predictable, and quality is guaranteed (Chong and Matthews, 2004). However, it is more difficult to manage and control because many physical cables and equipment are necessary. For the new generation equipment, most of the voice network's core switches are packet switches. However, these are in a closed system, which makes them relatively easy to manage. The Internet, which is the biggest public network, has millions of routers and switches. Thus, guaranteeing the QoS on the Internet is a very difficult task. A common method used to manage the quality of the packet transmission over the Internet is to measure the network conditions and change the codec and quality of the voice to match the situation (Akhshabi et al., 2011).

Voice quality measurement can be carried out using either subjective or objective methods. The Mean Opinion Score (MOS) is the most widely used subjective measurement of voice quality and is recommended by the International Telecommunication Union (ITU). A value is normally obtained as an average opinion of quality based on people's opinion to grade the quality of speech signals on a five-point scale under controlled conditions, as set out in the ITU-T standard P.800 (ITU-T P.800, 1996). The subjective test can be listening only (i.e., one way) or conversational. In the latter case, the voice quality scores are sometimes referred to as conversational "MOSc." There has been a paucity of published literature on modeling of the conversational quality of VoIP, which is the main contribution of this work.

The objective measurement of voice quality can be intrusive or nonintrusive. Intrusive methods are more accurate, but normally are unsuitable for monitoring live traffic because of the need for reference data and to utilize the network. Voice quality modeling is a method to predict the voice quality based on some parameters from the environment of the VoIP transmission. Network impairments such as delay, jitter, and loss of data packet directly affect the voice quality. The voice codec (Narbutt and Davis, 2005) encodes and compresses voice before it packetizes and sends the packets over the Internet. In addition, some prior research works have shown that gender and language also affect voice quality (Cai et al., 2010). The purpose of developing voice quality prediction models is to predict the voice communication quality on IP networks to help network administrators or applications to (automatically or manually) possibly fine-tune their system to improve the QoS. An accurate objective and nonintrusive measurement method is the ideal for a measurement model. The use of bio-inspired methods seems particular fitting for this purpose.

In this chapter, objective and nonintrusive voice quality measurement models are built and discussed using several techniques, such as regression, neural network, data mining classification, and so on. Several parameters include network impairments (packet loss and delay), codec, and language and gender of the speaker, and are used as features of the data set to model and investigate the performance of the voice measurement models.

The remaining parts of this chapter are organized as follows: Section 5.2 gives background on speech quality measurement, outlining the differences between subjective and objective techniques, as well as intrusive vs. nonintrusive means for the latter technique. Then, it will provide an overview of the bio-inspired techniques that have been used thus far. Section 5.3 describes the models used in the detailed comparative study. Section 5.4 elucidates the experimental testbed. Section 5.5 shows the results and provides discussion of the modeling and simulation outcome. Finally, Section 5.6 concludes this work.

5.2 SPEECH QUALITY MEASUREMENT BACKGROUND

In the past, speech quality assessment methods have been developed with either subjective or objective techniques. The subjective method is an assessment by captured data (voice or data packets) and mathematical expressions. The details of each method are described as follows.

5.2.1 SUBJECTIVE METHODS

Details of the subjective measurement are presented in ITU-T Recommendation P.800 (ITU-T P.800, 1996). The procedures involve asking the testers to rate the quality of hearing/listening from a set of speech samples under a controlled environment. The quality scores are arranged in a five-point scale. The most used five-point scale is Absolute Category Rating (ACR). The ACR rating scale and its opinion values are shown in Table 5.1.

The average quality score from each test condition is termed the MOS. The subjective test can be conducted by a listening or conversational test. In order to avoid misunderstanding and incorrect interpretation of MOS, ITU-T has published recommendation P.800.1 to define the terminology in terms of MOS, as to whether specific values of MOS are related to listening or conversational quality (ITU-T P.800.1, 2006). If the test is a listening test, MOS-LQ is used as the listening quality index. If the test is a conversational test, MOS-CQ is used as the conversational quality

Table 5.1 Absolute category rating (ACR) scale

ACR Scale	Opinion
5	Excellent
4	Good
3	Fair
2	Poor
1	Bad

index. It is well known that subjective testing is the most accurate and reliable method for assessing speech quality. However, a subjective test is time- and money-consuming. The results from the same subject may change if the tester is made to repeat the experiment. These drawbacks of subjective testing have led to the development of objective methods.

5.2.2 INTRUSIVE OBJECTIVE METHODS

Intrusive objective methods normally take two signals into consideration: one is the reference signal, whereas the second is the degraded signal. The reference signal is injected into the network under test and the (degraded) output signal is compared to the reference signal. The speech quality is computed from the difference between references and degraded signal by a perception-based model. Several techniques have been developed to measure voice quality intrusively: Perceptual Assessment of Speech Quality (PAMS), Perceptual Speech Quality Measure (PSQM), Measuring Normalizing Blocks (MNB), Enhanced Modified Bark Spectral Distortion (EMBSD), and Perceptual Evaluation of Speech Quality (PESQ) (Hammer et al., 2004). PESQ is the most popular intrusive objective method for speech quality prediction (ITU-T P.862, 2001). It was developed to measure the perception of speech quality in many kinds of network, such as Integrated Services Digital Network (ISDN), IP, and so on. It was developed by the cooperation between British Telecommunications and KPN Research in the Netherlands, by combining the ideas of PSQM+ and PAMS. PESQ uses a perception model to compare an original signal with a degraded signal received at the destination side. The basic structure of PESQ methods is illustrated in Figure 5.1.

PESQ takes into consideration not only codec but also network parameters such as packet loss, low bit-rate coding, and jitter. However, the effects of loudness loss, delay, side tone effect, echo, and other impairments related to two-way interaction are not reflected in the PESQ. The resulting PESQ score can be converted to MOS using a mapping function as described in ITU-T recommendation P.862.1 (ITU-T P.862.1, 2003). The wideband extension of the PESQ was later published in ITU-T recommendation P.862.2 (ITU-T P.862.2, 2007). The wideband extension version uses a different model to measure wideband quality and uses a different mapping function to the MOS scale. ITU launched some experiments to investigate the

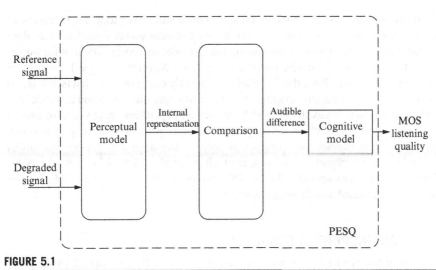

FIGURE 5.1

PESQ functional diagram.

accuracy of the PESQ method. The experiment covers a wide range of impairment conditions. The results show that PESQ has a high correlation with subjective testing, with a correlation coefficient of 0.935 (ITU-T P.862, 2001). Intrusive objective methods are more accurate to measure perceived speech quality. However, they are not appropriate to use in real-time traffic monitoring applications.

5.2.3 NONINTRUSIVE OBJECTIVE METHODS

There are two approaches of nonintrusive methods. One is the signal-based approach, which measures a speech quality from the degraded signal. ITU-T recommendation P.563 (3SQM) is a representative of this approach (ITU-T P.563, 2004). It is used for predicting the perceived speech quality of narrowband telephony by analyzing the output-degraded signal from the network under test. According to the recommendation, it is also used for live network monitoring. Another one is the parameter-based approach, which measures speech quality from impairment parameters in the network under test. The most popular parameter-based nonintrusive method is presented in ITU-T recommendation G.107, and it is known as the E-model (ITU-T G.107, 2011). The E-model is designed for transmission planning. It can calculate the speech quality with transmission impairment parameters such as packet loss, delay, and jitter. The speech qualities calculated from the E-model are presented in the form of R-factors, which lie in the range from 0 to 100. The speech quality with R equal to zero represents an extremely bad quality, whereas R equal to one hundred represents a very high quality. Generally, R less than 60 is considered poor quality.

In recent years, the E-model has become a most attractive model for nonintrusive speech quality prediction. The reason is that the E-model was designed to be used for network planning purposes, thus making the E-model an appropriate tool for implementation in VoIP live traffic monitoring systems. Nevertheless, the E-model still has some limitations. First, the E-model is a statically based model. That is, it is based on a set of constant parameters that are related to a specific codec and network conditions. This fact makes the E-model calculation dependent on the case of known network parameters such as delay, packet loss, and jitter. However, past research has shown that nonnetwork parameters such as language and gender also affect the accuracy of objective measurement (Ren et al., 2008; Ebem et al., 2011; Triyason and Kanthamanon, 2012). This makes the E-model not flexible for use in different cultural and language environments.

5.2.4 BIO-INSPIRED METHODS

The use of bio-inspired methods in speech quality modeling is inspired by the prevalence of bio-inspired computing in the artificial intelligence field. The main difference is that bio-inspired techniques tend to take more of an evolutionary approach to learning. The earlier works involve the use of artificial neural networks in a machine learning approach to model VoIP speech quality (Sun, 2004; Sun and Ifeachor, 2006). Then, Raja et al. (2007) recognized that the traditional models used have been empirical ones that involve mathematical functions such as logarithm and exponents. As a natural extension, they used the concept of genetic programming to search for the desirable combination of functions to be used in the empirical models. The 2007 pioneer work received honorable mention by Koza (2010) as being "human-competitive." This is currently an area of significant interest for various researchers, including the authors.

5.3 MODELING METHODS

5.3.1 METHODOLOGY FOR CONVERSATIONAL QUALITY PREDICTION (PESQ/E-MODEL)

The fundamental aspect of PESQ/E-model improvement is to use the PESQ MOS as the calibration parameter for the E-model. PESQ has a higher accuracy than the E-model, but it is an intrusive measurement method (Sun and Ifeachor, 2006). To use PESQ as a speech quality measurement tool, reference samples have to be injected into the network under test conditions. This makes PESQ inappropriate for real-time monitoring tasks. Therefore, the aim of the PESQ/E-model technique is to improve the accuracy of the nonintrusive method to have the same precision as an intrusive method. The advantage of this model is to provide a nonintrusive monitoring tool that predicts the MOS as well as PESQ. Figure 5.2 illustrates how PESQ scores are used to improve the accuracy of the E-model.

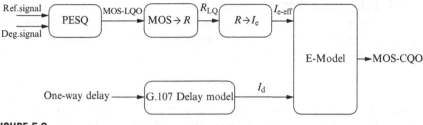

FIGURE 5.2

The PESQ/E-model diagram.

The E-model is a computation model developed for network planning. It was developed by a European Telecommunication Standards Institute (ETSI) work group of ITU-T. It was designed as a nonintrusive measurement technique that would not require the injection of a reference speech file. The model calculates a basic quality value determined by a network, later subtracted with the transmission impairment factors. The E-model combines the effect of each impairment factor into a rating factor scale "R," ranging from 0 (poor) to 100 (excellent). The R factor can be further transformed into the MOS scale with a mapping function from ITU-T. The R factor can be calculated by the following equation:

$$R_{\text{conversation}} = R_0 - I_s - I_d - I_{\text{e-eff}} + A \tag{5.1}$$

- R_0 represents the basic signal-to-noise ratio at 0 dB$_r$.
- I_s is the sum of all impairments that may occur in a voice transmission.
- I_d is the impairment factor due to delay.
- $I_{\text{e-eff}}$ is an effective equipment impairment factor (low bit-rate coding, packet loss)
- A is an advantage factor or expectation factor ($A = 10$ for a wired connection).

Most of the parameters for E-model calculation are defined in ITU-T recommendation G.113 (ITU-T G.113, 2007). According to Cole and Rosenbluth (2001), some parameters of the E-model are not related to packet-switched communication, such as I_s and A. In the case of VoIP, the original E-model can be reduced to the simplified version of AT&T labs as shown in Equation (5.2).

$$R_{\text{conversation}} = R_0 - I_d - I_{\text{e-eff}} \tag{5.2}$$

The method of calculation for each parameter is described as follows.

5.3.1.1 Basic signal-to-noise ratio, R₀

R_0 is the basic signal-to-noise ratio at 0 dB$_r$ point. According to recommendation G.107, the calculation details of R_0 are too complicated for direct calculation. If all parameters are set to the default values, the calculation result for R_0 is equal to 93.2 (ITU-T G.107, 2011). In the work shown in this chapter, the value of R_0 is set to the default as mentioned in recommendation G.107.

5.3.1.2 Delay impairment factor, I_d

According to recommendation G.107, I_d is the impairment factor representing all impairment from delay of voice signal. I_d is the combination of three delay factors, as shown in Equation (5.3).

$$I_d = I_{dte} + I_{dle} + I_{dd} \tag{5.3}$$

The factor I_{dte} gives an estimate for the impairment due to the talker's echo. The factor I_{dle} represents impairments due to the listener's echo, and the factor I_{dd} represents the impairment caused by excessive absolute delay. There are three different types of delay that are associated with I_d calculation. The average absolute delay T_a represents the total one-way delay between the send and receive sides and is used to estimate the impairment due to excessive delay. The average one-way delay T represents the delay between the receive side and the point in a connection where a signal coupling occurs as a source of echo. The average round-trip delay T_r represents the delay in a 4-wire loop that is related to the listener's echo. In the case of VoIP with no circuit-switched network interworking, the various measurement points become a single pair of points as mentioned in Cole and Rosenbluth (2001). The relation of these types of delay can be expressed in the form of one-way delay d as shown in Equation (5.4).

$$d = T_a = T = T_r/2 \tag{5.4}$$

Then, the I_d parameters can be calculated from one-way delay values, where the relation between one-way delay and I_d parameter is shown in Figure 5.3 (the calculation method of I_d can be conducted through the complicated set of equations as shown in recommendation G.107). According to Sun (2004), the values of I_d can be obtained through the more simplified Equation (5.5) by a curve-fitting method. This model is true for one-way delay of less than 400 ms.

$$I_d = 0.024d + 0.11(d - 177.3)H(d - 177.3)$$
$$\text{where} \begin{cases} H(x) = 0 & \text{if } x < 0 \\ H(x) = 1 & \text{if } x \geq 0 \end{cases} \tag{5.5}$$

Note that for one-way delay of $d < 100$ ms, the effect of the delay impairment factor is not significant and can be discarded (Voznak, 2011). Equation (5.2) can then be reduced to a more simplified form, as shown in Equation (5.6).

$$R_{conversation} \approx R_{listening} = R_0 - I_{e\text{-eff}} \tag{5.6}$$

This fact implies that in the case of a small one-way delay, the perceived listening and conversation quality have almost the same value. This agrees with the subjective test experiment by Gros and Chateau (2002), which reported that in the case of VoIP with only packet-loss impairment, the conversational subjective test can be substituted with the listening subjective test, which is less expensive in terms of both time and cost.

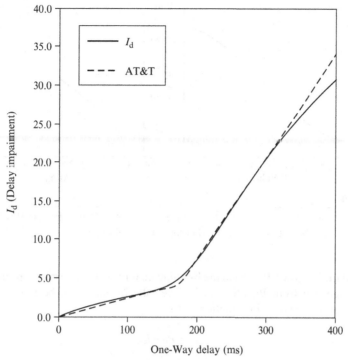

FIGURE 5.3

Relation between one-way delay and I_d from G.107 and AT&T simplified model.

5.3.1.3 MOS-to-R conversion function and effective equipment impairment factor, $I_{e\text{-}eff}$

ITU-T recommendation G.107 defines the conversion function between R and MOS as shown in Equation (5.7) and Figure 5.4 (left).

$$(R \leq 0) : \text{MOS} = 1$$
$$(0 < R < 100) : \text{MOS} = 1 + 0.035R + R(R - 60)(100 - R) \cdot (7 \times 10^{-6}) \qquad (5.7)$$
$$(R \geq 100) : \text{MOS} = 4.5$$

However, the step function given in Equation (5.7) cannot be converted back directly to obtain the R value in the form of MOS. The conversion from MOS to R-value can be done by using a more complicated set of equations as mentioned in Hoene et al. (2006). Different techniques have been developed for this purpose. In particular, the inverse function is created from the curve of G.107 and the curve-fitting technique is used. The inverse curve fits well with the third-order polynomial function as shown in Equation (5.8) and Figure 5.4 (right).

$$R = 3.113\,\text{MOS}^3 - 26.105\,\text{MOS}^2 + 89.31\,\text{MOS} - 59.293 \qquad (5.8)$$

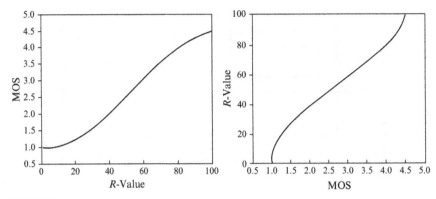

FIGURE 5.4

(Left) The conversions function of R to MOS from ITU-T Rec. G.107. (Right) The inverted graph of function MOS to R is fit by third-order polynomial regression.

The result of PESQ is reflected in the listening quality. The impairment from delay I_d can be neglected from Equation (5.2). Then the E-model can be reduced to the listening form, as shown in Equation (5.9).

$$R_{\text{listening}} = R_0 - I_{\text{e-eff}} \qquad (5.9)$$

Listening quality MOS from the PESQ method can be converted to $R_{\text{listening}}$ with Equation (5.8). The $R_{\text{listening}}$ can be converted to $I_{\text{e-eff}}$ using Equation (5.9).

5.3.2 NONLINEAR SURFACE REGRESSION MODEL

The results from the PESQ/E-model method show that the relation among levels of packet loss in the network, one-way delay, and MOS-CQO (conversational quality from an objective model) is nonlinear. Hence, a polynomial surface equation can be constructed to describe this relation. In the work shown in this chapter, the training data set can fit quite well with a third-order polynomial surface function (cubic function) with a goodness of fit (R^2) between 0.93 and 0.95. The general form of cubic function is shown in Equation (5.10) as follows.

$$\text{MOS} - \text{CQO} = a\rho^3 + b\delta^3 + c\rho^2 + d\delta^2 + e\rho^2\delta + f\rho\delta^2 + g\rho + h\delta + i\rho\delta + j \qquad (5.10)$$

where ρ is percent of packet loss in network and δ is one-way network delay in milliseconds, and the cubic equation coefficients ($a, b, c, d, e, f, g, h, i$, and j) are the set of constants of surface fitting, which varies in each experiment scenario. In this study, the experiment scenario is the set of codec, language, and gender. Figure 5.5 shows the plot of cubic equation surface fitting from the results of PESQ/E-model in case of G.711 codec/Thai language/male speech.

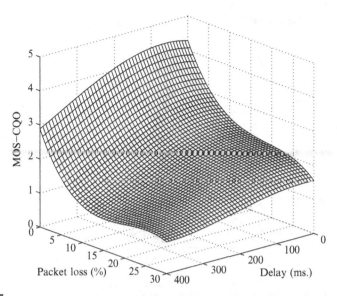

FIGURE 5.5

Nonlinear surface regression model of PESQ/E-model of G.711/Thai/male.

5.3.3 **NEURAL NETWORK MODEL**

A neural network model is represented by its architecture that shows how to transform two or more inputs into an output. The transformation is given in the form of a learning algorithm. In this work, the feed-forward architecture used is a multilayer perceptron (MLP) that utilizes back propagation as the learning technique. The difference from other mathematical models like E-model is that this model is able to retrain to learn a new relationship of voice quality and impairment factor (Sun, 2004). This is a great benefit for IP networks in which various parameter factors are not constant. In particular, a three-layer MLP neural network is created from the open-source Waikato Environment for Knowledge Analysis (Weka) software (Hall et al., 2009). The input nodes are the impairment parameters consisting of packet-loss rate, delay, codec types, gender, and language, while the output node is the predicted voice quality, MOS-CQO. The number of hidden nodes is optimized by trial-and-error analysis to be five nodes, and the model is created from the training data set using tenfold cross-validation. The structure of the neural network model used is shown in Figure 5.6.

5.3.4 **REPTree MODEL**

The decision tree is one of the most widely used classifiers for classifying problems. The REPTree classifier is based on the concept of the decision tree, but it calculates the information gain with entropy and uses it as the splitting criterion. It reduces error from variance by pruning. REPTree only sorts values for numeric attributes once.

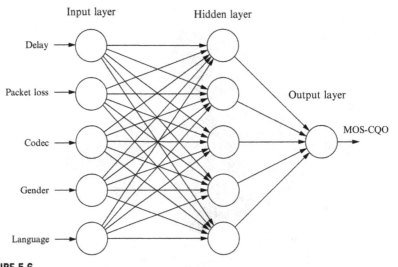

FIGURE 5.6

Feed-forward neural network with five hidden nodes structure.

Missing values are dealt with using the C4.5 method using fractional instances (Zhao and Zhang, 2008). To compare the accuracy of this classifier method, Weka software is used to create the REPTree model from the training data set.

5.4 EXPERIMENTAL TESTBED

To simulate a VoIP system and provide the training and testing data sets, an experimental testbed is set up. The details of the experimental testbed are shown in Figure 5.7 as follows.

Step 1:

The training and testing speech files are fed into the simulation of the VoIP system named Network-emulator (Imankulov, 2009). Network-emulator is a simple utility intended to test how network losses affect speech quality in VoIP-based applications. The training data set comes from ITU-T recommendation P.50 speech corpus (ITU-T P.50, 1999) and *Thai Speech Set for Telephonometry* (TSST) (Daengsi et al., 2010) corpus. All speech signals have lengths between 8 and 12 seconds. ITU-T P.50 speech corpus contains British English language with 16 sentences (8 female and 8 male speakers), whereas TSST speech corpus contains 8 Thai sentences from 4 male and 4 female speakers. The signal-sampling rate is revised to narrowband sampling. G.711, G.729, and Speex codec are used for the encoding task. For simplicity, the Bernoulli loss pattern (random loss) is set up with 0 to 30% (increment by 3%) by Network-emulator software. To remove the impact from bias of the data set, the testing data set uses the signal from a different speech corpus. The testing speech

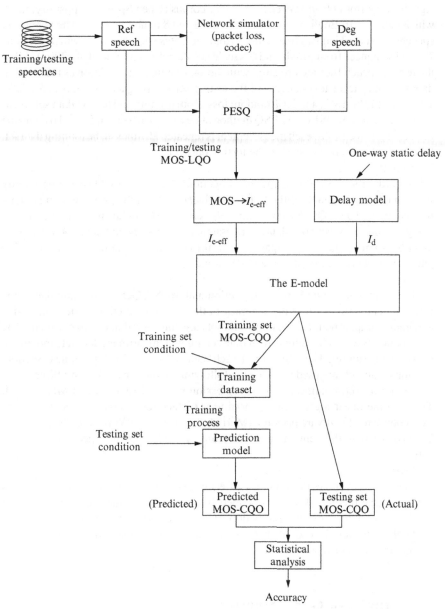

FIGURE 5.7

Experimental testbed diagram.

signals come from *Open Speech Repository* corpus (Open Speech Repository, n.d.), which contains 16 English sentences (8 female and 8 male speakers). The testing set speech signal for Thai language comes from the *Lotus* speech corpus, which contains 12 Thai sentences from 6 male and 6 female speakers (Patcharikra et al., 2005). All of the testing speech files are prepared with the same signaling condition as the training data set. The packet-loss patterns of the testing set are assigned to 0%, 3%, 5%, 8%, 10%, 13%, 15%, 18%, and 20% random loss. Both training and testing data sets measure the listening MOS by the PESQ method as described in Section 5.2.2. To avoid the impact of loss locations, each measurement is repeated 80 times in the training data set. Only a single instance is used in the testing data set.

Step 2:

According to the PESQ/E-model in Section5.3.1, MOS-LQO (listening quality from an objective model) of the training and testing set in the previous step is transformed into MOS-CQO by combining the effect from delay impairment. In this study, the one-way network delays are assigned to be between 0 and 400 ms (increment by 40 ms). The delay impairment factor (I_d) can be obtained by transforming one-way network delay using the delay model.

Step 3:

By combining the impairment condition and MOS-CQO, the training data set is formed. In order to evaluate the accuracy of the speech prediction model with codec and human impairment factor, three data sets are considered in this experiment. The three data sets are codec impairment data set, human impairment data set, and mixed impairment data set. The structure of each data set is described in the next section. The three data sets are used to create the following prediction models: multiple linear regression model, nonlinear surface regression model, REPTree, and MLP model. The multiple linear regression and nonlinear surface regression models are created by a commercial software package (MatLab 8.3, The MathWorks Inc., Natick, MA, 2014a), whereas the remaining models are created with the open-source Weka software.

Step 4:

In order to evaluate the accuracy of each prediction model, a testing data set is applied to the prediction model obtained from the training data set. The predicted MOS-CQO values from the prediction models are compared to the actual MOS-CQO of the testing data set. Pearson's correlation coefficient and several statistical errors' analysis are reported.

5.4.1 THE DATA SETS' STRUCTURE

5.4.1.1 Codec impairment data set

The objective of this data set is to measure the effect from codec impairment parameters commonly based on packet loss and network delay in each prediction model. The structure of the training and testing data sets from codec impairment is shown in Tables 5.2 and 5.3.

Table 5.2 The structure of training data for codec impairment set

Codec	Packet Loss	Delay	MOS-CQO
1, 2, 3	0-30%	0-400 ms	1-4.5

Codec: 1=G.711, 2=G.729, 3=Speex

Table 5.3 The structure of testing data for codec impairment set

Codec	Packet Loss	Delay	MOS-CQO
1, 2, 3	0%, 3%, 5%, 8%, 10%, 13%, 15%, 18%, 20%	0-400 ms	1-4.5

Codec: 1=G.711, 2=G.729, 3=Speex

5.4.1.2 Human impairment data set

The objective of the second data set is to consider the different characteristics of users. In this data set, a set of gender and language factors is used to investigate the accuracy of each speech prediction model. The structure of this data set is shown in Tables 5.4 and 5.5.

5.4.1.3 Mixed impairment data set

The mixed impairment data set combines the effect from codec impairment and human impairment together. The objective of this data set is to evaluate the accuracy of the speech prediction model in the case of more complex conditions. The structure of the mixed impairment data set is shown in Tables 5.6 and 5.7.

Table 5.4 The structure of training data for human impairment data set

Gender	Language	Packet Loss	Delay	MOS-CQO
1,2	1,2	0-30%	0-400 ms	1-4.5

Gender: 1=Male, 2=Female
Language: 1=Thai, 2=English

Table 5.5 The structure of testing data for human impairment data set

Gender	Language	Packet Loss	Delay	MOS-CQO
1,2	1,2	0%, 3%, 5%, 8%, 10%, 13%, 15%, 18%, 20%	0-400 ms	1-4.5

Gender: 1=Male, 2=Female
Language: 1=Thai, 2=English

Table 5.6 The structure of training data for mixed impairment data set

Codec	Gender	Language	Packet Loss	Delay	MOS-CQO
1,2,3	1,2	1,2	0-30%	0-400 ms	1-4.5

Codec: 1=G.711, 2=G.729, 3=Speex
Gender: 1=Male, 2=Female
Language: 1=Thai, 2=English

Table 5.7 The structure of testing data for mixed impairment data set

Codec	Gender	Language	Packet Loss	Delay	MOS-CQO
1,2,3	1,2	1,2	0-30%	0-400 ms	1-4.5

Codec: 1=G.711, 2=G.729, 3=Speex
Gender: 1=Male, 2=Female
Language: 1=Thai, 2=English

5.4.2 THE PERFORMANCE MEASURES

The accuracy of each speech prediction model is analyzed by the Pearson's correlation coefficient, the mean absolute error, the root mean squared error, and the relative absolute error. The results are shown in the next section.

5.5 RESULTS AND DISCUSSION

According to the last section, more than 1400 test conditions have been conducted for each modeling method, while each test is represented with an IP network environment obtained from variation of packet-loss rate, one-way delay, codecs, gender, and language as the test condition. In this section, the experimental results for each speech quality prediction model and their performance comparison are discussed in more detail.

5.5.1 CORRELATION COMPARISON

To investigate the prediction performance of each speech quality prediction model, the testing set MOS-CQO from four prediction models, namely, nonlinear surface (SF) regression model, multiple linear regression (LR) model, neural network (NN) model, and REPTree (RT) are compared with the results of MOS-CQO from the PESQ/E-model. Pearson's correlation coefficient and other statistical measures are calculated. The statistical details are shown in Tables 5.8–5.10. The correlation coefficient should be as close to 1 as possible, whereas the error measures should be minimized for improved accuracy.

Linear regression yielded the lowest correlation coefficient in all data sets, thus making it unreliable to use for prediction. Also, from the IP network condition, there

Table 5.8 Performance comparison of speech prediction model in codec impairment data set

Statistical Analysis	LR	RT	NN	SF
Correlation coefficient	0.8951	0.9859	0.9890	0.9826
Mean absolute error	0.3290	0.1051	0.0944	0.1149
Root mean squared error	0.4213	0.1364	0.1268	0.1514
Relative absolute error	51.57%	16.47%	14.80%	17.64%

Table 5.9 Performance comparison of speech prediction model in human impairment data set

Statistical Analysis	LR	RT	NN	SF
Correlation coefficient	0.8692	0.9759	0.9730	0.9645
Mean absolute error	0.4193	0.1603	0.2426	0.1054
Root mean squared error	0.5142	0.2130	0.3098	0.2536
Relative absolute error	59.95%	22.92%	34.69%	27.56%

Table 5.10 Performance comparison of speech prediction model in mixed impairment data set

Statistical Analysis	LR	RT	NN	SF
Correlation coefficient	0.8637	0.9765	0.9726	0.9698
Mean absolute error	0.4090	0.1595	0.2569	0.1799
Root mean squared error	0.4914	0.2108	0.3134	0.2334
Relative absolute error	63.60%	24.81%	39.95%	27.46%

is a nonlinear relationship with MOS. The model providing the highest accuracy for the first data set was neural network, while REPTree showed the best accuracy in the other data sets. Neural network and REPTree both showed a high correlation coefficient, but neural network tends to have more errors than REPTree and nonlinear surface regression. This is because the correlation coefficient does not take into account the accuracy of individual components in the calculation; instead, it only measures the similarity of the curve shape (ITU-T P.862, 2001).

5.5.2 RESIDUAL ANALYSIS

The error of prediction can be further analyzed by using a residual plot, which is the difference between the predicted values and the actual values. Figures 5.8–5.10 show the residual plots of each model, with actual data in codec, language, and gender

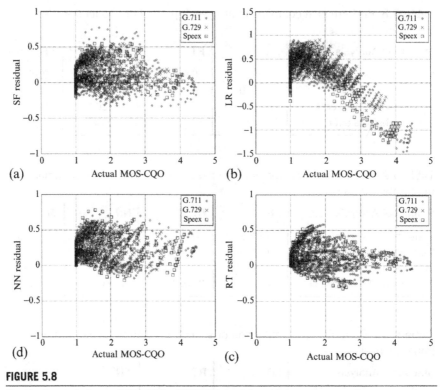

FIGURE 5.8

Residual error between MOS-CQO of PESQ/E-model and (a) nonlinear surface regression model, (b) multiple linear regression model, (c) neural network model, and (d) REPTree model in codec dimension.

dimension from all data sets. According to the residual analysis, if a model can predict the quality score accurately, the residual error will be close to zero and all residual plots will fall around the middle horizontal axis. From the results, it shows that all prediction models tend to have an overprediction of MOS between ranges 1-3, whereas linear regression shows an underprediction of MOS in ranges 4-5. There is no significant difference in the residual error for the codec dimension, except G.711 tends to have more error in the surface regression model (extreme overprediction and underprediction). It can be seen that both language and gender have a clearly significant effect on the performance of the speech quality prediction model. MOS of Thai language are overpredicted in all models, and MOS of female speech tend to have more errors under all conditions. Overall, correlation and residual analyses show that nonlinear surface regression, neural network, and REPTree are reliable models for the speech quality evaluation task, whereas the multiple linear regression model provided the lowest correlation coefficient and the highest errors, thus making it unreliable to use for speech quality prediction.

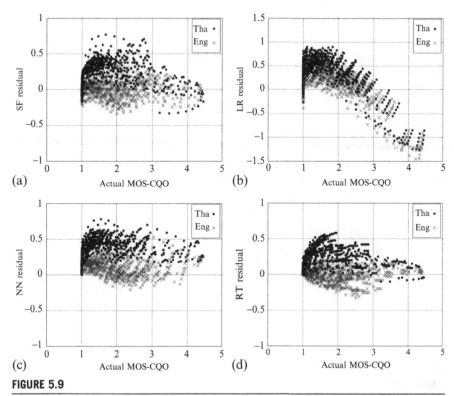

FIGURE 5.9

Residual error between MOS-CQO of PESQ/E-model and (a) nonlinear surface regression model, (b) multiple linear regression model, (c) neural network model, and (d) REPTree model in language dimension.

5.6 CONCLUSIONS

In this work, an overview of the use of bio-inspired algorithms in modeling of real-time nonintrusive VoIP performance is given. Then, an in-depth analysis is provided on the performance of the bio-inspired method of neural network and decision-tree-based REPTree in comparison to nonlinear surface and multiple linear regressions. The main objective of the work is to analyze the reliability of different classifier methods for nonintrusive speech quality prediction tasks in the conversational mode. The results show that neural network and REPTree are reliable tools for real-time quality monitoring tasks under both network and human impairments. The experimental results show a good correlation of classifier methods with ITU-T PESQ/E-model objective tests. Generally, there are both advantages and disadvantages for nonlinear regression and bio-inspired classifier models for speech quality prediction. The nonlinear regression model is simple

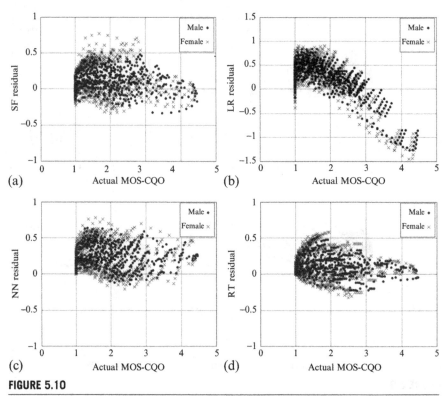

FIGURE 5.10

Residual error between MOS-CQO of PESQ/E-model and (a) nonlinear surface regression model, (b) multiple linear regression model, (c) neural network model, and (d) REPTree model in gender dimension.

and provides high accuracy, but it is specific to a set of features. The bio-inspired model has a learning ability with high adaptability, but requires more effort for fine-tuning before being applicable to a particular condition. The performance investigation in codec dimension does not show any significant bias in the prediction accuracy of the models investigated. Language and gender investigations show that Thai language is overpredicted in all models and female speech shows more errors than male speech in every condition. This shows that the accuracy of the objective speech quality prediction model is not only dependent on codec and network impairments, but human impairments such as gender and language also have a significant impact. Future work may consider more human impairments such as age and more languages (e.g., tonal vs. Latin-based), and other neural network architectures. In addition, we have started some preliminary work on implementing genetic programming models, gender prediction based on MOS-CQO, and QoS and traffic shaping to improve the speech quality over VoIP.

REFERENCES

Akhshabi, S., Begen, A.C., Dovrolis, C., 2011. An experimental evaluation of rate-adaptation algorithms in adaptive streaming over HTTP. In: Proceedings of the Second Annual ACM Conference on Multimedia Systems, MMSys'11. ACM, New York, NY, USA, pp. 157–168.

Cai, Z., Kitawaki, N., Yamada, T., Makino, S., 2010. Comparison of MOS evaluation characteristics for Chinese, Japanese, and English in IP telephony. In: Universal Communication Symposium (IUCS), 2010 4th International. Presented at the Universal Communication Symposium (IUCS), 2010 4th International, pp. 112–115.

Chong, H.M., Matthews, H.S., 2004. Comparative analysis of traditional telephone and voice-over-Internet protocol (VoIP) systems. In: 2004 IEEE International Symposium on Electronics and the Environment, 2004. Conference Record. Presented at the 2004 IEEE International Symposium on Electronics and the Environment, 2004. Conference Record, pp. 106–111.

Cole, R.G., Rosenbluth, J.H., 2001. Voice over IP performance monitoring. SIGCOMM Comput. Commun. Rev. 31, 9–24.

Daengsi, T., Preechayasomboon, A., Sukparungsee, S., Chootrakool, P., Wutiwiwatchai, C., 2010. The development of a Thai speech set for telephonometry. In: The International Committee for The Co-Ordination and Standardization of Speech Databases and Assessment Techniques 2010. http://desceco.org/O-COCOSDA2010/proceedings/paper_53.pdf, Available at (Accessed on 17 August 2014).

Ebem, D.U., Beerends, J.G., Van Vugt, J., Schmidmer, C., Kooij, R.E., Uguru, J.O., 2011. The impact of tone language and non-native language listening on measuring speech quality. J. Audio Eng. Soc. 59, 647–655.

Gros, L., Chateau, N., 2002. The impact of listening and conversational situations on speech perceived quality for time-varying impairments. Presented at the International Conference on Measurement of Speech and Audio Quality in Networks, Prague, Czech Republic, pp. 17–19.

Hall, M., Frank, E., Holmes, G., Pfahringer, B., Reutemann, P., Witten, I.H., 2009. The WEKA data mining software: an update. ACM SIGKDD Explor. Newsl. 11 (1), 10–18.

Hammer, F., Reichl, P., Ziegler, T., 2004. Where packet traces meet speech samples: an instrumental approach to perceptual QoS evaluation of VoIP. In: Twelfth IEEE International Workshop on Quality of Service, 2004. IWQOS 2004. Presented at the Twelfth IEEE International Workshop on Quality of Service, 2004. IWQOS 2004, pp. 273–280.

Hoene, C., Karl, H., Wolisz, A., 2006. A perceptual quality model intended for adaptive voip applications: research articles. Int. J. Commun. Syst. 19, 299–316.

Imankulov, R., 2009. Network-Emulator [Computer Program]. http://imankulov.github.io/network-emulator, Available at (Accessed on 17 August 2014).

ITU-T Recommendation G.107, 2011. The E-model: A Computational Model for Use in Transmission Planning. International Telecommunication Union, Geneva, Switzerland.

ITU-T Recommendation G.113, 2007. Transmission Impairments Due to Speech Processing. International Telecommunication Union, Geneva, Switzerland.

ITU-T Recommendation P.50, 1999. Artificial Voices. International Telecommunication Union, Geneva, Switzerland.

ITU-T Recommendation P.563, 2004. Single-Ended Method for Objective Speech Quality Assessment in Narrow-Band Telephony Applications. Switzerland, Geneva.

ITU-T Recommendation P.800, 1996. Methods for Subjective Determination of Transmission Quality. International Telecommunication Union, Geneva, Switzerland.

ITU-T Recommendation P.800.1, 2006. Mean Opinion Score (MOS) terminology. International Telecommunication Union, Geneva, Switzerland.

ITU-T Recommendation P.862, 2001. Perceptual Evaluation of Speech Quality (PESQ): An Objective Method for End-to-End Speech Quality Assessment of Narrow-Band Telephone Networks and Speech Codecs. International Telecommunication Union, Geneva, Switzerland.

ITU-T Recommendation P.862.2, 2007. Wideband Extension to Recommendation P.862 for The Assessment of Wideband Telephone Networks and Speech Codecs. International Telecommunication Union, Geneva, Switzerland.

ITU-T Recommendation P862.1, 2003. Mapping Function for Transforming P.862 Raw Result Scores to MOS-LQO. International Telecommunication Union, Geneva, Switzerland.

Koza, J.R., 2010. Human-Competitive Results Produced by Genetic Programming. Genet. Program Evolv. Mach. 11, 251–284.

Narbutt, M., Davis, M., 2005. An assessment of the audio codec performance in Voice over WLAN (VoWLAN) systems. In: Proceedings of The Second Annual International Conference on Mobile and Ubiquitous Systems: Networking and Services, MOBIQUITOUS'05. IEEE Computer Society, Washington, DC, USA, pp. 461–470.

Open Speech Repository, n.d. Open Speech Repository [Online]. Available at http://www.voiptroubleshooter.com/open_speech/british.html (Accessed on 9 September 2014).

Patcharikra, C., Treepop, S., Sawit, K., Nattanun, T., Wutiwiwatchai, C., 2005. LOTUS: large vocabulary Thai continuous speech recognition corpus. In: NSTDA Annual Conference S&T in Thailand: Towards the Molecular Economy.

Raja, A., Azad, R.M.A., Flanagan, C., Ryan, C., 2007. Real-time, non-intrusive evaluation of VoIP. In: Ebner, M., O'Neill, M., Ekárt, A., Vanneschi, L., Esparcia-Alcázar, A.I. (Eds.), Genetic Programming, Lecture Notes in Computer Science. Springer, Berlin Heidelberg, pp. 217–228.

Ren, J., Zhang, H., Zhu, Y., Gao, C., 2008. Assessment of effects of different language in VOIP. In: International Conference on Audio, Language and Image Processing, 2008. ICALIP 2008. Presented at the International Conference on Audio, Language and Image Processing, 2008. ICALIP 2008, pp. 1624–1628.

Sun, L., 2004. Speech Quality Prediction for Voice Over Internet Protocol Networks. Ph.D. thesis, School of Computing, Communications and Electronics, Faculty of Technology, Plymouth University, UK.

Sun, L., Ifeachor, E.C., 2006. Voice quality prediction models and their application in VoIP networks. IEEE Trans. Multimed. 8, 809–820.

Triyason, T., Kanthamanon, P., 2012. Perceptual evaluation of speech quality measurement on Speex codec VoIP with tonal language Thai. In: Papasratorn, B., Charoenkitkarn, N., Lavangnananda, K., Chutimaskul, W., Vanijja, V. (Eds.), Advances in Information Technology, Communications in Computer and Information Science. Springer, Berlin Heidelberg, pp. 181–190.

Voznak, M., 2011. E-model modification for case of cascade codec arrangement. Int. J. Math Mod. Meth Appl. Sci. 5, 1439–1447.

Zhao, Y., Zhang, Y., 2008. Comparison of decision tree methods for finding active objects. Adv. Space Res. 41, 1955–1959.

On the Impact of the Differential Evolution Parameters in the Solution of the Survivable Virtual Topology-Mapping Problem in IP-Over-WDM Networks

Fernando Lezama[1], Gerardo Castañón[1], and Ana Maria Sarmiento[2]

[1]Department of Electrical and Computer Engineering, Tecnológico de Monterrey, Monterrey, Mexico
[2]Department of Industrial Engineering, Tecnológico de Monterrey, Monterrey, Mexico

CHAPTER CONTENTS

6.1 INTRODUCTION

Nowadays, it is evident that telecommunications are developing almost exponentially worldwide in response to the ever-increasing bandwidth demand and transmission distances required in communication networks. The systems fit to cope with this exponential growth are led by optical systems using wavelength-division multiplexing (WDM) technology.

In WDM networks, the high bandwidth capacity can be divided into some non-overlapping channels operating at different wavelengths, through which the upper layers (IP, Ethernet, etc.) can transmit data (Kaldirim et al., 2009). Because every physical fiber link carries a huge amount of data, a simple link failure or a node failure can cause a significant loss of information.

Considering this, optical networks must be designed to ensuring resilience in case of failure, in order to avoid this loss of information. In this scenario, survivability, that is, the ability of a network to withstand and recover from failures or attacks, is one of the most important requirements in today's networks. Its importance is magnified in fiber-optic networks with throughputs in the order of gigabits and terabits per second (Zhou and Subramaniam, 2000).

More specifically, in IP-over-WDM networks, a virtual topology is placed over the physical topology of the optical network. The virtual topology consists of virtual links that are in fact the lightpaths in the physical topology. There are two main approaches to protect the IP-over-WDM networks: WDM protection level and IP restoration level (Sahasrabuddhe et al., 2002). In WDM protection level, backup paths for the virtual topology are reserved; this provides faster recovery for time-critical applications but is less resource-efficient because the resources are reserved without knowledge of the failure. In IP restoration level, failures are detected by the IP routers, which adapt their routing tables, and therefore no action is taken at the optical layer.

Considering the IP restoration-level scenario, an important challenge is to make the routing of the virtual topology onto the physical topology survivable. Each virtual link (IP link) should be mapped on the physical topology as a lightpath, and usually a fiber physical link is used for more than one lightpath. Therefore, a single failure of a physical link can disconnect more than one IP link in the virtual topology. To achieve the IP restoration level, the virtual topology needs to remain connected after a failure occurs. Failures can be of many types: node failure, link failure, or multiple link failures. In order to call a mapping survivable, we need to specify the type of failure that it has to survive. A single link failure is one of the most common failures in optical networks (Mukherjee, 2006). The problem of routing virtual links into a physical topology in such a way that the virtual topology (lightpaths set up on the physical network) remains connected after a physical link failure occurs is known as the survivable virtual topology mapping (SVTM) problem. This combinatorial problem is NP-complete (Modiano and Narula-Tam, 2002).

The importance of the SVTM problem can be assessed by a number of approaches proposed in the literature to obtain suboptimal solutions. Integer linear programming (ILP) models (Modiano and Narula-Tam, 2002; Todimala and Ramamurthy, 2007) have been successfully used to solve the SVTM problem in small optical networks. However, as the network's size increases, so does the dimension of the ILP model, whose solution typically requires an extensive computation effort (and execution time), which renders the model impractical for medium- to large-scale networks.

Therefore, different heuristic-based algorithms have been proposed. Among them are the evolutionary algorithms (EAs), ant colony optimization (ACO), and so on. Ducatelle and Gambardella (2004) present a local search algorithm that can provide survivable routing in the presence of physical link failures. Their algorithm can easily be extended for the cases of node failures and multiple simultaneous link failures. Another approach is presented by Kurant and Thiran (2004), in which the problem or task is divided into a set of independent and simple subtasks. The combination of solutions of these subtasks is a survivable mapping. Finally, nature-inspired heuristics have been used to the SVTM. Ergin et al. (2009) and Kaldirim et al. (2009) present an efficient EA and a suitable ACO algorithm, respectively, to find a survivable mapping of a given virtual topology while minimizing resource usage. In Ergin et al. (2010), a comparison between both algorithms, ACO and EA, is presented.

Differential evolution (DE) is a very simple but very powerful stochastic global optimizer for a continuous search domain. It was proposed by Storn and Price (1997) to represent a very complex process of evolution. Intelligently using the differences between populations (a population is a set of candidate solutions), and with the manipulation of few control parameters, they created a simple but fast linear operator called differentiation, which makes DE unique. Additionally, studies show that DE in many instances outperforms other EAs (Das and Suganthan 2011; Lezama et al., 2012; Vesterstrom and Thomsen, 2004). More specifically, DE exploits a population of potential solutions to effectively probe the search space. The algorithm is initialized with a population of random candidate solutions, conceptualized as individuals. For each individual in the population, a descendant is created from three parents. One parent, the main, is disturbed by the vector of the difference of the other two parents. If the descendant has a better performance resulting in the objective function, it replaces the individual. Otherwise, the individual is retained and is passed on to the next generation of the algorithm. This process is repeated until it reaches the termination condition. A complete theoretical analysis of the algorithm is presented in Storn and Price (1997).

In this chapter, we study the application of a DE algorithm to the SVTM problem. Two algorithms, named basic DE-VTM (B-DE-VTM) and enhanced DE-VTM (E-DE-VTM), are proposed. An illustrative example is presented to clearly understand the method. Moreover, a detailed analysis on the impact of the DE parameters (population size (NP), mutation constant (M), and recombination constant (RC)) on the quality of solutions is presented. The application of DE to the VTM problem was presented in Lezama et al. (2014). Nevertheless, the variability of the quality of solutions due to the DE parameters is an issue that must be studied, because of its importance in the VTM problem in optical networks.

The chapter layout is as follows: in Section 6.2, the problem formulation is presented. The proposed search algorithm is discussed in Section 6.3, and an illustrative example is presented in Section 6.4. The results are explained in Section 6.5, and conclusions are addressed in Section 6.6.

6.2 PROBLEM FORMULATION

The physical network is modeled as an undirected graph $G_p = (V_p, E_p)$, where V_p is the set of physical nodes numbered $\{1, 2, ..., |V_p|\}$, and E_p is the set of physical links $\left\{ e_{i,j}^p, i, j \in V_p \right\}$ with cardinality $|E_p|$. In a similar way, the virtual topology is modeled as an undirected graph $G_v = (V_v, E_v)$ in which V_v is a subset of V_p, and E_v is a set of virtual links representing lightpaths on the physical topology.

In the physical topology, $e_{i,j}^p$ is in E_p if there is a link between nodes i and j. On the other hand, the virtual topology has a set of edges (lightpaths) E_v, where an edge $e_{i,j}^v$ exists in E_v if both nodes s and d are in V_v and there is a lightpath between them. Figure 6.1a and b presents a physical and virtual topology, respectively.

There is no SVTM that protects the network against all component failures; therefore, we need to define what kind of failure our network is able to survive, for example, single link failures, node failure, multiple link failures, and so on. Our analysis considers single link failure, which is one of the most common in optical networks (Mukherjee, 2006; Grover, 2004), and is also less cumbersome to analyze.

To illustrate the SVTM problem, consider the physical topology (WDM network) presented in Figure 6.1a. Figure 6.1b presents a virtual topology with virtual links

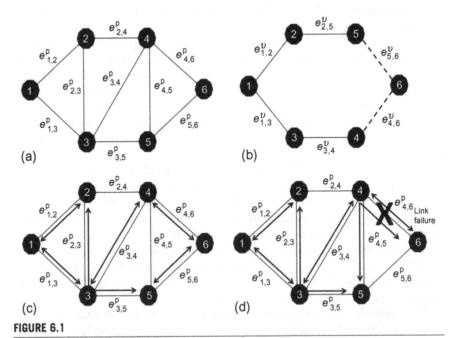

FIGURE 6.1

Illustrative survivable and unsurvivable virtual topology mapping for a simple 6-node network. (a) Physical topology. (b) Virtual topology. (c) SVTM. (d) Unsurvivable virtual topology mapping. Solid arrows in (c) and (d) represent virtual links mapped onto the physical topology.

$e_{1,2}^v$, $e_{2,5}^v$, $e_{5,6}^v$, $e_{4,6}^v$, $e_{3,4}^v$, and $e_{1,3}^v$, which in fact are ligthpaths (IP links) that need to be mapped onto the WDM network. Usually, a physical fiber link is used by more than one lightpath. Therefore, a single failure of a physical link can disconnect more than one IP link in the virtual topology. To achieve the IP restoration level, the virtual topology needs to remain connected after a failure. Figure 6.1c presents an SVTM against a single link failure. Observe that a single link failure disconnects at most one virtual link of the virtual topology, so the remaining virtual links remain connected, achieving the IP restoration level. Obtaining survivable mappings is relatively easy for small networks. However, as the network's complexity increases, so does the difficulty of determining them. To show what an unsurvivable mapping would look like, in Figure 6.1d we have routed the lightpath $e_{5,6}^v$ through the physical links $e_{4,5}^p$ and $e_{4,6}^p$. It can be seen that a failure in the physical link $e_{4,6}^p$ disconnects the virtual links $e_{5,6}^v$ and $e_{4,6}^v$ of the virtual topology (dashed lines in Figure 6.1b), leaving the virtual node 6 isolated in the virtual topology, which clearly indicates that the mapping is not survivable.

Lightpaths can be mapped taking into account different metrics along with the requirement of survivable mapping, such as the reduction of wavelengths to connect the lightpaths, or the number of wavelength links (*NWLs*). Based on the ILP formulations previously done in Modiano and Narula-Tam (2002), we have chosen to reduce the *NWL*, because this metric gives a better idea of the use of the resources in the network. The primary impact of the reduction of the *NWL* is on the delays and transmission impairments of the signal; it also helps reduce the network's resource wastage.

A wavelength link is defined as a physical link that a lightpath traverses in the virtual topology. The *NWL* is defined as

$$\text{NWL} = \sum \text{lp}_{s,d} \quad \forall s \neq d \in V_v, \tag{6.1}$$

where $\text{lp}_{s,d}$ is the length of the path in number of hops that connects nodes (s,d) in the virtual topology. To illustrate the *NWL*, in Figure 6.1c and d, each solid arrow represents a virtual link mapped onto the physical topology. In fact, every virtual link is a route that connects a pair of nodes in the physical topology. Each route has a cost associated to its length based on the number of hops. In Figure 6.1c, the number of hops of all mapped virtual links is equal to 7 (i.e., *NWL* = 7). Notice from Figure 6.1c and d that the only difference between them is that in Figure 6.1d, the virtual link $e_{5,6}^v$ has been mapped through the physical links $e_{4,5}^p$ and $e_{4,6}^p$, resulting in an unsurvivable mapping with an *NWL* = 8 (i.e., the length of the routes in number of hops is equal to 8), which is a worse solution in both metrics (survivability and *NWL*).

6.3 DE ALGORITHM

We applied the DE algorithm to find an SVTM in the network. As mentioned before, a DE algorithm uses a population (*Pop*) of individuals and iterates by creating new populations until an optimal or near optimal solution is obtained. The individuals represent specific solutions to the problem, so that an encoding that is well suited to it is necessary. Our approach, inspired in Banerjee and Sharan (2004), assumes that a set of paths to meet demand is available. Therefore, in an initialization step, we calculate k paths between all pairs of nodes, using the k-shortest paths algorithm (Yen, 1971). The precalculation of paths could be based on hops or distance. For simplicity, the paths are precalculated based on hops and are considered bidirectional.

To encode our individuals, we generate a vector of dimension D, for example,

$$X = [k_{1 \to 2}, k_{1 \to 3}, k_{1 \to 4}, \ldots, k_{N-1 \to N}], \tag{6.2}$$

where D is equal to the number of virtual links (lightpath connections) between virtual nodes in the virtual network, N is the number of virtual nodes (i.e., $N = V_v$), and $k_{i \to j}$ represents the selection of an available path between nodes i and j. Each virtual link is assigned a position in the individual, starting with virtual link $e^v_{1,2}$ (if it exists), and proceeds (in a row major order) to virtual link $e^v_{N-1,N}$.

The DE algorithm is run for a limited number of iterations (generations). At the beginning of the algorithm, assuming we do not have information about the optimum, the initial population is created randomly. DE employs repeated cycles of recombination and selection to guide the population toward the vicinity of the global optimum. We apply the probability operators—crossing and mutation—to each individual in a population to obtain new individuals (children). These new individuals have some properties of their ancestors. These ancestors are kept or deleted by selection. The term "generation" is used to designate the conversion of all individuals into new ones, that is, to move from one population to another.

DE has three crucial control parameters: the mutation constant (M), which controls the mutation strength; the recombination constant (RC) to increase the diversity in the mutation process; and the population size (NP). Throughout the execution process, the user defines M and RC values in the range of $[0,1]$, and the Pop size NP.

At each generation, all individuals in the Pop are evaluated in turn. The individual being evaluated is called the target vector. For each target vector x^i, where $i = 1, \ldots, NP$, a mutant individual m^i is generated according to

$$m^i = x^{r1} + M * \left(x^{r2} - x^{r3} \right), \tag{6.3}$$

where $x^{r1}, x^{r2}, x^{r3} \in Pop$: $x^{r1} \neq x^{r2} \neq x^{r3} \neq x^i$. x^{r1}, x^{r2}, x^{r3} are three random individuals from the Pop, mutually different and also different from the current target vector x^i, and the mutation constant M is a scaling factor in the range of $[0,1]$. The M operator is used to control the magnitude of the difference between two individuals in Equation (6.3). This operator allows us to manage the trade-off between exploitation and exploration in the search process.

Then, the recombination operator RC is applied to increase the diversity in the mutation process. This operator is the last step in the creation of the trial vector t^i. To create the trial vector, the mutant individual, m^i, is combined with the current target vector x^i. Particularly, for each component j, where $j = \{1, 2, \ldots D\}$, we choose the jth element of the m^i with probability RC, otherwise from the x^i. Moreover, a random integer value Rnd is chosen from the interval $[1,D]$ to guarantee that at least one element is taken from m^i. Choosing a random number rand in the $[0,1]$ interval for each jth component, then the t^i is created as follows:

$$t^{i,j} \begin{cases} m^{i,j} & \text{if } (rand < RC \lor Rnd = \text{j}) \\ x^{i,j} & \text{otherwise} \end{cases} \quad \forall j. \tag{6.4}$$

After we create the trial vector t^i, it is necessary to verify the boundary constraints of each element of t^i to avoid creating an infeasible solution. This could happen because any jth element created by Equation (6.3) that is not in the allowed range of the specification of a problem has an RC probability of being selected. If any element of the trial vector violates the constraints, it is replaced with a random number in the allowed range.

Finally, the selection operator is applied; this operator is a simple rule of elitist selection of the vectors that improve the objective function. This is done by comparing the fitness between the trial vector and the target vector in the objective function, using

$$Pop_{next} \begin{cases} t^i & \text{if } f(t^i) < f(x^i) \\ x^i & \text{otherwise} \end{cases}, \tag{6.5}$$

where Pop_{next} is the population of the next generation that changes by accepting or rejecting new individuals. The best individual in the population and the global best individual are kept at the end of each generation to keep track of the best solution found so far.

6.3.1 FITNESS OF AN INDIVIDUAL

Our objective is to minimize the NWL (Equation 6.1), which means that the length of the chosen paths must be as short as possible while considering the survivability requirement. This objective is used to evaluate the fitness of an individual. We include penalties in the objective function when the survivability requirement is not met. To count the penalties added to a solution, we erase each physical link one by one and count how many of these removed links disconnect the virtual topology. According to this, the fitness of an individual is given by

$$\text{Fitness} = NWL + w * p, \tag{6.6}$$

where NWL is the sum of the paths' length, p is the number of links that disconnect the virtual network, and w is a weighted factor multiplying p.

6.3.2 PSEUDOCODE OF THE DE ALGORITHM

Under these considerations, a pseudocode of the basic DE algorithm B-DE-VTM is presented in Figure 6.2.

Set the control parameters *M*, *RC*, and *NP*.
Create an initial *Pop*.
Evaluate the fitness of every individual in *Pop*.
repeat
   for each individual $x^i \in Pop$ **do**
      Select three individuals from *Pop* (i.e., $x^{r1}, x^{r2}, x^{r3} \in Pop: x^{r1} \neq x^{r2} \neq x^{r3} \neq x^i$).
      Apply mutation Equation (6.3).
      Apply recombination Equation (6.4).
      Verify boundary constraints.
      if boundary constraints are violated **then**
         Modify the infeasible elements.
      end if
      Apply selection operator Equation (6.5).
      Update *Pop*.
   end for
until a satisfactory solution is obtained or a computational limit is exceeded.

FIGURE 6.2

B-DE-VTM pseudocode.

6.3.3 ENHANCED DE-VTM ALGORITHM

We develop a modification of the DE-VTM algorithm that makes it more efficient and robust. We refer to the application of the algorithm without any modification as basic DE algorithm B-DE-VTM. The algorithm that we propose, which is a modification of the B-DE-VTM, is called enhanced DE algorithm E-DE-VTM.

In this algorithm, we define a variation of the DE algorithm's main operator that better suits our purpose. In Equation (6.3), a new mutant individual is created over a continuous space based only on the addition of the scaled difference between two individuals to another one $x^{r1} + M * (x^{r2} - x^{r3})$. We observed that the "arithmetic operation" may lead to infeasible solutions and the single operator "+" limits the diversity of the new possible mutant vector. So we redefine Equation (6.3) as follows:

$$m^i = \left\lfloor x^{r1} \pm M * \left(x^{r2} - x^{r3} \right) \right\rfloor. \tag{6.7}$$

The "+" or the "−" operator is decided randomly for simplicity. The floor function generates integer elements only, which are needed for this particular problem. This simple modification creates more diverse new individuals.

6.4 ILLUSTRATIVE EXAMPLE

We present an illustrative example of our E-DE-VTM algorithm, using the simple physical network of five nodes shown in Figure 6.1a and the virtual topology from Figure 6.1b.

As mentioned in Section 6.3, our approach assumes that a set of precalculated paths is available. In an initialization step, we calculate k shortest paths between all pairs of nodes. Table 6.1 shows $k = 5$ shortest paths for the 6-node network.

Table 6.1 k-shortest paths for the 6-node network. The selected paths for the individual in the example are in bold

Node	1	2	3	4	5	6
1	[ ]	[1,2] [1,3,2] **[1,3,4,2]** [1,3,5,4,2] [1,3,5,6,4,2]	**[1,3]** [1,2,3] [1,2,4,3] [1,2,4,5,3] [1,2,4,6,5,3]	[1,2,4] [1,3,4] [1,2,3,4] [1,3,2,4] [1,3,5,4]	[1,3,5] [1,2,3,5] [1,3,4,5] [1,2,4,5] [1,2,3,4,5]	[1,2,4,6] [1,3,4,6] [1,3,5,6] [1,2,3,4,6] [1,2,4,5,6]
2		[ ]	[2,3] [2,1,3] [2,4,3] [2,4,5,3] [2,4,6,5,3]	[2,4] [2,3,4] [2,1,3,4] [2,3,5,4] [2,1,3,5,4]	**[2,3,5]** [2,4,5] [2,3,4,5] [2,1,3,5] [2,4,3,5]	[2,4,6] [2,3,4,6] [2,4,5,6] [2,3,5,6] [2,1,3,4,6]
3			[ ]	[3,4] [3,2,4] [3,5,4] **[3,1,2,4]** [3,5,6,4]	[3,5] [3,4,5] [3,2,4,5] [3,4,6,5] [3,1,2,4,5]	[3,4,6] [3,5,6] [3,4,5,6] [3,2,4,6] [3,5,4,6]
4				[ ]	[4,5] [4,3,5] [4,5,6] [4,2,3,5] [4,2,1,3,5]	[4,6] **[4,5,6]** [4,3,5,6] [4,2,3,5,6] [4,2,1,3,5,6]
5					[ ]	**[5,6]** [5,4,6] [5,3,4,6] [5,3,2,4,6] [5,3,1,2,4,6]
6						[ ]

Now, each virtual link from Figure 6.1b must be represented with an element of an individual in DE. Using a row major order, for this illustrative example, an individual will have the form of $X = [k_{1\to2}, k_{1\to3}, k_{2\to5}, k_{3\to4}, k_{4\to6}, k_{5\to6}]$, representing selected routes for the virtual topology. As stated in the pseudoalgorithm in Figure 6.2, we first set the control parameters. For this illustrative example, the values are set to $NP = 5$, $M = 0.2$, and $CR = 0.5$. Then, an initial random population is created, and the fitness of each particle is calculated. To explain how the fitness of each individual is calculated, consider the individual X:

$$X = [3, 1, 1, 4, 2, 1].$$
(6.8)

The elements in the vector of individual X indicate which routes are selected from Table 6.1 (the ones marked in bold). For instance, the first element in the vector, which is a 4, indicates that for the virtual link $e^y_{1,2}$, the third route (i.e., $k_{1\to2} = [1, 3, 4, 2]$) is selected. The second element in the vector, which is a 1, indicates that for the virtual link $e^y_{1,3}$, the first route (i.e., $k_{1\to2} = [1, 3]$) is selected, and so on. Figure 6.3a presents the individual representation, the virtual links, and the selected route from this example.

For this solution, considering Equation (6.1), we get an $NWL = 12$, which is the total length of the six selected routes. The mapping of the selected routes is presented in Figure 6.3b. Moreover, Figure 6.3b shows that with this solution, if a failure occurs in physical links $e_{1,3}^p$ (failure 1), $e_{2,4}^p$ (failure 2), and $e_{5,6}^p$ (failure 3), respectively, the virtual topology becomes disconnected. The virtual links disconnected due to these three failures are presented in Figure 6.3c–e (dashed lines). Because the failures of three physical links led to a disconnected virtual topology, a penalty of 3 has to be considered for this example. Finally, the fitness of the individual using Equation (6.6), with an arbitrary weight factor of $w = 50$, is given by

$$\text{Fitness} = NWL + w * p = 12 + 50 * 3 = 162. \qquad (6.9)$$

We illustrate the application of the mutation, recombination, and selection operators of DE. As stated in Section 6.3, from the initial Pop, we select a target individual x^i. Suppose that our initial population is

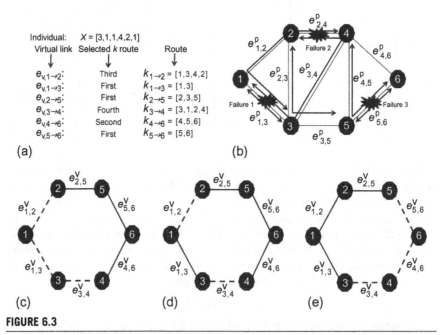

(a)   (b)   (c)   (d)   (e)

FIGURE 6.3

Illustrative individual representation. (a) Individual X and selected routes that represent to perform the mapping. (b) Physical topology and the virtual mapping of individual X. (c) Disconnected virtual links due to failure 1. (d) Disconnected virtual links due to failure 2. (e) Disconnected virtual links due to failure 3.

$$Pop = \begin{cases} x^1 = [3, 4, 3, 5, 4, 2] & \text{fitness} = 369 \\ x^2 = [5, 1, 5, 2, 4, 1] & \text{fitness} = 266 \\ x^3 = [4, 1, 3, 2, 4, 1] & \text{fitness} = 315 \\ x^4 = [2, 1, 2, 5, 5, 3] & \text{fitness} = 266 \\ x^5 = [2, 1, 1, 4, 4, 2] & \text{fitness} = 214 \end{cases}$$ (6.10)

We select a target vector in the first generation, that is, $x^1 = [3, 4, 3, 5, 4, 2]$. Then, three random individuals x^{r1}, x^{r2}, and x^{r3} mutually different and also different from the current target vector are selected from the *Pop* as well. For this example, let $r1 - 4$, $r2 - 3$, and $r3 - 5$. Figure 6.4 shows the application of the mutation operator (Equation 6.7) to get the mutant individual m^i.

Once we have the mutant individual m^i, we apply the recombination operator (Equation 6.4) to combine the target individual and the mutant individual. Figure 6.5 illustrates how the trial individual t^i is formed with elements chosen from the target individual x^1 and the mutant individual m^i. The elements are chosen with probability *RC*.

Finally, the selection operator (Equation 6.5) is applied between the target individual x^1 and the trial individual t^i. This operator is a simple rule of elitist selection. The individual with the best fitness value will survive to the next generation. In this example, the trial vector t^i will replace x^1 in the next generation because $f(t^i) = 268 < f(x^1) = 369$.

$$\lfloor x^4 \pm M * (x^3 - x^5) \rfloor = m^i$$

$$\lfloor 2 \pm 0.2 * (4 - 2) \rfloor = 1$$
$$\lfloor 1 \pm 0.2 * (1 - 1) \rfloor = 1$$
$$\lfloor 2 \pm 0.2 * (3 - 1) \rfloor = 2$$
$$\lfloor 5 \pm 0.2 * (2 - 4) \rfloor = 5$$
$$\lfloor 5 \pm 0.2 * (4 - 4) \rfloor = 5$$
$$\lfloor 3 \pm 0.2 * (1 - 2) \rfloor = 2$$

FIGURE 6.4

Mutant individual generation.

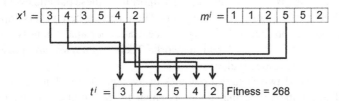

FIGURE 6.5

Trial individual generation. Elements to create t^i are taken from the target individual and the mutant individual with probability *RC*.

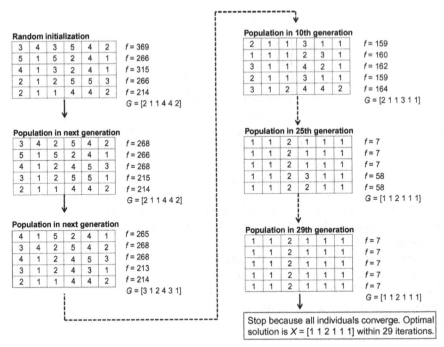

FIGURE 6.6

Illustration of E-DE-VTM simulation.

The process is repeated for each individual in *Pop* to form the next generation's population. The algorithm stops when a termination condition is met.

The capability of the applied E-DE-VTM is illustrated in Figure 6.6. The E-DE-VTM algorithm finds the optimal SVTM within 29 iterations.

6.5 RESULTS AND DISCUSSION

It is worth noting that the DE-VTM algorithms require the manipulation of few control parameters (NP, M, RC). In the proposed algorithms, these parameters have a significant impact on the quality of the solutions. Therefore, we believe it is important to present an analysis on their effects. We concentrated our analysis in two real-size networks, the NSF and the Japan network. Their topologies are presented in Figures 6.7 and 6.8, respectively.

We have created two virtual topologies of average degree 3 for each physical network. The number of virtual nodes is equal to the number of physical links (i.e., $|V_p| = |V_v|$). Using a set of default parameter values showed in Table 6.2, we use a sweeping technique to assess their impact on the solutions, one at a time. We have limited the algorithms to run for 500 and 1000 generations to appreciate the

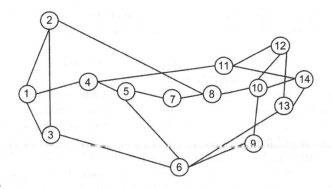

FIGURE 6.7

NSF network topology. 14 nodes, 21 links.

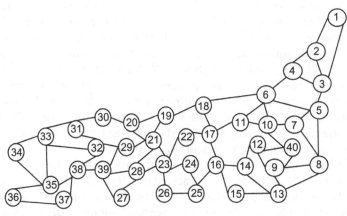

FIGURE 6.8

Japan network topology. 40 nodes, 65 links.

Table 6.2 Default parameter values for the test

Topology	NSFnet	Japan
NP	15	30
M	0.5	0.5
RC	0.5	0.5
k	6	9
GEN	500	1000

convergence rate. To calculate the fitness of Equation (6.6), we set the weight value to $w = 50$. Our experimentation was carried out on a 2.35 GHz, Pentium R Dual-Core PC.

From Equation (6.6), our fitness value is formed by the sum of two objectives. In order of priority, the first requirement that has to be fulfilled is the survivability requirement, and when this is met, it is also expected that the *NWL* be reduced as much as possible, because this reflects a better distribution of resources. Our parameter analysis is therefore presented for the convergence rate regarding these two objectives instead of the total fitness.

The parameter analysis is performed for both the B-DE-VTM and the E-DE-VTM algorithms, because the impact of the parameter's selection is slightly different on them. The main difference is that the E-DE-VTM has a better speed of convergence and is more robust. At the end of the section, a comparison of the two algorithms is presented.

Figure 6.9 shows the convergence rate of the penalties and the *NWL* when varying the *NP* value for the NSFnet. It is clear that a low value of *NP* ($NP = 5$) leads to stagnation in both objectives. However, a big value ($NP = 25$), even when guaranteeing survivability (Figure 6.9a and c), has a poor convergence rate regarding the *NWL* objective (Figure 6.9b and d). So a value within this range is recommendable for this particular case. For instance, from Figure 6.9b and d, it can be observed that values of $NP = 20$ and $NP = 15$ achieve the best *NWL* for B-DE-VTM and E-DE-VTM, respectively. Moreover, these values guarantee survivability for both algorithms.

Figure 6.10 shows the convergence rate of the penalties and the *NWL* when varying the *M* value for the NSFnet. We can observe in Figure 6.10a that the best convergence rate for the B-DE-VTM is obtained when $M = 0.1$, achieving the survivability requirement within just 10 generations. Nevertheless, in Figure 6.10b it is observed that with $M = 0.1$, the *NWL* is high. Because the fitness is a combination of both metrics, we conclude that $M = 0.5$ is the best value for the B-DE-VTM, because at this value the best results for survivability and *NWL* are obtained (Figure 6.10a and b). On the other hand, for the E-DE-VTM, it is clear from Figure 6.10c and d that $M = 0.1$ is the best value, because it renders the best convergence rate and *NWL*.

Figure 6.11 shows the convergence rate of the penalties and the *NWL* when varying the *RC* parameter for the NSFnet. The simulation was run for the B-DE-VTM and the E-DE-VTM, finding that the main difference was the convergence rate. For the E-DE-VTM, Figure 6.11a and b shows that $RC = 0.1$ presents an excellent convergence rate, achieving also the survivability requirement. The same value ($RC = 0.1$) can be recommended for the B-DE-VTM. *RC* values in the range of 0.5-0.9 convey the worst performance for both algorithms. Also, note in Figure 6.11b that the E-DE-VTM algorithm is very robust to variations of the *RC* parameter.

We also studied the effect of the *k* parameter, which determines the number of paths between nodes that are obtained at the beginning of both algorithms (with the *k*-shortest paths algorithm). This parameter determines the size of the solution space, becoming larger as *k* increases, even when the dimension of

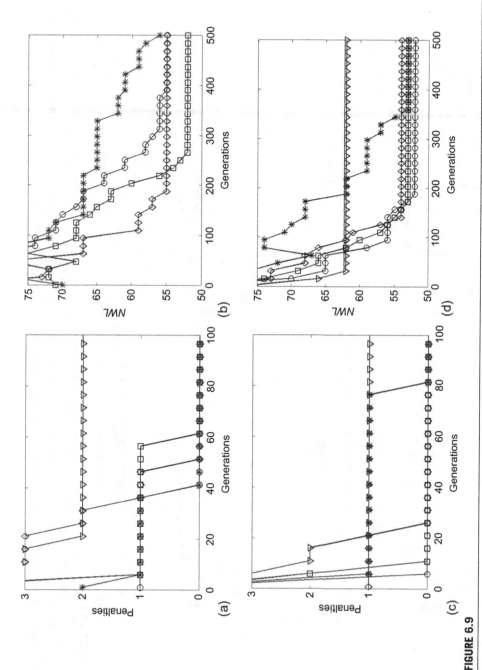

FIGURE 6.9

Convergence rate when varying *NP* parameter with $M=0.5$, $RC=0.5$, and $k=6$ for the NSF network. ∇: $NP=5$ $\diamond$: $NP=10$, $\circ$: $NP=15$, $\square$: $NP=20$, $*$: $NP=25$. (a) B-DE-VTM: Penalties due to unsurvivable mapping vs. Generations. (b) B-DE-VTM: *VWL* vs. Generations. (c) E-DE-VTM: Penalties due to unsurvivable mapping vs. Generations. (d) E-DE-VTM: *NWL* vs. Generations.

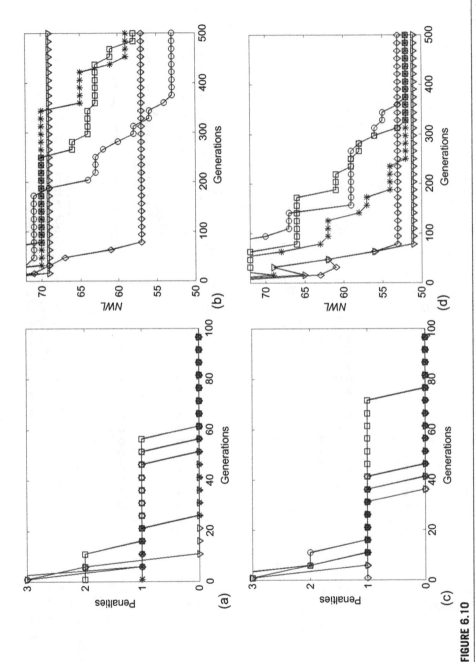

FIGURE 6.10

Convergence rate when varying M parameter with $NP=15$, $RC=0.5$, and $k=6$ for the NSF network. ∇: $M=0.1$, ◇: $M=0.3$, ○: $M=0.5$, □: $M=0.7$, *: $M=0.9$. (a) B-DE-VTM: Penalties due to unsurvivable mapping vs. Generations. (b) B-DE-VTM: NWL vs. Generations. (c) E-DE-VTM: Penalties due to unsurvivable mapping vs. Generations. (d) E-DE-VTM: NWL vs. Generations.

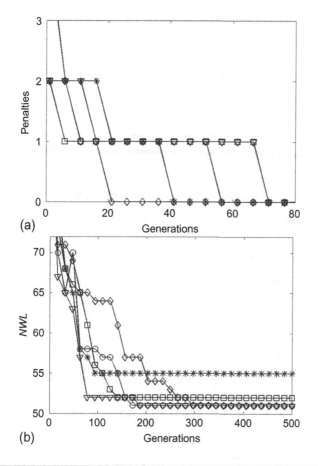

FIGURE 6.11

Convergence rate when varying RC parameter with $NP=15$, $M=0.5$, and $k=6$ for the NSF network. ∇: $RC=0.1$, $\Diamond$: $RC=0.3$, $\circ$: $RC=0.5$, $\Box$: $RC=0.7$, *: $RC=0.9$. (a) E-DE-VTM: Penalties due to unsurvivable mapping vs. Generations. (b) E-DE-VTM: NWL vs. Generations.

the problem, D, remains the same. Figure 6.12a shows the variation of penalties and NWL when the parameter k is modified. It is clear from the figure that a small value ($k=2$) leads to stagnation for the survivability requirement. This is because having a low number of paths limits the capacity of the algorithm to find a survivable mapping. On the contrary, Figure 6.12b shows that the larger k becomes, the lower the convergence rate that is obtained, because the solution space increases and the search for the global optimum requires more computational effort. A set of preliminary tests is required to find an acceptable value for k. For the NSF network, a value of $k=4$ is enough to meet the survivability requirement with a low NWL value as well. The performance of both algorithms was very similar, the only difference being the convergence rate. That is why we only present the graphs for the E-DE-VTM: the same conclusions apply for the B-DE-VTM.

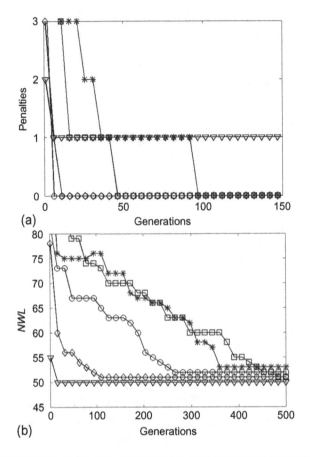

FIGURE 6.12

Convergence rate when varying k parameter with $NP=15$, $M=0.5$, and $RC=0.5$ for the NSF network. ∇: $k=2$, $\diamond$: $k=4$, $\circ$: $k=6$, $\square$: $k=8$, *: $k=10$. (a) E-DE-VTM: Penalties due to unsurvivable mapping vs. Generations. (b) E-DE-VTM: NWL vs. Generations.

We tested our algorithms in the more realistic Japanese network, with 40 nodes, shown in Figure 6.8. It is important to mention that this is a more complex network, and in our results, even when we reduce the number of penalties, we were not able to find a survivable topology mapping for the virtual topology analyzed. This is due to the stop criterion of 1000 generations of simulation. It is possible to find an SVTM by letting the algorithm run for a higher number of generations. However, the main purpose of this chapter is to analyze the impact of the parameters in the quality of the solutions, and 1000 generations are enough to show their effect.

Figure 6.13 shows the performance of the algorithms when varying the NP parameter for the B-DE-VTM. From Figure 6.13a, note that as the value of NP increases, the performance worsens. Also, Figure 6.13b shows that the NWL is not the dominant metric, due to the weight value (i.e., $w=50$) in the survivability

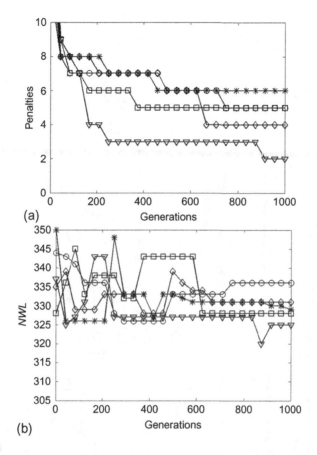

FIGURE 6.13

Convergence rate when varying NP parameter with $M=0.5$, $RC=0.5$, and $k=9$ for the Japan network. ∇: $NP=10$, $\diamond$: $NP=20$, $\circ$: $NP=30$, $\square$: $NP=40$, *: $NP=50$. (a) B-DE-VTM: Penalties due to unsurvivable mapping vs. Generations. (b) B-DE-VTM: NWL vs. Generations.

metric of Equation (6.6). That makes the NWL metric oscillate, not showing any apparent tendency. However, even when having these fluctuations, it is evident that a small value of NP is preferred, because in Figure 6.13b the best NWL value is obtained with a small population. The experiment was repeated for the E-DE-VTM algorithm, having a similar performance. Both algorithms (B-DE-VTM and E-DE-VTM) require a small value of NP (i.e., $NP=10$) to render the best results.

Figure 6.14 shows the convergence rate of the penalties and the NWL when varying the M value for the Japanese network. We can observe in Figure 6.14a a very similar tendency to the one of the NSF network. According to this figure, an $M=0.1$ gives the best performance for the B-DE-VTM, considering the penalties for unsurvivability. For the E-DE-VTM, we found the same behavior ($M=0.1$ as the best value) with a better convergence rate. Figure 6.14b shows that the NWL

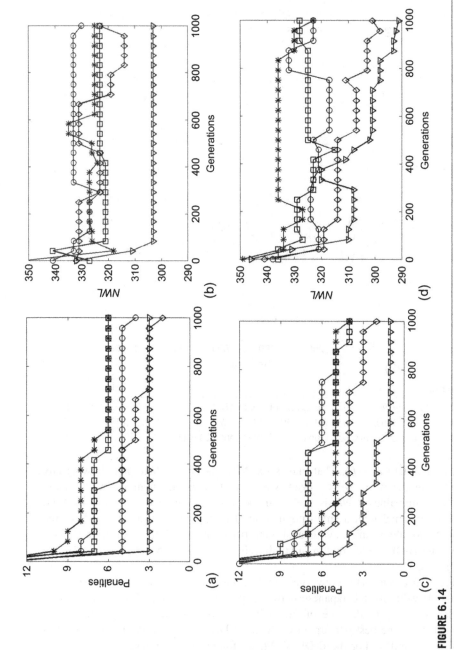

FIGURE 6.14

Convergence rate when varying M parameter with $NP = 30$, $RC = 0.5$, and $k = 9$ for the Japan network. ∇: $M = 0.1$, $\diamondsuit$: $M = 0.3$, $\circ$: $M = 0.5$, $\square$: $M = 0.7$, $*$: $M = 0.9$. (a) B-DE-VTM: Penalties due to unsurvivable mapping vs. Generations. (b) B-DE-VTM: NWL vs. Generations. (c) E-DE-VTM: Penalties due to unsurvivable mapping vs. Generations. (d) E-DE-VTM: NWL vs. Generations.

value does not change when the M value varies from 0.3 to 0.9. Also, the value obtained for the NWL when $M=0.1$ is the best in both algorithms.

Regarding the RC parameter, the behavior is very similar to that of the M parameter. Both algorithms present a similar behavior. Considering Figure 6.15a and b, it is evident that a value of $RC=0.1$ is preferred for the E-DE-VTM, obtaining the best values of penalties and NWL (for both algorithms). Notice that for the NWL, the value obtained with $RC=0.1$ is much better.

Finally, Figure 6.16 shows the performance of the k parameter in the Japanese network for the E-DE-VTM. The plots for the B-DE-VTM were not included because the tendency was very similar to the enhanced DE version. In priority order, Figure 6.16a shows an interesting behavior regarding the penalties, because of unsurvivability. It can be noticed that a small value of k (i.e., $k=3$) gives the worst

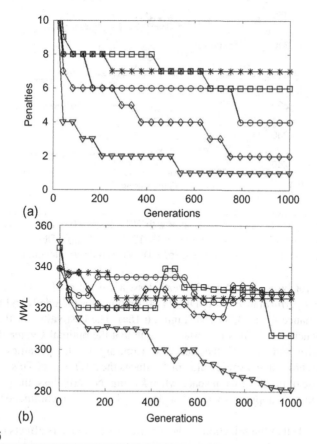

(a)

(b)

FIGURE 6.15

Convergence rate when varying RC parameter with $NP=30$, $M=0.5$, and $k=9$ for the Japan network. $\triangledown$: $RC=0.1$, $\Diamond$: $RC=0.3$, $\circ$: $RC=0.5$, $\square$: $RC=0.7$, *: $RC=0.9$. (a) E-DE-VTM: Penalties due to unsurvivable mapping vs. Generations. (b) E-DE-VTM: NWL vs. Generations.

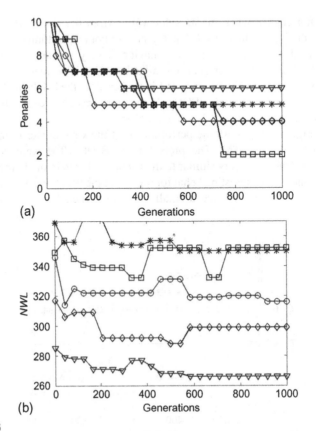

FIGURE 6.16

Convergence rate when varying k parameter with $NP=30$, $M=0.5$, and $RC=0.5$ for the Japan network. ∇: $k=3$, $\diamond$: $k=6$, $\circ$: $k=9$, $\square$: $k=12$, *: $k=15$. (a) E-DE-VTM: Penalties due to unsurvivable mapping vs. Generations. (b) E-DE-VTM: NWL vs. Generations.

performance for both algorithms. Nevertheless, a big value of k (i.e., $k=15$) also give the worst performance for B-DE-VTM (graphic not included) and the second worst performance for E-DE-VTM (Figure 6.16a). The best value for the penalties is obtained when $k=12$. It is important to note that the optimal k value depends on the complexity and size of the network. Contrary to the penalties' behavior, Figure 6.16b shows that a small value of k reduces the NWL value. This is explained by the increase of the dimension space when k is big. Nevertheless, the priority is on the survivability mapping, so a value around $k=12$ is recommended for this network.

We conclude that the selection of the algorithm's parameters affects the solution procedure in a critical way. A careful selection of these values should be considered, rendering an analysis of these parameters as very important. After a careful analysis of the DE parameters, Table 6.3 summarizes the recommended values for the analyzed networks and proposed algorithms.

Table 6.3 Parameter values recommended for the proposed DE-VTM algorithms

Parameter	NSFnet		Japan	
	B-DE-VTM	E-DE-VTM	B-DE-VTM	E-DE-VTM
NP	20	15	10	10
M	0.5	0.1	0.1	0.1
RC	0.1	0.1	0.1	0.1
k	4	4	12	12

6.6 CONCLUSIONS

In this chapter, we proposed the application of DE algorithms to the SVTM problem. We presented two algorithms based in DE, named B-DE-VTM and E-DE-VTM. We confirmed the effective capabilities in terms of convergence speed and quality of the solutions obtained, minimizing the *NWL* and reducing the penalties because of unsurvivability. We presented an illustrative experiment to demonstrate the methodology of our DE-VTM algorithm and show the effectiveness and efficiency of the proposed EA. Despite the fact that the solutions are somewhat sensitive to variations of the DE-VTM algorithm's parameters, the computed results showed that a good combination of these parameters leads to a system performance improvement and to a superior convergence rate. The analysis was performed in networks of different size to appreciate the different behavior of the algorithms when complexity increases. We have provided a practical DE model to solve the SVTM problem that is simple to implement and could also be extended by adding other features into the objective function. As further work, we propose to investigate the use of the DE-VTM algorithms in different scenarios, or to include other metrics to measure the performance and quality of the solutions.

REFERENCES

Banerjee, N., Sharan, S., 2004. An evolutionary algorithm for solving the single objective static routing and wavelength assignment problem in WDM networks. In: Proceedings of the International Conference on Intelligent Sensing and Information Processing (ICISIP), pp. 13–18.

Das, S., Suganthan, P., 2011. Differential evolution: a survey of the state-of-the-art. IEEE Trans. Evol. Comput. 15 (1), 4–31.

Ducatelle, F., Gambardella, L., 2004. A scalable algorithm for survivable routing in IP-over-WDM networks. In: Proceedings of the International Conference on Broadband, Networks (BroadNets), pp. 54–63.

Ergin, F., Yayimli, A., Uyar, A., 2009. An evolutionary algorithm for survivable virtual topology mapping in optical WDM networks. In: Applications of Evolutionary Computing. Lecture Notes in Computer Science, vol. 5484. Springer, Berlin/Heidelberg, pp. 31–40.

Ergin, F., Kaldirim, E., Yayimli, A., Uyar, A., 2010. Ensuring resilience in optical WDM networks with nature-inspired heuristics. IEEE/OSA J. Opt. Commun. Netw. 2 (8), 642–652.

Grover, W.D., 2004. Mesh-based Survivable Networks. Prentice Hall, Englewood Cliffs, NJ.

Kaldirim, E., Ergin, F., Uyar, S., Yayimli, A., 2009. Ant colony optimization for survivable virtual topology mapping in optical WDM networks. In: Symposium on Computer and, Information Sciences (ISCIS), pp. 334–339.

Kurant, M., Thiran, P., 2004. Survivable mapping algorithm by ring trimming (SMART) for large IP-over-WDM networks. In: Proceedings of the International Conference on Broadband, Networks (BroadNets), pp. 44–53.

Lezama, F., Castañón, G., Sarmiento, A., 2012. Differential evolution optimization applied to the wavelength converters placement problem in all optical networks. Comput. Netw. 56 (9), 2262–2275.

Lezama, F., Castañón, G., Sarmiento, A., Callegati, F., Cerroni, W., 2014. Survivable virtual topology mapping in IP-over-WDM networks using differential evolution optimization. Photon. Netw. Commun. 28, 306–319. http://dx.doi.org/10.1007/s11107-014-0455-1.

Modiano, E., Narula-Tam, A., 2002. Survivable lightpath routing: a new approach to the design of WDM-based networks. IEEE J. Sel. Areas Commun. 20 (4), 800–809.

Mukherjee, B., 2006. Optical WDM Networks. Springer, New York.

Sahasrabuddhe, L., Ramamurthy, S., Mukherjee, B., 2002. Fault management in IP-over-WDM networks: WDM protection versus IP restoration. IEEE J. Sel. Areas Commun. 20 (1), 21–33.

Storn, R., Price, K., 1997. Differential evolution – a simple and efficient heuristic for global optimization over continuous spaces. J. Glob. Optim. 11 (4), 341–359.

Todimala, A., Ramamurthy, B., 2007. A scalable approach for survivable virtual topology routing in optical WDM networks. IEEE J. Sel. Areas Commun. 25 (6), 63–69.

Vesterstrom, J., Thomsen, R., 2004. A comparative study of differential evolution, particle swarm optimization, and evolutionary algorithms on numerical benchmark problems. In: Congress on Evolutionary Computation (CEC), vol. 2, pp. 1980–1987.

Yen, J.Y., 1971. Finding the k shortest loopless paths in a network. Manag. Sci. 17 (11), 712–716.

Zhou, D., Subramaniam, S., 2000. Survivability in optical networks. IEEE Netw. 14 (6), 16–23.

Radio Resource Management by Evolutionary Algorithms for 4G LTE-Advanced Networks

7

M. Jawad Alam and Maode Ma

School of Electrical and Electronic Engineering, Nanyang Technological University,

Singapore

CHAPTER CONTENTS

7.1 INTRODUCTION TO RADIO RESOURCE MANAGEMENT

Long-Term Evolution (LTE) is an emerging radio access network technology, an evolution of Universal Mobile Telecommunications System standardized in the third Generation Partnership Project (3GPP). LTE-Advanced (LTE-A) is an enhancement of LTE. LTE/LTE-A is scheduled to provide support for all Internet Protocol (IP)-based traffic. In light of technological advances, the media offered through social networks and other applications is becoming increasingly content rich, with traffic volume per subscriber accelerating at a fast rate. Another factor contributing to the increase of data content is the development of new services, including mobile TV, e-learning solutions, and online gaming. Quality of service (QoS) is crucial for providing IP-based services. These services also have varying QoS requirements. To fulfill different traffic performance requirements and improve the end user quality experience, mechanisms for optimizing service and user experience will play an important role in future fourth generation (4G) LTE networks. The further evolution of LTE will lead to LTE-A, promising downlink (DL) data rates up to 3 Gbps and uplink data rates up to 1 Gbps. To achieve these high data rates, LTE-A brings new technologies in LTE. Carrier aggregation (CA) is one of the most significant, with spectrum in CA aggregated across different bands to achieve higher bandwidth. Multiple-input and multiple-output (MIMO) enhancements have already helped increase throughput in LTE-A, and network coverage in LTE-A is enhanced using relay nodes and femtocells. The introduction of a mechanism for machine-to-machine (M2M) communication and interworking with heterogeneous networks will also increase traffic load in 4G LTE-A networks. With the introduction of these new technologies, there is a need to support QoS mechanisms for LTE-A. Radio resources management (RRM) and optimization is one of the key elements to achieve and provide QoS in LTE/LTE-A. The significance of QoS is also important from a range of views, including telecom vendors' and operators'. From the service provider's point of view, service differentiation and subscriber differentiation are important factors in networks. As such, network operators and service providers are pushing toward the development of an enhanced resource management mechanism to support this differentiation and demand.

Optimization of radio resources is one of the major research areas in LTE-A. By improving RRM techniques, the quality of the network can subsequently be improved. Some major techniques for RRM are link adaption, hybrid automatic repeat request, admission control, handover, packet scheduling, and load balancing. In LTE-A, some new technologies are currently in the process of being introduced. Research work for RRM has been carried out on different layers, and work has also been undertaken on crosslayers, so that improvement in one layer can deliver some improvements for another layer, with the combination improving overall resource utilization.

In LTE-A, some work has been done in relays and M2M. It is estimated that when compared with the current cellular users, the number of machines using mobile networks would be significantly greater. New mechanisms to optimize such network resources would be required for M2M communication, as would QoS arrangements. Support of self-organizing networks and heterogeneous networks are also some of the established research areas in LTE-A. A brief overview of the main technologies introduced in LTE-A is given below. There are several procedures for RRM, some of which are briefly summarized. The radio interface of LTE consists of three layers (layers 1, 2, and 3). Layer 1 is also known as the physical layer, and deals with the modulation and demodulation of physical channels. Link adaption, error correction, power control, and some other services are also provided by the physical layer. Layer 2 is the media access control (MAC) layer, and deals with packet scheduling and related activities. Layer 3 is the radio resources control layer, handling connection control activities such as admission control and mobility control.

7.1.1 FRAME STRUCTURE

A brief description of LTE/LTE-A physical layer features is necessary to discuss the RRM techniques. To understand RRM in LTE, a basic understanding of the physical layer concepts of LTE is required. Structures of frame, subframe, and slot are shown in Figure 7.1. LTE frames are 10 ms in duration, with each frame divided into 10 subframes that each measure 1 ms. A subframe is further divided into two slots each of 0.5 ms. Each slot consists of six or seven orthogonal frequency-division multiplexing (OFDM) symbols. A resource block (RB) is defined as a block over 0.5 ms duration (1 slot) for 12 consecutive subcarriers.

A RB is the unit of block used in scheduling and scheduling decisions that are done in the base station, with each RB consisting of a maximum of resource elements. Orthogonal frequency-division multiple access (OFDMA) has been chosen for DL due to its properties, and single-carrier frequency-division multiple access (SC-FDMA) is chosen for uplink mainly due to its peak-to-average-power ratio (PAPR) property.

7.1.2 DL AND UPLINK

OFDMA is used in the DL, and LTE-A uses multiuser (MU)-MIMO in DL to increase capacity. In OFDM, the available bandwidth is divided into subcarriers and data is transmitted on them in parallel. Each of the subcarriers can be modulated using a different modulation technique.

SC-FDMA is used in the uplink. SC-FDMA can help reduce PAPR and offers better power efficiency than OFDMA. Power consumption is the principal consideration in the uplink. In SC-FDMA, subcarriers are not independently modulated, resulting in lower PAPR for OFDM transmissions.

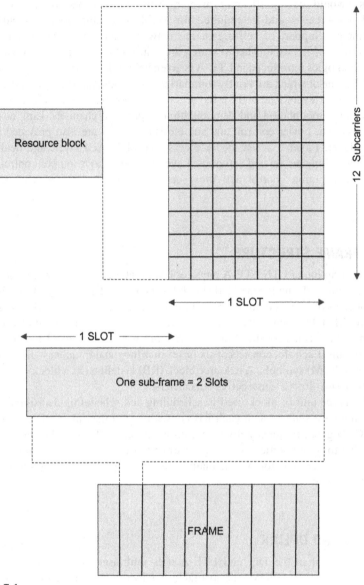

FIGURE 7.1

Resource block and frame structure.

7.2 **LTE-A TECHNOLOGIES**

A brief introduction to the new technologies introduced in 4G LTE-A is provided below.

7.2.1 **CARRIER AGGREGATION**

3GPP specifies 20 MHz bandwidth for LTE systems. Support for wider transmission bandwidth is required in LTE-A. CA is the method suggested to expand the bandwidth, with multiple component carriers (CCs) aggregated to form an overall larger bandwidth. Two or more (up to 5) CCs can be aggregated to support up to 100 MHz in LTE-A. Using CA 3 Gbps in DL is possible. There are two scenarios for CA: contiguous and noncontiguous. In the first case, the carrier deployment is contiguous, but it is possible that contiguous bandwidth is not available. In that case, noncontiguous components can be used. Noncontiguous bandwidth aggregation can be from the same band, or different bands (Figure 7.2).

7.2.2 **RELAY NODES**

In LTE-A, the concept of relay nodes has been proposed. Relay is a type of communication in which the relay station helps send messages from base station to user equipment (UE) that may not ordinarily be inside the range of the base station. This way, the signal and service of the base station is extended (Figure 7.3)

There are two terminologies used for relays: Type 1 and Type 2. In Type 1, the main purpose is to extend the service and signal so that it gives the ability to UE, which is outside the range of the base station, to communicate through the relay. Type 2 helps improve the service quality and link capacity. In this type, the relay node and the UE are both within the base station coverage and can communicate directly with the base station. By using the multipath diversity and transmission gain, overall system capacity can be improved for these types of relays.

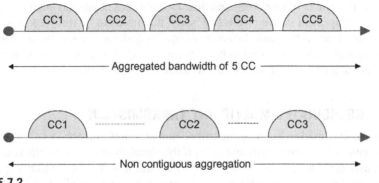

FIGURE 7.2

Contiguous and noncontiguous aggregation.

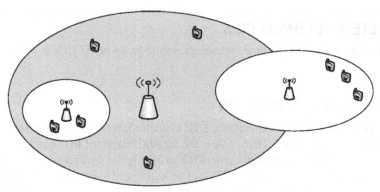

FIGURE 7.3

Relay nodes.

Relays create additional interference challenges. Handoff decisions need to take new requirements, such as backhaul availability and load balancing across base stations, into consideration. Dynamic and adaptive resource partitioning is needed, and the number of relays and their impact on the performance of the network also require investigation. QoS support for relay network needs to guarantee both delay and bandwidth in the relay network.

7.2.3 FEMTOCELL

Femtocell technology has been included in the standardization of LTE-A. Femtocell is made of a low-cost base station or evolved node B (eNB). It serves the subscriber within a short range of 10-30 m. High capacity gains can be achieved from the reduced cell size. Some of the benefits of femtocell are improved indoor coverage, boosted spectral efficiency, offloading of the macrocell network, and the removal of dead spots. There are some challenges associated with its use: as high numbers of femtocells are deployed, overlaying networks will have cross-tier interference. Because the number of femtocells and their locations in a network is unknown, interference management is one of the key issues that needs to be resolved in 4G networks. Interference between different femtocells and the macrocells are some of the problematic areas that need to be addressed for QoS guarantee. There has been some research undertaken in this area to address these concerns.

7.2.4 COORDINATED MULTIPOINT TRANSMISSION

This method is used to improve the cell edge user's performance. A cell edge user can receive or transmit to different cells. If the signal from different cells is coordinated, user performance can be improved. This will also improve system capacity. Coordinated multipoint (CoMP) transmission can be categorized into intersite and intrasite, based on the base station coordination method.

7.3 SELF-ORGANIZATION USING EVOLUTIONARY ALGORITHMS

LTE-A networks are designed to be self-organized based on 3GPP specification, with the network able to self-configure and self-optimize to improve its performance. If there is a service outage, the network has, to an extent, the ability to self-heal. Self-organization includes automated network planning, management, optimization, healing, and configuration. A summary of studies carried out in network self-organization has been carried out (Hu et al., 2010; Zhongshan et al., 2014). The main advantage of a self-organization network (SON) is overall network deployment and maintenance cost reduction. The complexity and scale of the problem resembles some of the natural systems. Self-organization in LTE-A can be broadly categorized into three categories.

1. *Self-configuration*: In this configuration step, a new base station (eNodeB) is added to the network. With the help of self-organization, eNodeB is able to auto-configure with minimal outside support. This reduces network deployment time and cost. Some network planning is still required, but the initial setup time and cost is significantly reduced.
2. *Self-optimization*: Network performance can be measured using different key performance indicators. Some of the factors to consider are capacity, load balance, QoS, and quality of experience. Self-optimization will help improve different parameters and adapt them according to the network dynamics. Self-optimization techniques also optimize the RRM, increase network capacity, improve QoS by load-balancing network traffic, and perform intercell interference coordination (ICIC).
3. *Self-healing*: Network self-healing is used to fix any network failure or loss of coverage. Techniques developed can be used to detect failures and assist in recovery of the network.

4G LTE-A networks will be SONs. The algorithms to provide self-organization are vendor-specific and another important research area. Nature-inspired algorithms can be adapted to work on LTE-A SON: a brief introduction to some popular algorithms that can be adapted is provided here, with popular algorithms discussed using a layered approach

7.3.1 SON PHYSICAL LAYER

In self-organizing networks, one of the main features required is self-configuration. At the physical layer, network synchronization is performed and timing synchronization is required to ensure the whole network is synchronized and working together. The two broad categories of synchronization are centralized and distributed synchronization. For LTE-A SON, a distributed solution seems more appropriate as there is no centralized authority. In nature, a firefly exhibits the phenomenon of self-synchronization. Fireflies are able to synchronize their flashing during the night (Buck, 1988). They have been used as a symbol for synchronization in wireless systems (Tyrrell et al.,

2006), and the mathematical model of Mirollo and Strogatz is used as a reference point for this particular model (Mirollo and Strogatz, 1990). The pulse-coupled oscillation (PCO)-based synchronization model can be used in firefly-inspired LTE SON, in place of packets. Two major types of PCO, identical and nonidentical synchronization models, can be used to achieve network synchronization. Difficulties with PCO models arise when there is a delay in the network. In an identical PCO, all eNodeB will try to synchronize to the same phase. Nonidentical PCO models have also been used in wireless networks to achieve network synchronization. Another technique for synchronization inspired from nature is cricket synchronization. Crickets are able to synchronize their chirping (Walker, 1969). This natural phenomenon has inspired some synchronization techniques that can be used in SON. The main challenges in network synchronization are network scalability: as the network increases synchronization, performance decreases. Achieving quick synchronization in a small number of cycles is also a significant challenge, as is adapting to coupling strength.

7.3.2 SON MAC LAYER

Spectrum efficiency, resource management/allocation, and scheduling are the main functions of the MAC layer. In self-organizing networks, techniques for cooperation, resource allocation, and automated load balancing are deployed in the network. The need for such a system in an LTE network has been highlighted (Zhang et al., 2010), although with no definitive solution, this is still under research. Drawing once again on the natural world, evolutionary cooperation techniques used by ants for task allocation can be applied to self-organized networks. Load balancing can be performed by optimizing routing algorithms. Additional methods inspired by the natural world in routing and load balancing include the bee hive method (Farooq, 2006) for routing, freshwater crab-capture process for traffic congestion problems (Tonguz, 2011), and task allocation by division of labor among ants (Robson and Traniello, 1998). The main challenges in SON MAC are to reduce the communication overhead for large networks and the speed of convergence for task allocation. Game theory can also be utilized to optimize the resource allocation in LTE-A SON, although finding equilibrium for such a game model presents a range of additional operational challenges.

7.3.3 SON NETWORK LAYER

Routing, load balancing, and security are some of the issues handled in the SON network layer. From nature, the mammalian immune system can be used to model an artificial immune system (De Castro and Timmis, 2002). Intrusion and misbehavior detection capabilities can be modeled based on pattern recognition, self-learning, and memorization characteristics inspired by nature. Another class of solutions for these problems is inspired from intercellular communication (Dressler and Krüger, 2004). Foraging is the process of searching for food resources in nature. For routing purposes, the principle of ant foraging can be used. In wireless communications, an ant colony optimization (ACO) mechanism is usually used to find optimal solutions (Bonabeau et al., 1999). In this method, the laying and sensing of pheromones are modeled as data structures that record time and nodes, with the pheromone concentration deciding the most appropriate routing path. AntNet (Di Caro and Dorigo, 2011) and AntHocNet (Di Caro et al., 2005) are two of the most popular algorithms found in packet-based networks. In AntNet, an ant randomly searches for its food. Once the food is found, the traversal path will be followed back to the origin. During this process, all nodes are updated with the latest information. Returning ants will help to reach a conclusion on the forward path. Genetic algorithms (GAs) have also been added to improve the AntNet (White et al., 1998). AntHocNet, on the other hand, improves upon AntNet by reducing the overhead for relying on repeated paths, and the setup stage sends some additional messages to gather information from neighbors. Another nature-inspired algorithm is that of the bee hive (Farooq, 2006), with the added advantage of this method being the reduced size of the routing table. The selection method is inspired from the honeybee colony method (Seeley and Towne, 1992). The routing of a packet from one node to another depends on the quality of neighbor, and this method also helps achieve load balancing in the network. The multiple ACO method has also been used to model routing in some complex networks (Sim and Sun, 2002), and can also be used for load-balancing purposes. The main difference from the single-colony method is that different types of pheromones can be detected by ants, meaning different types of ants can communicate and cooperate through interaction with these pheromones. High concentrations of pheromones from a different colony will result in an ant avoiding that path. This way, load balancing is also taken care of to an extent. Some of the main challenges in this area are the need to improve the mapping of a nature-inspired system to artificial systems. Load-balancing convergence, fault tolerance, and support for heterogeneous networks are some of the main challenges that need to be addressed.

7.3.4 LTE-A OPEN RESEARCH ISSUES AND CHALLENGES

LTE-A specifications are under development, with self-organizing technologies poised to be a major part of future networks. Most of the methods explained in the earlier section are related to communication networks in general. For LTE-A, a solution still remains to be found for self-organization. In the next section, we will take a look into evolutionary algorithms (EAs) that can be applied in LTE-A. Nature-inspired EAs are good candidate solutions, and research is ongoing to model these for the LTE-A network. For the physical layers, network scalability and synchronization performance is the main challenge. For the MAC layer, the challenges are in reducing communication overhead, resource allocation improvement, and reducing energy consumption. For network layers, routing path optimization and network security pose the greatest barriers. The biggest common challenge across all methods is to model the nature-based system into LTE-A.

7.4 EAs IN LTE-A

Optimization of network resources is one of the most challenging problems in wireless communication. When the optimization solution is too difficult or complex to find, or takes too long to solve, heuristic methods for optimization can be employed as opposed to traditional methods. One of the major categories for heuristic algorithms is nature-inspired algorithms. Nature-inspired algorithms are gaining popularity as a result of their ability to provide good optimization solutions within a reasonable time frame for complex, dynamic problems. Figure 7.4 shows some categories of metaheuristic algorithms that have been inspired by natural phenomena.

A brief introduction of the various categories of nature-inspired algorithms is given here. One of the main categories is EAs. In EAs, the design and implementation of the solution is based on evolutionary processes and can be classified into several subcategories (De Castro, 2006). One of the primary categories is genetic programming, which uses biological evolution and natural selection as inspiration for solving programming problems. The performance of the algorithm is measured in term of fitness functions, such as inheritance, mutation, selection, and crossover, which are used to inform solutions for optimization problems (Das et al., 2006). Game theory can be utilized to employ strategies and mathematical criteria, which could also help to reach an optimal solution. Classical game theory consists of a set of players and a set of strategies, with game-theory algorithms finding optimal solutions with a wide set of strategies and players. Particle swarm optimization (PSO) is a popular optimization technique inspired by nature (Bonabeau et al., 1999). PSO is inspired by the flocking pattern of birds and fish, and ACO algorithms (Dorigo et al., 2006) are also used for routing in SON. Another method for optimization is

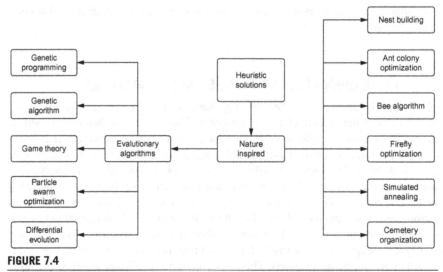

FIGURE 7.4

Nature-inspired algorithms.

differential evolution; it performs iterations to optimize the solution from the current population. EAs aside, there are also some techniques inspired by nature that can be used to solve the optimization problem. Nest-building inspiration is used to develop self-assembly algorithms, ant foraging behavior has been studied to develop optimization techniques for routing algorithms in communication networks, and cemetery organization has been used for data analysis and graph partitioning. Bee algorithms can also be used to solve multiobjective optimization problems. In this chapter, we will take a look at the use of these algorithms in the context of 4G LTE-A RRM.

7.4.1 NETWORK PLANNING

Network planning problems can also be solved using EA. In evolution theory, the population is defined as a candidate solution for the optimization problem. Fitness is the ability to survive and reproduce; as such, the fitness of an individual is the value of the cost function of that candidate. During the selection process, the best candidates are selected as parents of the next generation. A multiobjective network planning problem is defined as maximizing the number of users and minimizing the number of sites. Ali et al. (2011) adapt the nature of the problem as quantum-inspired evolutionary algorithm (QIEA) to achieve their optimized solution. QIEA has historically been used to solve combinatorial optimization problems. Each individual is represented by a string of quantum bits (qubit), where qubit is a probabilistic representation concept from quantum computing. A comparison study of the proposed algorithm is done with GAs.

Network planning problems can be modeled as a facility location problem. The goal is to maximize the number of UE and minimize the number of eNB. The objective function can be written as

$$\max\left(w_1 f_1(x) + w_2 f_2(y)\right) \tag{7.1}$$

subject to

$$g(i,j) > \delta, \quad i \in [1, 2, \ldots, X], j \in [1, 2, \ldots, Y] \tag{7.2}$$

$$\sum_i c(i,j) \leq \propto, \quad i \in [1, 2, \ldots, X], j \in [1, 2, \ldots, Y] \tag{7.3}$$

where w_1 and w_2 are weights of the two objective functions such that $w_1 + w_2 = 1$. f_1 is the ratio of the number of users in service to the total number. f_2 is the ratio of the number of free eNB to the total number of eNB. g represents the channel gain and c is the capacity of the eNB. In quantum computing, the state of qubit $[\propto \beta]^T$ can be written as

$$|\psi = \propto |0 + \beta |1^1 \tag{7.4}$$

where $\propto$ and β represent state of the qubit and are complex numbers. Typical QIEA pseudocode is given below:

```
1: l = 0,
2: Initialize the Q − bit individuals in the population Q(l)
3: while convergence criterion not satisfied
4:       Randomly generate a population, P(l),
          of binary strings from Q − bit individuals in Q(l),
5:       Evaluate the fitness of P(l)
6:       Store the best individuals from P(l) into B(l)
7:       Update Q(l + 1) with the fitness values computed in 5
8:       l = l + 1
9: end while
```

In the study by Ali et al. (2011), Ali models this problem using QIEA. The study compares QIEA with GA, with the findings showing that QIEA outperforms GA after the 40th iteration. Further research is required to add more factors into consideration.

7.4.2 NETWORK SCHEDULING

Scheduling is one of the most important areas for optimizing RRM, and EAs can be used to solve some complex optimization problems in wireless network scheduling. The study conducted by Thompson et al. (2010) proposes an adaptive scheduler, which focuses on minimizing intercell interference. An EA is used to optimize the scheduler for network parameters. ICIC can control the transmission power of some RBs to manage interference levels, and they have two major categories: static and semi-static. In static schemes, the communication is done during the initiation stage. In semi-static, frequent communication is done to reconfigure the network depending on the network environment. A popular example of static ICIC is soft frequency reuse (SFR). In SFR, orthogonal subbands are allocated to avoid interference, with this method limiting frequency utilization for cell edge users. In semi-static, there is more flexibility to allocate the subbands. A popular semi-static scheme is to allow cell center users to use the entire available spectrum, and allocate subbands to edge users depending on their load. The study proposes the adaption of softer frequency reuse. The entire frequency is available to the whole cell. Priority value for RB scheduling is calculated. A proportional fair algorithm is adapted to add a factor γ. γ controls the decision whether RB is part of the full-power frequency subband and whether or not the center or edge user is being scheduled. Reference signal received quality is used to determine the number of edge users and center users. A strict parameter is used to determine whether full-power or low-power RB can be scheduled.

An EA chromosome represents a vector of parameter choice for the adaptive scheduler. The study (Thompson et al., 2010) experimented with two EAs to search for ideal parameterization. The EAs are steady states, with two new children produced in each generation. A roulette-wheel selection operator is used to select

Algorithm	Average UE Throughput	Sector Throughput	5th % UE Throughput
SFR	1	1	1.0
SerFR	0.685	0.7777	1.069

FIGURE 7.5

Existing algorithm at 100% load (Thompson et al., 2010).

Strict	Power	RSRQ Threshold	Edge Threshold	UE Throughput	Sector Throughput	5th % UE Throughput
False	0.85	38.1	93.9	0.659	0.751	1.341
True	0.85	22.2	93.9	0.642	0.741	1.474
False	0.85	78.9	83.6	0.645	0.743	1.353

FIGURE 7.6

EA results from adaptive scheduler with 100% load (Thompson et al., 2010).

two parents, and uniform crossover is applied to produce two children. Each child is mutated with single gene mutation. If the selected child from the population is better than the current worse than, it would be integrated into the main population. In multi-objective EA, the relative fitness of each chromosome is determined by calculating the domination count across the three fitness measures.

The results from the study show a trade-off between fifth percentile throughput and average throughput. The cell edge user throughput can be increased by this method. MO-EA results also conclude a similar trend (Figures 7.5 and 7.6).

7.4.3 ENERGY EFFICIENCY

One of the most challenging problems in wireless communications is the reduction of overall energy consumption. The complexity of the problems resembles that of natural phenomena. GAs can be used to find an optimal solution for energy reduction. In Feng-Seng and Kwang-Cheng (2011), a GA is used to model the problem, and iterative tracking is used to minimize the entire system energy consumption. The study assumes that the OFDMA transmitter consists of a control processor, a specific circuit for encoding/decoding, and an analog front end. The problem to minimize entire system energy consumption can be formulated as follows:

$$\min\nolimits_{N_{in},\,c_k,\,\pi_{(n,\,m)},\,b_{(n,\,m)}} \left(E_{cp} + E_{sc} + E_{AFE}\right) \tag{7.5}$$

subject to

$$c_k \in C = \{(r_1, g_1),\, \ldots\ldots,\, (r_\theta, g_\theta)\} \quad \forall\, k \tag{7.6}$$

$$\pi_{n,m} \in \{0, \ldots k \ldots K\},\ b_{n,m} \in \{0, 2, \ldots \omega\} \,\forall\, n,m \tag{7.7}$$

$$\sum_{m=1}^{M}\sum_{n=1}^{N}\delta(k-\pi_{n,m})b_{n,m}=MTd_k r_k^{-1} \quad \forall n,m \tag{7.8}$$

$$\{C_k,\pi_{n,m}, b_{n,m}\}=g(N_{in}) \tag{7.9}$$

where E_{cp} is the energy consumption in the control processor, E_{sc} is the energy consumption of the specific circuit, and E_{AFE} is the energy consumption of the analog front end. The first constraint is on the coding scheme. The coding gain and rate will be selected from set C. The second constraint is one selection on UE from 0 up to K and modulation rate. The third constraint regards the QoS of the system. The last model shows that coding rate, UE, and modulation order are decided by an optimization algorithm.

In the initialization phase, randomly select feasible solutions are chosen that satisfy the constraints. This will be the first generation. Then, we need to compute the total energy consumption and order the solutions according to energy consumption. During the execution phase, iterations are performed from generation 1 to G. Parent solutions are selected to reproduce children solutions. After iterations, solutions in the latest generation are expected to show lower energy consumption.

7.4.4 LOAD BALANCING

4G LTE-A formulates the concept of M2M communication in the 3GPP standard specification. The most important challenge faced in this new technology is resource management for M2M. The number of M2M devices is expected to be huge and, if not properly handled, could lead to severe congestion across the network. It would have a similar effect on the network users. One of the methods used to handle the random access by these devices is access class barring (ACB). EAs can be used to model this problem in order to find an optimal solution. In Tingsong et al., an evolutionary game-based model is used to solve this particular problem.

Machine-type communication (MTC) devices can be modeled as players in game theory. Strategies will be user choice of eNB, and payoff will be defined with utility function. Two important factors in evolutionary game theory are replicator dynamics and equilibrium. The replicator dynamic is the rate of strategy change, and equilibrium is the stable point where there is no incentive for individual players to change their choice or strategy. The utility function for a user to connect to eNB can be expressed as

$$\pi_i(x)=U\left(\frac{C_i}{\sum_{a\in\varphi}N^a x_i^a}\right)-P\left(\sum_{a\in\varphi}N^a x_i^a\right) \tag{7.10}$$

where C_i is the channel capacity for coverage area φ. N^a is the number of users in area a. x_i^a is the proportion of users in the area who select base station i as service provider. Using the generalized expression, utility functions for microcell and picocell can be defined in the network. Once equilibrium is reached, the payoff for individual players cannot be improved further.

In the proposed solution, using evolutionary game-based model, MTC devices are going to perform random access. MTC devices will act in accordance with evolutionary game theory until equilibrium is achieved. Once equilibrium is achieved, the ACB factor will be broadcasted. Next, each MTC device will draw a random number from the allowed range and compare it with the ACB factor. If it is less than the factor, the device will proceed to random access procedure. If not, it will be barred from access until the next turn. The ACB factor will be updated after random access.

7.4.5 RESOURCE ALLOCATION

7.4.5.1 Genetic algorithm

GAs can be used to optimize some parameters in LTE-A relays. In Mao et al. (2012), the study investigates resource allocation in an LTE-A MU cooperative system. The problem is defined to maximize the system throughput with a total power constraint. To solve the problem, a GA is proposed. GAs have strong global search ability, and they will provide the best resource allocation through an iterative method. The comparison study done by Ali et al. (2011) is conducted with an Equal Bit Allocation (EBA) algorithm. The problem with the equal-bit allocation algorithm is that it cannot change allocation according to the channel conditions. This leads to wastage of power and reduced throughput. In the study, a centralized scheduler has been proposed. The channel information is gathered in the central controller, and decisions about the routing and modulation scheme to be used are made.

The objective is to maximize the total data rate under total power constraint. Assuming that P_{total} is the total power and R is total data, the object function can be described as

$$R = \text{arg max}_{b_n \in \{0 \sim B_{\max}\}} \sum_{n=1}^{N} b_n \tag{7.11}$$

subject to

$$\sum_{n=1}^{N} \frac{P_{\text{req}}(b_b)}{G_{\text{eq}}(n)} \leq P_{\text{total}} \tag{7.12}$$

$$\beta(n) + \sum_{k=1}^{K} \sum_{j=1}^{N} \alpha_k(n,j) = 1 \tag{7.13}$$

$$\beta(n) \leq 1 \tag{7.14}$$

$$\sum_{k=1}^{K} \sum_{j=1}^{N} \alpha_k(n,j) \leq 1 \tag{7.15}$$

where the number of bits carried by subcarrier n is $\beta(n)$. Matrix $\beta(n)$ shows whether or not subcarrier n has a direct link to the destination node. Matrix $\alpha_k(n,j)$ is used to indicate which subcarrier will be chosen by relay node to transmit to the destination node. $G_{eq}(n)$ is the equivalent gain. The value of α and β will be computed using an iterative method. In the proposed GA, first, the initial parameter is generated. The initial population is randomly generated using these parameters. The object value of the initial population is computed using object function. Following this, the GA will perform iterations until the maximum number allowed is reached during the selection step. At this point, the individual with the highest fitness will be the output. According to crossover probability, new populations will be updated through crossover operation. The next step will be to update the population with mutation probability. The process will move into the next loop after this (Figure 7.7).

Simulation results from the study show that the total throughput increases when relay is used with the proposed GA. When the study compares the GA with EBA using multiple relays, the results show that the GA provides better throughput. The reason for this is that the GA can adaptively allocate bits for each subcarrier. It shows that the GA can be adapted to solve some of the 4G network problems. The focus of study was relays and throughput increase; when we add in more factors, such as number of users, fair allocation, delay, QoS, CA, and CoMP, the problem becomes much more complex. Further research to model and solve these problems is ongoing.

7.4.5.2 Game theory

Matching game comes under cooperative game theory. Cooperative games are games where agents cooperate in the decision-making process. Matching game has many applications, and the topic has been extensively researched. Different

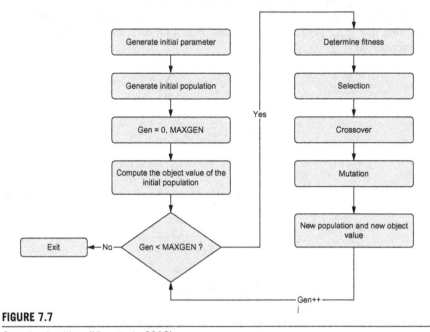

FIGURE 7.7

Genetic algorithm (Mao et al., 2012).

variants of the theory have been applied in areas such as college admissions, kidney donor matching, job matching in hospitals, and stable marriage problems. These problems can be modeled as matching problems, and resource allocation in 4G LTE-A can be modeled as a matching game.

To model a game, we must first define the players. In the case of LTE-A, this would be the UE devices in the network and eNB. The next step is to define the actions that players can take: this could be requesting resources or resource allocation, admission into the network, leaving the network, and so on. The next step is to define the payoff in the game: what each player stands to gain from taking an action. The payoffs are defined in term of utility functions based on actions.

LTE/LTE-A uses the SC-FDMA in the uplink transmission. Scheduling is done on a subframe basis in both uplink and DL. Resource allocation is done in terms of RBs. RB is the smallest and most basic unit of resource allocation in the network. For LTE-A devices that support multicluster allocation, the smallest allocation unit will be in terms of RB group. The size of the group can be seen from 3GPP specs. The chapter has summarized few of the methods in LTE-A RRM. Evolutionary models currently used in LTE-A have been briefly explained. We denote the collection of RBs, that is, RBG, as R. The total number of uplink RBG can be computed as

$$N_{RBG}^{UL} = N_{RB}^{UL} / R \tag{7.16}$$

In LTE-A, CA/MU-MIMO and multicluster adds a new dimension to the problem. In 4G networks, both LTE and LTE-A devices need to be scheduled. As such, assigning RBs in those bands that are supported by a particular device also needs to be addressed. Existing LTE-scheduling algorithms cannot be used in such scenarios, and there is a need to develop and enhance mechanisms for carrier-aggregated networks with backward compatibility of QoS. For uplink scheduling, power consideration is the main concern for resource allocation. The UE uplink power control formula in 3GPP specs is given as

$$P = \min \left\{ PC_{MAX}, 10\log_{10}(M, C^i) + P^o + \propto * PL_c + TF + F_c \right\} \text{ (dbm)} \tag{7.17}$$

In the above equation, PC_{MAX} stands for maximum transmit power of the UE in cell c in subframe i. PL is path loss and M is the number of RBs allocated to UE. P^o is the cell specific parameter and α is path-loss compensation factor. Resource allocation in uplink is limited by power. The optimization problem is to maximize the utility

$$U^i = \sum_{cc=1}^{CC_{max}} \sum_{k=1}^{RB_{max}} r_{k,cc}^i \tag{7.18}$$

$$\text{Max} \sum_{i=1}^{Max\ UE} U^i \tag{7.19}$$

$$\text{Subject to } f_i(\gamma) \geq \gamma^{Thresh} \tag{7.20}$$

$$N_i \leq RB_{max}, \quad N_i \geq 0 \tag{7.21}$$

C_{max} is the maximum CC available and RB_{max} is the max RB available in the CC. γ^{Thresh} is the QoS threshold for user flow and f_i is the QoS function for user i. N_i is the total PRB assigned to user i and RB_{max} is the total RB available in system. The problem is to maximize the system throughput while fulfilling QoS requirements for all users. The exhaustive search for optimal solutions would be impractical, so a heuristic solution can be achieved by modeling the problem as a matching game and achieving a Pareto-efficient matching. Stable matching problems match two sets of agents with preference lists in such a way that the outcome is stable. With a stable outcome, there is no incentive for agents on either side to form a new pair. The definition of matching market and matching game for the system is given below.

Definition I

A matching market can be defined in the form (UE, RB, P), where

$$UE = \{UE_i, UE_{i+1}, UE_{i+2}, UE_{i+3}, \dots, UE_{i+n}\} \tag{7.22}$$

$$RB = \{RB_j, RB_{j+1}, RB_{j+2}RB_{j+3}, \dots, RB_{j+n}\} \tag{7.23}$$

$$P = (P_k)_{k \in (UE \cup RB)} \tag{7.24}$$

$$UE \cap RB = \emptyset \tag{7.25}$$

UE is the first set of agents, which is a set of UE devices. RB is the second set of agents, which is a set of system RBs. Each of the agents has preference over other agents. P donates the list of preference of agents, P_{ue} denotes the preference of agent UE, and P_{rb} denotes the preference list of agent RB. The preference list can be described in form

$$P_{ue} = \{\varphi(ue^1), \varphi(ue^2), \varphi(ue^3), \dots, \varphi(ue^k)\} \tag{7.26}$$

$$P_{rb} = \{\gamma(rb^1), \gamma(rb^2), \gamma(rb^3), \dots, \gamma(rb^k)\} \tag{7.27}$$

where $\varphi(ue)$ and $\gamma(rb)$ are the preference functions of agents ue and rb.

Definition II

A matching is a function

$$\mu : RB \cup UE \rightarrow RB \cup UE, \text{ such that} \tag{7.28}$$

$$\mu(ue) = rb \leftrightarrow \mu(rb) = ue \text{ for all } rb \in RB, \tag{7.29}$$

$$ue \in UE, \tag{7.30}$$

$$\mu(rb) \in UE \rightarrow \mu(rb) = rb \text{ for all } rb \in RB, \tag{7.31}$$

$$\mu(\text{ue}) \in \text{RB} \rightarrow \mu(\text{ue}) \quad \text{for all ue} \in \text{UE} \tag{7.32}$$

That is, an outcome matches agents on one side to agents on the other side, or to themselves. If ue is matched to rb, then rb is matched to ue. Agents' preferences over outcomes are determined by their preferences for their own mates at those outcomes. Let $\omega(m,w)$ be the utility function of (rb,ue) belonging to RB × UE; then a matching game can be expressed as (RB,UE,ω).

A heuristic solution to the problem would be to solve the matching game using the Gale Shapley algorithm (Gale and Shapley, 1962). The proposed algorithm performs rounds similar to Gale Shapley and allocates resources requested by the agent. A preference matrix will be defined for both sets of agents. In the algorithm, the aim is to match devices that support a particular band to those RBs that belong to the required band. For devices that can support multiple bands, allocation can be done over different bands. The algorithm takes into consideration channel quality indicator and allocates the RB that has the best continuous quality improvement (CQI) for the current channel, taking carrier load into consideration. For stable matching, a list of preferences for agents needs to be formed. From a UE perspective, it will prefer those RBs that offer a greater data rate. The preferences of UE i over RB j can be defined by the following equation:

$$U_{i,j}^{\text{dl}} = (C_i + Q_c) * \alpha \tag{7.33}$$

For eNB, the preference list for RB will be created. RB prefers UE on the basis of the QoS profile. eNB has to ensure QoS for UE, so agents with higher QoS requirements get preference over agents with low QoS requirements. The preference V_i of RB i over UE j can be defined as

$$V_{i,j}^{\text{dl}} = (\text{QoS}_j - \beta) \tag{7.34}$$

This problem can be solved by applying the Gale Shapley algorithm. An optimal matching would be the one that could maximize/minimize the corresponding utility function for matching μ (Definition II).

$$\text{Maximize} \sum_{i,j} U_{i,j}^{\text{dl}} \tag{7.35}$$

$$\text{Minimize} \sum_{i,j} V_{i,j}^{\text{dl}} \tag{7.36}$$

The algorithm starts with both agents creating their preference list. In the first step, proposals are done by UE. This is followed by the RB selecting a candidate from the proposals. After this step, each RB that receives some proposal will have at least one pair formed. Some of the UEs will be rejected, which will follow in the next phase. This continues until all pairs are formed or no more proposals can be done. It can be proved that the matching obtained from this algorithm is Pareto efficient. In each of the iterations, a UE proposes to the best available candidate, and the proposal is only rejected if a better pair can be formed. A better pair can only be formed if the QoS requirement for the other proposal is higher than the current one. In this way, the best QoS requirement proposal always gets preference (Figure 7.8).

In the simulation scenario, two noncontiguous bands are considered with 4CCs. LTE devices are supported by one band, and LTE-A is supported on both bands. The algorithm provides a better matching than the round robin method. Throughput difference between using two algorithms can be seen in the results. As LTE-A devices can use two bands, they will provide better utilization. LTE devices can support only one band, so the second band resources are unutilized. We take a

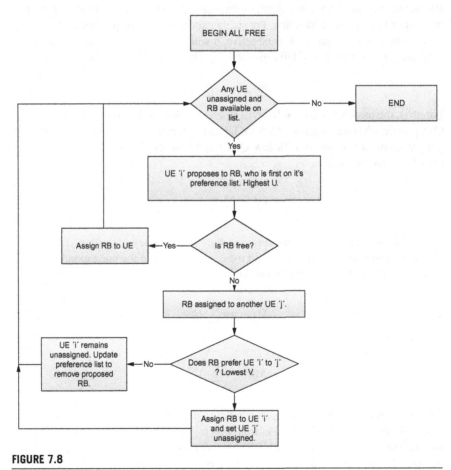

FIGURE 7.8

The heuristic solution using a game model.

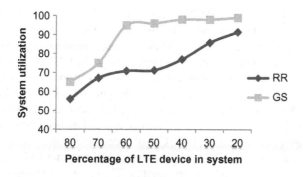

FIGURE 7.9

System utilization comparison of RR and GS in two bands (4CC).

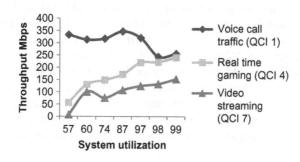

FIGURE 7.10

QoS differentiation for different traffic classes using GS in two bands (4CC).

mix of LTE/LTE-A devices. At the start, 80% of the devices are LTE and 20% are LTE-A. Over simulation time, LTE-A will enter the eNB area and LTE will leave (Figures 7.9 and 7.10).

GS gives a better throughput because of better CQI mapping. Due to QoS differentiation, we find that when utilization of the system is low, high-priority traffic is scheduled. Once the utilization increases, other traffics are also scheduled. This algorithm provides a mechanism to support QoS differentiation. The result shows that GS can provide QoS differentiation for LTE-A (Figures 7.9 and 7.10).

7.5 CONCLUSION

This chapter has summarized several of the methods used in LTE-A RRM. Evolutionary models currently used in LTE-A have been briefly explained. Modeling of the LTE-A system as an evolutionary system remains the biggest challenge. There has been limited work carried out in this area due to the difficulty involved in mapping a natural system against its artificial counterpart. The main technologies introduced in LTE-A are CA, M2M, relays, CoMP, femtocells, and SON. With the introduction of these technologies, models that were developed earlier for cellular systems are no longer valid. A lot of different aspects need to be covered, and there is further research required in creating such models. In LTE-A network planning, facility location is the biggest challenge. The current proposed solution, based on a QIEA, addresses some concerns; however, the parameters need to be fine-tuned. The research focuses on base station placing, but more is required to focus on LTE-A relays. In network scheduling, an EA model has been proposed for LTE. There is research required for LTE-A. CoMP access and femtocells introduce some challenges in LTE-A scheduling. The existing model may not address all issues, so there is a need for further study on new models. The energy consumption model presented was for LTE. The same model can be used in LTE-A; however, because the circuitry of devices will be different, there is room to make it more efficient. Resource allocation is one of the most important aspects of RRM. With resource allocation, the main aim is to improve the QoS of the system. QoS on M2M and femtocells are some of the challenging issues in this area. Load balancing also has a significant research gap. M2M device growth is going to be explosive over the coming years, with the networks subsequently receiving a heavy load. New methods are required to effectively deal with this. LTE-A M2M is one of the most sought-after research areas, and scalability of M2M networks remains a huge challenge. Due to the scale of the problem, EAs present a potentially promising solution, and modeling an M2M system to evolutionary systems is going to be a main research focus in the future.

REFERENCES

Ali, H.M., Ashrafinia, S., et al., 2011. Wireless mesh network planning using quantum inspired evolutionary algorithm. In: Vehicular Technology Conference (VTC Fall), 2011 IEEE.

Bonabeau, E., Dorigo, M., et al., 1999. Swarm Intelligence: From Natural to Artificial Systems. Oxford University Press, New York.

Buck, J., 1988. Synchronous rhythmic flashing of fireflies. II. Q. Rev. Biol. 63, 265–289.

Das, S.K., Banerjee, N., et al., 2006. Solving optimization problems in wireless networks using genetic algorithms. In: Olariu, S., Zomaya, A. (Eds.), Handbook of Bioinspired Algorithms and Applications. Chapman and Hall/CRC, Boca Raton, p. 219.

De Castro, L.N., 2006. Fundamentals of Natural Computing: Basic Concepts, Algorithms, and Applications. CRC Press, Boca Raton.

De Castro, L.N., Timmis, J., 2002. Artificial Immune Systems: A New Computational Intelligence Approach. Springer.

Di Caro, G., Dorigo, M., 2011. AntNet: distributed stigmergetic control for communications networks. arXiv preprint arXiv:1105.5449.

Di Caro, G., Ducatelle, F., et al., 2005. AntHocNet: an adaptive nature-inspired algorithm for routing in mobile ad hoc networks. Eur. Trans. Telecommun. 16 (5), 443–455.

Dorigo, M., Birattari, M., et al., 2006. Ant colony optimization. Comput. Intell. Mag. IEEE 1 (4), 28–39.

Dressler, F., Krüger, B., 2004. Cell biology as a key to computer networking. In: German Conference on Bioinformatics.

Farooq, M., 2006. From the Wisdom of the Hive to Intelligent Routing in Telecommunication Networks: A Step Towards Intelligent Network Management Through Natural Engineering. PhD Thesis, Dortmund University of Technology, Dortmund.

Feng-Seng, C., Kwang-Cheng, C., 2011. Iterative tracking the minimum of overall energy consumption in OFDMA systems. In: 22nd International Symposium on Personal Indoor and Mobile Radio Communications (PIMRC), 2011. IEEE.

Gale, D., Shapley, L.S., 1962. College admission and the stability of marriage. Am. Math. Mon. 69, 9–15.

Hu, H., Zhang, J., et al., 2010. Self-configuration and self-optimization for LTE networks. Commun. Mag. IEEE 48 (2), 94–100.

Mao, S., Zu, Y., et al., 2012. Resource allocation based on genetic algorithm in the multiuser cooperative OFDM system. In: 2012 Eighth International Conference on Wireless Communications, Networking and Mobile Computing (WiCOM).

Mirollo, R.E., Strogatz, S.H., 1990. Synchronization of pulse-coupled biological oscillators. SIAM J. Appl. Math. 50 (6), 1645–1662.

Robson, S.K., Traniello, J.F., 1998. Resource assessment, recruitment behavior, and organization of cooperative prey retrieval in the ant *Formica schaufussi* (Hymenoptera: Formicidae). J. Insect Behav. 11 (1), 1–22.

Seeley, T.D., Towne, W.F., 1992. Tactics of dance choice in honey bees: Do foragers compare dances? Behav. Ecol. Sociobiol. 30 (1), 59–69.

Sim, K.M., Sun, W.H., 2002. Multiple ant-colony optimization for network routing. In: Proceedings of the First International Symposium on Cyber Worlds, 2002. IEEE.

Thompson, A., Corne, D., et al., 2010. Optimization of LTE scheduling using multi-objectiveevolutionary algorithms. In: UK Workshop on Computational Intelligence (UKCI), 2010.

Tingsong, J., Xu, T., et al., 2014. Evolutionary game based access class barring for machine-to-machine communications. In: 16th International Conference on Advanced Communication Technology (ICACT), 2014.

Tonguz, O.K., 2011. Biologically inspired solutions to fundamental transportation problems. IEEE Commun. Mag. 49 (11), 106–115.

Tyrrell, A., Auer, G., et al., 2006. Fireflies as role models for synchronization in ad hoc networks. In: Proceedings of the First International Conference on Bio Inspired Models of Network, Information and Computing Systems. ACM.

Walker, T.J., 1969. Acoustic synchrony: two mechanisms in the snowy tree cricket. Science 166 (3907), 891–894.

White, T., Pagurek, B., et al., 1998. ASGA: improving the ant system by integration with genetic algorithms. In: Genetic Programming, pp. 610–617.

Zhang, H., Qiu, X.-S., et al., 2010. Achieving distributed load balancing in self-organizing LTE radio access network with autonomic network management. In: GLOBECOM Workshops (GC Wkshps), 2010 IEEE, pp. 454–459.

Zhongshan, Z., Keping, L., et al., 2014. On swarm intelligence inspired self-organized networking: its bionic mechanisms, designing principles and optimization approaches. IEEE Commun. Surv. Tutorials 16 (1), 513–537.

Robust Transmission for Heterogeneous Networks with Cognitive Small Cells

8

Carrson C. Fung

Department of Electronics Engineering, National Chiao Tung University,
Hsinchu, Taiwan

CHAPTER CONTENTS

8.1 INTRODUCTION

Technologies such as Internet-of-Things (IoTs) and cloud computing will put severe strain on existing wireless communication infrastructure, as the number of operating devices is envisioned to grow exponentially. This makes existing strategies such as increasing spectrum efficiency no longer feasible. One of the solutions that has been proposed recently to tackle this problem, while sustaining the future goal of increasing capacity by 1000 times, is to increase the number of cells per square kilometer. This can be achieved by deploying more cells of different types/technologies in a given area within the existing macrocells, thus creating a two-tier network. In such a network, the cells can be serviced by high-power macro base stations (MBSs), relay nodes, or low-power small cell (femtocell, picocell, microcell) base stations (SBSs), resulting in a highly dense heterogeneous network (HetNet). Unfortunately, it is expected that many of these cells will be randomly deployed; this will undoubtedly increase the amount of intercell, or cross-tier, interference, making a cross-tier spectrum sharing strategy a viable option for HetNet.

In a HetNet with cognitive small cells, the SBS and small cell user (SCU) have the capability to autonomously adapt their parameters based on surrounding conditions. The small cells can dynamically sense spectrum usage by the macrocells and adapt their transmission to optimize overall usage of the spectrum. Cognitive radio (Haykin, 2005) has shown to be effective in providing the intelligence needed for implementing cognitive small cells. Interference among different cells can be mitigated by using cognitive small cells that can transmit opportunistically by first sensing the status of the spectrum, and then using the channel without interrupting the macrocell users. This chapter will illustrate cognitive transmission strategy for two-tier HetNet, where the network only consists of two types of base stations: the MBS and SBS, in which the cognitive transmitters can transmit opportunistically, using primarily underlay spectrum sharing. This allows the primary and secondary users to transmit simultaneously, under the constraint that the transmit power of the secondary transmission is below a certain noise floor when they reach the primary receivers. This requires the secondary users to first utilize an opportunity identification phase, commonly known as spectrum sensing, to see whether or not transmission is possible. Because this is a vital part of all cognitive transmission schemes, Section 8.2 will briefly overview some of the commonly used sensing techniques.

Notations:
Upper (lower) boldface letters indicate matrices (column vectors).
Superscript ^{H} denotes Hermitian, ^{T} denotes transposition.
$[\mathbf{A}]_{ij}$ denotes the $(i,j)^{\text{th}}$ element of $\mathbf{A}$.
$[\mathbb{E}\cdot]$ stands for statistical expectation of the entity inside the square bracket.
$\mathbf{A} \succ 0$ and $\mathbf{A} \succeq 0$ designate $\mathbf{A}$ as a symmetric positive definite and symmetric semi-definite matrix, respectively.
$\mathbf{I}_N$ denotes an $N \times N$ identity matrix. $\mathbf{1}_{M \times N}$ and $\mathbf{1}_M$ denote an $M \times N$ matrix and $M \times 1$ vector, respectively, containing 1 in all of their entries.
$\otimes$ and $\circ$ denote the Kronecker and Hadamard products, respectively.
$\text{vec}(\cdot)$ is the vectorization operator.
$\text{tr}(\mathbf{A})$ denotes the trace of the matrix $\mathbf{A}$.
$\|\cdot\|_F$ denotes Frobenius norm.
$\lambda_{\max}(\mathbf{A})$ denotes the maximum eigenvalue of matrix $\mathbf{A}$.
$|S|$ denotes the cardinality of the set S.

8.2 SPECTRUM SENSING FOR COGNITIVE RADIO

Spectrum sensing has been considered the most important task in the establishment of cognitive radio. Spectrum sensing includes awareness about the interference temperature and presence of existing users. Although spectrum is traditionally viewed as measuring the spectral content around the cognitive node, or measuring the interference temperature over the spectrum, it has become more accepted in practice that spectrum sensing should involve the attainment of spectrum usage characteristics across not only frequency, but also time, space, and code. It can also include determining what type of signal is occupying the spectrum.

Sensing algorithms can roughly be categorized into five techniques (Arslan, 2007). *Matched filtering* has been known to render optimal performance when the transmitted signal is known. Its relative short time for detection allows it to quickly determine whether or not the primary user is in the proximity of the cognitive user. However, implementation of such a technique requires perfect knowledge of the primary user's signaling features such as bandwidth, operating frequency, modulation type, pulse shaping, and so on. In *waveform-based sensing*, the cognitive user exploits a priori known sequence embedded inside the primary user's data stream to detect the presence of the primary user. Similar to synchronization, the cognitive user correlates the receive signal from the primary user with the known sequence pattern. A peak in the resulting signal signifies the presence of the primary user. The disadvantage of this technique is the requirement that the cognitive user must have a priori knowledge about which sequences are available to exploit in the primary user signal. The third type of method is based on exploiting cyclostationarity embedded inside modulated signals, which is not present in noise signals, making them wide-sense stationary in nature. Assuming the received signal at the secondary user is $y(n) = s(n) + \eta(n)$, where $s(n)$ and $\eta(n)$ are the modulated and noise signal, respectively, then the cyclic spectral density function of $y(n)$ is (Gardner, 1991)

$$S_{yy}(f, \alpha) = \sum_{\tau} r_{yy}^{\alpha}(\tau) e^{-j2\pi f \tau}, \quad \text{where } r_{yy}^{\alpha}(\tau) = E[y(n+\tau)y^*(n-\tau)e^{j2\pi\alpha n}] \text{ is the cyclic}$$

autocorrelation function, and α is the cyclic frequency. Because the cyclic autocorrelation function will output peak values when α is equal to the fundamental frequencies of $s(n)$, it can be used to detect the primary user, assuming that the secondary user has knowledge of α. *Wavelet transform-based techniques* transform the received signal's power spectrum density (PSD) into the wavelet domain and attempt to detect edges in the PSD of a wideband channel. The edges correspond to transitions from occupied band to empty band, or vice versa. Once the edges have been detected, the power within the bands between two edges are estimated, which then can be used to indicate the presence or absence of the primary user. *Energy detector-based techniques* are the least computational expensive techniques among all sensing techniques, as the secondary user only needs to compute the energy of the received signal. The energy is then used to compare to a certain threshold, which can be determined based on the noise signal level. However, not only is it difficult to have knowledge of the noise level, but this technique will incur a high failure rate when the signal-to-noise ratio (SNR) of the received signal is low.

8.3 UNDERLAY SPECTRUM SHARING

Recently, spectrum underlay techniques using single or multiple antennas have been proposed as a viable option to implement cognitive radio systems. The relationship between the primary and secondary users in a cognitive radio system is analogous to the one between the macrocell and small cell networks. The secondary users in a cognitive radio system can be regarded as the aggressors (e.g., in the macrocell), while the primary users can be regarded as the victims. The interference can be managed by constraining the transmit power emitted by the aggressor-transmitter (A-Tx) toward its victims so that the interference toward the victims is unnoticeable. In this case, the communicating link models the channel between the A-Tx and A-Rx while the interfering link models the channel between the A-Tx and victims. Both the multiple-input and multiple-output (MIMO) and doubly selective fading single-input and single-output (SISO) channel scenarios will be considered, with a single A-Tx and A-Rx with multiple victims. However, performance of such a system is hindered by inaccurate CSI between the A-Tx and its victims, and between the A-Tx and A-Rx, with channel estimation error in the former usually greater due to possible lack of hand shaking.

Stochastic and deterministic mismatch models to model mismatch between actual and estimated channel-state information (CSI) have been incorporated in transmitter (Zheng et al., 2009, 2010; Islam et al., 2009; Zhang et al., 2009; Gharavol et al., 2010; Phan et al., 2009; Huang et al., 2012) and transceiver (Gharavol et al., 2011; Du et al., 2012) designs to guard against performance loss caused by channel estimation noise. The stochastic model attempts to model the mismatch in the form of confidence level. However, using the stochastic model requires the knowledge of the distribution function, or the first- and/or second-order statistics of the function of the channel estimate. The optimality of the design will be affected if the assumed probability density function and/or corresponding parameters are incorrect or inaccurately estimated.

In contrast, the deterministic mismatch model, or worst-case performance model, assumes a priori knowledge about the maximum error bound and models the error using norm ball. This will be known hereafter as norm ball mismatch model (NBMM), which can be generalized to an ellipsoid by assuming a priori knowledge about the directionality of the mismatch. The deterministic mismatch model has been applied to a wide array of designs, from robust precoder design with single A-Tx and single victim (Islam et al., 2009; Zhang et al., 2009) and robust precoder design with multiple A-Txs and multiple victims (Gharavol et al., 2010; Zheng et al., 2009), to robust precoder design for multicast signal transmission (Phan et al., 2009; Huang et al., 2012) and robust transceiver design (Gharavol et al., 2011; Du et al., 2012). Unfortunately, the error bound, i.e., the radius of the norm ball, that is assumed to bound the energy of the mismatch is often chosen arbitrarily, thus affecting optimality of the design. By assuming the channel coefficients and estimation noise are jointly Gaussian, a closed-form equation for the error bound was given in Pascual-Iserte et al. (2006), in which estimation of the Gaussian parameters is required.

8.3.1 UNDERLAY SPECTRUM SHARING FOR HETEROGENEOUS NETWORKS WITH MIMO CHANNELS

A new deterministic mismatch model, called *sparsity enhanced mismatch model* (SEMM), was proposed in Chang and Fung (2014a) to better account for the CSI-estimate mismatch compared to the NBMM, which can take advantage of the potential offered by MIMO-interference channels. The heterogeneous system considered includes a single A-Tx and A-Rx, with multiple victims. In the context of precoder design using the SEMM, the proposed model allows the A-Tx to utilize higher transmit power without violating the interference constraint placed at the victims, resulting in enhanced performance in the communicating link. This is achievable because the additional transmit power has been "absorbed," or allocated, to the sparse elements of the interfering channel. This is made possible by judiciously choosing a basis expansion model (BEM) to represent the MIMO channel.

8.3.2 UNDERLAY SPECTRUM SHARING FOR HETEROGENEOUS NETWORKS WITH DOUBLY SELECTIVE FADING SISO CHANNELS

The problem of channel-estimate mismatch is exasperated when the channel is time-varying. The Gauss-Markov autoregressive order-one (AR1) model is often used to capture temporal variation when the channel is varying slowly. However, as the coherence time of the channel decreases, the effectiveness of the AR1 model diminishes. Similar to the case of MIMO channel, a BEM has been proposed in Ma and Giannakis (2003) to combat such a problem in which the channel can be represented as a linear combination of basis functions of time with their corresponding coefficients. Note that BEM coefficients remain invariant within one temporal block, and only vary from block to block. Fourier basis based on short-time Fourier transform was previously proposed to model time-selective fading channels, but suffers from spectrum leakage (Zemen and Mecklenbrauker, 2005).

A new mismatch model, called *sparsity-enhanced mismatch model—reverse discrete prolate spheroidal sequence* (SEMMR), was proposed in Chang and Fung (2014b) to better account for the CSI-estimate mismatch compared to the NBMM. Using the proposed SEMMR transceiver, which includes a two-stage precoder and decoder, the A-Tx is able to utilize higher transmit power without violating the interference constraint placed at the victims, resulting in enhanced performance in the communicating link, similar to the SEMM. Discrete prolate spheroidal sequence (DPSS) has recently been proposed as an alternative to represent time-selective fading channels (Zemen and Mecklenbrauker, 2005). DPSS has proven to be the optimal solution to represent band- and also time-limited signals by providing the maximal concentration of energy in the time and frequency domains, respectively (Slepian, 1978). Given that the channel is Doppler-limited, i.e., Doppler-shift-limited, a few DPS coefficients (the principal components or PCs) are sufficient to represent the channel, implying that DPS basis has a good energy compaction property. DPSS has also been used to model frequency- and doubly selective fading (DSF) SISO channels (Carrasco-Alvarez et al., 2012). Unfortunately, the PCs of the communicating and interfering channels are naturally aligned, which hinders performance. The SEMMR-based two-stage transceiver scheme is able to bypass such problems by reversing the order of the DPS coefficients of all the interfering channels, thus maximizing the orthogonality between the DPS coefficients of the communicating and interfering channels in subspace, spanned by the DPS basis.

Regardless of which basis is employed, the use of the SEMM and SEMMR offers better performance than the NBMM, brought about by the existence of dual parameters in the mismatch model, which allows both mismatch models to have fine-grained control of the mismatch energy bound.

8.4 SYSTEM MODEL
8.4.1 SYSTEM MODEL WITH MIMO CHANNEL

A system consisting of a single aggressor transmitter and receiver equipped with multiple antennas is considered. All victims have n_V antennas, while the A-Tx and A-Rx have n_T and n_R antennas, respectively. The channel between the A-Tx and A-Rx is referred to as the communicating link hereafter, and is denoted as $\mathbf{H} \in \mathbb{C}^{n_R \times n_T}$. The channel between the A-Tx and the kth victim is referred to as the interfering link hereafter, and is denoted as $\mathbf{G}_k \in \mathbb{C}^{n_V \times n_T}$. Orthogonal space-time block code (OSTBC) is used to encode the data signal vector $\mathbf{s} \in \mathbb{C}^{B \times 1}$ at the A-Tx containing 4-QAM modulated symbol with equal probability and equal power, i.e., $|s_i|^2 = 1$. Therefore, the receive signal matrix at the A-Rx is $\mathbf{Y} = \mathbf{HFC} + \mathbf{N} \in \mathbb{C}^{n_R \times T}$, where $\mathbf{F} \in \mathbb{C}^{n_T \times B}$ is the precoder matrix and $\mathbf{C} \in \mathbb{C}^{B \times T}$ is an orthogonal space-time codeword matrix such that $\mathbf{CC}^H = T\mathbf{I}_B$, with T denoting the number of symbols encoded in time and B denoting the number of data streams during T. $\mathbf{N} \in \mathbb{C}^{n_R \times T}$ is the noise matrix containing zero-mean, independent Gaussian-distributed samples with known variance σ_n^2. It is assumed that the noise is uncorrelated with the data signal. Finally, a maximum likelihood (ML) decoder is employed at the A-Rx to recover an estimate of the data signal $\hat{\mathbf{s}} \in \mathbb{C}^{B \times 1}$.

8.4.2 SYSTEM MODEL WITH DOUBLY FADING SELECTIVE SISO CHANNEL

Because time- and frequency-selective fading channels are special cases of DSF channels, the signal and system notations adopted will include all three scenarios. OFDM transmission will be adopted for all frequency-selective (and doubly selective) fading channels for ease of decoding. Hence, the data frame (matrix) $\mathbf{S} \in \mathbb{C}^{M \times N}$ can be used to describe data transmission across DSF channels. In other words, it is assumed that OFDM signaling is used, where a resource block contains N OFDM symbols with M subcarriers.[1] The data frame can be serialized with $[\mathbf{s}]_i$ as data sample for transmission, where $i = nM + m$ is the serial index, with $n \in \{1, \cdots, N\}$ and $m \in \{1, \cdots, M\}$. Thus, the received signal can be written in vector form as $\mathbf{y} = \mathbf{h} \circ \mathbf{f} \circ \mathbf{s} + \mathbf{n} = \mathbf{H}\mathbf{F}\mathbf{s} + \mathbf{n}$, where $\mathbf{h}$, $\mathbf{f}$, $\mathbf{s}$, and $\mathbf{n}$ denote the channel, precoding, data, and channel-noise vector with zero mean, independent, and Gaussian-distributed elements, respectively. $\mathbf{H} \triangleq \mathrm{Diag}(\mathbf{h})$ and $\mathbf{F} \triangleq \mathrm{Diag}(\mathbf{f})$ define the corresponding channel and per-time and/or per-tone precoder matrix, respectively. Depending on the type of channel under consideration, $\mathbf{h}$ (and associated precoding, data, and noise vectors) will have different sizes. Specifically, $\mathbf{h}$ can be

$$\mathbf{h} = \begin{cases} \mathrm{vec}(\mathbf{H}_d) \in \mathbb{C}^{MN} & \text{doubly-selective} \\ \mathbf{h}_t \in \mathbb{C}^N & \text{time-selective} \\ \mathbf{h}_f \in \mathbb{C}^M & \text{frequency-selective} \end{cases} , \text{ where } [\mathbf{H}_d]_{mn} \text{ is the } (m,n)^{\text{th}} \text{ element}$$

of the (sampled) two-dimensional (2D) DSF channel function $H(f,t)$, $[\mathbf{h}_t]_n$ is the nth element of the (sampled) time-varying channel impulse function $h(t)$, and $[\mathbf{h}_f]_m$ is the mth element of the (sampled) frequency-response function $H(f)$. It should be obvious that $h(t)$ and $H(f)$ are special cases of the two-dimensional function $H(f,t)$. It shall be assumed that all transmissions are free of interblock interference and have negligible intercarrier interference.

8.5 PROBLEM FORMULATION

The design metric for the SEMM precoder and SEMMR first-stage precoder aims to maximize the worst-case SNR, subject to the interference leakage and transmit power constraints. That is, the design problem is formulated as

$$\begin{aligned} &\max \quad \text{worst-case SNR} \\ &s.t. \quad \text{interference leakage power} \leq I\text{th} \\ &\qquad \text{transmit power} \leq P\text{th}, \end{aligned}$$

where Ith is the interference leakage threshold and Pth is the transmit power threshold. Even though the single A-Tx scenario is considered herein, the mismatch model does not preclude the inclusion of multiple A-Txs, in which the problem formulation and methodology similar to the ones considered in Zheng et al. (2009) can be used. For both MIMO and DSF SISO channels, the receive SNR is proportional to transmit signal power $\|\mathbf{H}\mathbf{F}\|_F^2 = \left\| (\hat{\mathbf{H}} + \mathbf{\Delta}_H)\mathbf{F} \right\|_F^2$, where $\hat{\mathbf{H}}$ and $\mathbf{\Delta}_H$ denote the estimate of $\mathbf{H}$ and error (or mismatch) matrix of $\hat{\mathbf{H}}$, respectively, such that $\mathbf{H} = \hat{\mathbf{H}} + \mathbf{\Delta}_H$. Similarly, for the SEMM precoder, the interference power toward the kth victim can be expressed as $\|\mathbf{G}_k\mathbf{F}\|_F^2 = \left\| \left(\hat{\mathbf{G}}_k + \mathbf{\Delta}_{G,k} \right)\mathbf{F} \right\|_F^2$, where $\hat{\mathbf{G}}_k$ represents the estimate of $\mathbf{G}_k$ and $\mathbf{\Delta}_{G,k}$ denotes its corresponding error matrix. For the SEMMR transceiver, the interference

---

[1]The maximum Doppler shift is assumed fixed within the resource block.

power is expressed as $\left\|\mathbf{G}_k \overline{\mathbf{W}}\mathbf{F}\right\|_{\mathrm{F}}^2 = \left\|\left(\hat{\mathbf{G}}_k + \mathbf{\Delta}_{G,k}\right)\overline{\mathbf{W}}\mathbf{F}\right\|_{\mathrm{F}}^2$, where $\overline{\mathbf{W}}$ denotes the second-stage precoder matrix, the design of which will be described in the sequel. The additive mismatch matrices, $\mathbf{\Delta}_{\mathrm{H}}$ and $\mathbf{\Delta}_{G,k}$, are assumed to be bounded inside a deterministic region. Deterministic regions, such as ellipsoid, can be used, which can be described by $\mathcal{H}_{\mathrm{E}} \triangleq \left\{\mathbf{\delta}_{\mathrm{H}} | \mathbf{\delta}_{\mathrm{H}}^{\mathrm{H}} \mathbf{B} \mathbf{\delta}_{\mathrm{H}} \le 1\right\}$ and $\mathcal{G}_{\mathrm{E}} \triangleq \left\{\mathbf{\delta}_{G,k} | \mathbf{\delta}_{G,k}^{\mathrm{H}} \mathbf{D} \mathbf{\delta}_{G,k} \le 1\right\}$, where $\mathbf{\delta}_{\mathrm{H}} \triangleq \mathrm{vec}(\mathbf{\Delta}_{\mathrm{H}})$ and $\mathbf{\delta}_{G,k} \triangleq \mathrm{vec}(\mathbf{\Delta}_{G,k})$. Positive definite matrices, $\mathbf{B}$ and $\mathbf{D}$, define the shape and direction of ellipsoids that bound $\mathbf{\delta}_{\mathrm{H}}$ and $\mathbf{\delta}_{G,k}$, respectively. Unfortunately, without prior knowledge of $\mathbf{\delta}_{\mathrm{H}}$ and $\mathbf{\delta}_{G,k}$, $\mathbf{B}$ and $\mathbf{D}$ are usually assumed to be scaled identity matrices, i.e., $\mathbf{B} = \dfrac{1}{\varepsilon_{\mathrm{H}}^2}\mathbf{I}_{n_T n_R}$ and $\mathbf{D} = \dfrac{1}{\varepsilon_G^2}\mathbf{I}_{n_T n_R}$. $\mathcal{H}_{\mathrm{E}}$ and $\mathcal{G}_{\mathrm{E}}$ will then degenerate to the NBMMs $\mathcal{H}_{\mathrm{NBMM}} \triangleq \left\{\mathbf{\delta}_{\mathrm{H}} \big| \|\mathbf{\delta}_{\mathrm{H}}\|_2^2 \le \varepsilon_{\mathrm{H}}^2\right\}$ and $\mathcal{G}_{\mathrm{NBMM}} \triangleq \left\{\mathbf{\delta}_{G,k} \big| \|\mathbf{\delta}_{G,k}\|_2^2 \le \varepsilon_G^2\right\}$. Using vectorized mismatch vectors $\mathbf{\delta}_{\mathrm{H}}$ and $\mathbf{\delta}_{G,k}$, the receive signal and interference power for MIMO and DSF channels can be written respectively as

$$\left\|\left(\hat{\mathbf{H}} + \mathbf{\Delta}_{\mathrm{H}}\right)\mathbf{F}\right\|_{\mathrm{F}}^2 = \left(\hat{\mathbf{h}} + \mathbf{\delta}_{\mathrm{H}}\right)^{\mathrm{H}} \widetilde{\mathbf{Q}}_S \left(\hat{\mathbf{h}} + \mathbf{\delta}_{\mathrm{H}}\right) \quad \text{and} \quad \left\|\left(\hat{\mathbf{G}}_k + \mathbf{\Delta}_{G,k}\right)\mathbf{F}\right\|_{\mathrm{F}}^2 = \left(\hat{\mathbf{g}}_k + \mathbf{\delta}_{g,k}\right)^{\mathrm{H}} \widetilde{\mathbf{Q}}_I$$

$\left(\hat{\mathbf{g}}_k + \mathbf{\delta}_{g,k}\right)$, where $\widetilde{\mathbf{Q}}_S = \begin{cases} \widetilde{\mathbf{Q}}_R \triangleq \mathbf{Q}^T \otimes \mathbf{I}_{n_R} & \mathbf{MIMO} \\ \mathbf{Q} & \mathbf{DFS} \end{cases}$,

and $\quad \widetilde{\mathbf{Q}}_I = \begin{cases} \widetilde{\mathbf{Q}}_V \triangleq \mathbf{Q}^T \otimes \mathbf{I}_{n_v}, & \mathbf{MIMO}, \\ \overline{\mathbf{W}}\mathbf{Q}\overline{\mathbf{W}}^{\mathrm{H}}, & \mathbf{DFS}, \end{cases}$ with $\hat{\mathbf{h}} \triangleq \mathrm{vec}(\hat{\mathbf{H}})$, $\hat{\mathbf{g}}_k \triangleq \mathrm{vec}(\hat{\mathbf{G}}_k)$, and

$$\mathbf{Q} \triangleq \mathbf{F}\mathbf{F}^{\mathrm{H}} \in \begin{cases} \mathbb{C}^{n_T \times n_T} & \mathbf{MIMO} \\ \mathbb{C}^{MN \times MN} & \mathbf{doubly\text{-}selective} \\ \mathbb{C}^{N \times N} & \mathbf{time\text{-}selective} \\ \mathbb{C}^{M \times M} & \mathbf{frequency\text{-}selective} \end{cases} \quad . \quad \text{Note that the property}$$

$\mathrm{vec}(\mathbf{YXZ}) = \left(\mathbf{Z}^T \otimes \mathbf{Y}\right)\mathrm{vec}(\mathbf{X})$ is used to obtain the expressions for the MIMO case. However, this property is not necessary for the case of the DSF because $\mathbf{F} = \mathrm{Diag}(\mathbf{f})$ is a per-time and/or per-tone precoder.

Using the NBMM, the robust precoder can be designed to maximize the minimum SNR such that the instantaneous interference power does not exceed a certain threshold. The design problem is formulated in Chang and Fung (2014a). With the use of $\mathcal{G}_{\mathrm{NBMM}}$, all the mismatch elements in $\mathbf{\delta}_{G,k}$ are treated *equally*. This consequently hinders the designed precoder's performance in terms of maximizing transmit power performance. Such performance loss is mitigated in the SEMM and SEMMR, using the concept of basis expansion in which judicious choice of BEM for the channel allows the precoder to allocate power into components of the interference channels. These components are chosen on the basis that they do not dramatically increase the interference perceived by the victims, while effectively raising the received power at the A-Rx. This is possible because elements of $\mathbf{\delta}_{G,k}$ are bounded differently in the SEMM and SEMMR in terms of how much interference they will contribute to their corresponding victims.

For improved esthetics, without loss of generality, the subscript k in $\mathbf{G}_k$, its associated estimate, and mismatch matrix will be ignored in the derivation in the following sections, and will only appear in the final problem formulation.

8.6 SPARSITY-ENHANCED MISMATCH MODEL (SEMM)

It was shown in Chang and Fung (2012) that higher receive SNR at the A-Rx can be achieved using the structural mismatch model (SMM) by exploiting the sparse elements induced by the SMM. It is posited that if inherent sparse elements, or near-sparse elements, in $\mathbf{G}$ can be extracted and exploited, that will allow higher receive power at the A-Rx – not by constraining the average interference power, but the instantaneous interference power. This can be accomplished by modeling the mismatch error in $\hat{\mathbf{G}}$ instead of modeling the mismatch error in the spatial covariance matrix $\hat{\mathbf{R}}_G \triangleq \mathbb{E}[\hat{\mathbf{g}}\hat{\mathbf{g}}^H]$ as was done in the SMM.

The inherent sparsity of $\mathbf{G}$ (and $\mathbf{H}$) can be extracted by representing $\mathbf{G}$ using an appropriate BEM that provides high disparity between principal and nonprincipal components (PCs and NPCs). Such disparity is defined to be the ratio of the magnitude of the smallest PCs to the highest NPC. The SEMM is formulated by incorporating a BEM to expose the inherent sparsity of $\mathbf{G}$ and using a dual parameter concentric norm ball to judiciously bound the mismatch energy corresponding to the PCs and NPCs. In addition, unlike the SMM, not only is the amount of transmit power that can be allocated to the interfering link dictated by n_T, but it is also dependent on the energy level of the NPCs, the location of the NPCs, and the "correlation" of the PCs and NPCs between the communicating and the interfering links. The correlation indicates the relative location of the PCs and NPCs in the transform domain, and its precise definition will be given in the sequel. Simulation results demonstrating its effect on the performance of the precoder using the SEMM is given in Chang and Fung (2014a). Two BEMs were proposed in Chang and Fung (2014a), illustrating two extremes in terms of the amount of a priori channel knowledge assumed (and thus the level of parameterization). The first one, known as the angular domain representation (ADR) (Tse and Viswanath, 2005), uses complex exponential functions as the basis in which the MIMO channel can be represented in the angular domain as discrete angular points by using a superposition of complex exponential basis functions parameterized by the number of transmit and receive antennas. Unfortunately, the low level of parameterization can lead to large modeling error. To alleviate this problem, a second BEM using the eigenmode of the channel is incorporated into the SEMM, resulting in SEMM—eigenmode (SEMME). Sparse representation of MIMO channels using the eigenmode was shown in Weichselberger (2003) and Weichselberger et al. (2006), which is a basis of the eigenspace of the one-sided transmit and receive correlation matrix. Hence, it requires more parameterization than the ADR. The correlation matrices are estimated by sample averaging several channel realizations during a learning period. Because the learning period of the correlation matrices increases latency and may be relatively significant depending on the channel coherence time, an alternative parametric method based on the power azimuth spectrum (PAS) to obtain the correlation matrices is proposed in Chang and Fung (2014a). The decreased level of parameterization compared to using sample averaging, as will be shown in Section 8.9, does not significantly affect performance. The resulting mismatch model will be referred to as parametric SEMME (pSEMME).

The interfering link, $\mathbf{G}$, can be in general represented using the linear transformation

$$\mathbf{G} = \mathbf{U}_V \mathbf{G}_\times \mathbf{U}_T^T \in \mathbb{C}^{n_V \times n_T} \tag{8.1}$$

Where $\mathbf{U}_V$ and $\mathbf{U}_T$ are in general unitary matrices of size n_V and n_T, respectively. $\mathbf{G}_\times \in \mathbb{C}^{n_V \times n_T}$ is a coefficient matrix in some domain $\times$, which in this work is either angular (a) or eigenmode (e), although these are not the only choices. Using the same additive mismatch model that was used for $\mathbf{H}$ on $\mathbf{G}$, and applying (eq. 8.1), then $\mathbf{G} = \hat{\mathbf{G}} + \boldsymbol{\Delta}_G = \mathbf{U}_V \hat{\mathbf{G}}_\times \mathbf{U}_T^T + \mathbf{U}_V \boldsymbol{\Delta}_{G,\times} \mathbf{U}_T^T$, where $\hat{\mathbf{G}}_\times$ is the estimate of $\mathbf{G}_\times$, and $\boldsymbol{\Delta}_{G,\times}$ is the corresponding mismatch matrix. In other words, it is assumed that the error is only modeled in the coefficient matrix. Despite this assumption, simulation results will show that better SNR performance is achieved by employing the SEMM in precoder design than by using the NBMM, thus proving the efficacy of the model.

$\mathbf{G}$ can be vectorized as $\mathbf{g} \triangleq \mathrm{vec}(\mathbf{G}) = \hat{\mathbf{g}} + \boldsymbol{\delta}_G = \tilde{\mathbf{U}}\hat{\mathbf{g}}_\times + \tilde{\mathbf{U}}\boldsymbol{\delta}_{G,\times}$, where $\hat{\mathbf{g}} \triangleq \mathrm{vec}\left(\hat{\mathbf{G}}\right)$, $\boldsymbol{\delta}_G \triangleq \mathrm{vec}(\boldsymbol{\Delta}_G)$, $\hat{\mathbf{g}}_\times \triangleq \tilde{\mathbf{U}}^H \hat{\mathbf{g}}$, and $\boldsymbol{\delta}_{G,\times} \triangleq \tilde{\mathbf{U}}^H \boldsymbol{\delta}_G$, with $\tilde{\mathbf{U}} \triangleq (\mathbf{U}_T \otimes \mathbf{U}_V) \in \mathbb{C}^{n_V n_T \times n_V n_T}$. It is clear that $\boldsymbol{\delta}_{G,\times}$ is the coefficient vector of $\boldsymbol{\delta}_G$ inside the domain $\times$.

To improve the SNR performance of the NBMM precoder, it is advantageous to exploit the high disparity that exists between the PCs and NPCs of $\hat{\mathbf{G}}$. This can be achieved by properly selecting the basis matrices $\mathbf{U}_V$ and $\mathbf{U}_T$. This allows the mismatch corresponding to the PCs and NPCs to be treated *differently*. To accomplish this, define the SEMM as

SEMM (concentric norm ball) :
$$\mathcal{G}_{\mathrm{SEMM}} \triangleq \left\{ \boldsymbol{\delta}_{G,\times} \Big| \|\boldsymbol{\delta}_{G\times\mathrm{pc}}\|_2^2 \le \varepsilon_{G\times\mathrm{pc}}^2, \|\boldsymbol{\delta}_{G\times\mathrm{npc}}\|_2^2 \le \varepsilon_{G\times\mathrm{npc}}^2 \right\}, \tag{8.2}$$

which is depicted as two concentric norm balls in Figure 8.1(b) for $n_V n_T = 4$ and $n_{\mathrm{pc}} = 2$. In other words, $\mathcal{G}_{\mathrm{SEMM}}$ is parameterized by two parameters as compared

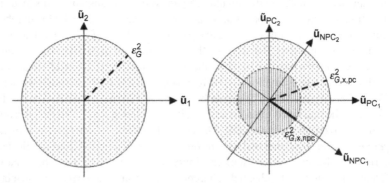

FIGURE 8.1

Illustration of uncertainty region for the NBMM and SEMM with $n_V n_T = 4$ and $n_{\mathrm{pc}} = 2$ (implying that $n_{\mathrm{npc}} = 2$). Note that the basis vectors, $\tilde{\mathbf{u}}_{\mathrm{PC}_i}$ and $\tilde{\mathbf{u}}_{\mathrm{NPC}_i}$, for $i = 1, 2$, are orthonormal to each other; hence, the picture in (b) is a bit misleading, as it is difficult to draw a space of dimension 4 on a 2-dimensional piece of paper. (a) NBMM, (b) SEMM.

to one in the NBMM (Figure 8.1(a)). The dual parameters in $\mathcal{G}_{\mathrm{SEMM}}$ allow it to have finer control over the mismatch compared to the single parameter $\mathcal{G}_{\mathrm{NBMM}}$. This is accomplished by careful selection of $\varepsilon_{G\times\mathrm{pc}}^2$ and $\varepsilon_{G\times\mathrm{npc}}^2$, which can be done parametrically using the method proposed in Pascual-Iserte et al. (2006), or nonparametrically using statistical learning (Chang and Fung, 2014a).

Further define the matrix $\prod_{\hat{G}}$ to identify the location of the PCs, i.e.,

$$[\Pi_{\hat{G}}]_{ij} = \begin{cases} 1 & \text{if } \left[\hat{\mathbf{G}}_\times\right]_{ij} \text{ is a PC} \\ 0 & \text{otherwise} \end{cases}, \text{ and } \mathbf{i}_{\mathrm{pc}} \triangleq \mathrm{vec}\left(\Pi_{\hat{G}}\right). \text{ Then } \boldsymbol{\delta}_{G\times\mathrm{pc}} \triangleq \mathbf{i}_{\mathrm{pc}} \circ \boldsymbol{\delta}_{G,\times} \text{ and }$$

$\boldsymbol{\delta}_{G\times\mathrm{npc}} \triangleq \left(\mathbf{1}_{n_Tn_V} - \mathbf{i}_{\mathrm{pc}}\right) \circ \boldsymbol{\delta}_{G,\times}$ denote the mismatch terms for the PC and NPC, respectively. Due to the orthogonality of $\boldsymbol{\delta}_{G\times\mathrm{pc}}$ and $\boldsymbol{\delta}_{G\times\mathrm{npc}}$, $\boldsymbol{\delta}_{G\times} = \boldsymbol{\delta}_{G\times\mathrm{pc}} + \boldsymbol{\delta}_{G\times\mathrm{npc}}$. Furthermore, define $\mathcal{S}_{\mathrm{pc}}$ and $\mathcal{S}_{\mathrm{npc}}$ as the set of PC and NPC, respectively, and $|\mathcal{S}_{\mathrm{pc}}| = n_{\mathrm{pc}}$,

$|\mathcal{S}_{\mathrm{npc}}| = n_{\mathrm{npc}}$, $\mathcal{S}_{\mathrm{pc}} \cap \mathcal{S}_{\mathrm{npc}} = \varnothing$ and $\min\limits_{[\hat{G}_\times]_{ij}\in\mathcal{S}_{\mathrm{pc}}} \left|\left[\hat{\mathbf{G}}_\times\right]_{ij}\right| > \min\limits_{[\hat{G}_\times]_{ij}\in\mathcal{S}_{\mathrm{npc}}} \left|\left[\hat{\mathbf{G}}_\times\right]_{ij}\right|$. Note that determination of the PCs is based on estimated channel matrix $\hat{\mathbf{G}}$, but misdetection of the PCs is possible when the disparity is not sufficient. In the simulations in Section 8.9, the PCs are determined by first sorting the coefficients in $\hat{\mathbf{G}}_\times$ in descending order, and then collecting the least number of coefficients that captures 90% of channel energy, starting with the largest PC.

With the definition of the SEMM, the amount of correlation between the PCs and NPCs in the angular domain of $\mathbf{H}$ and $\mathbf{G}$ can be defined as $r_{\mathrm{HGa}} \triangleq tr\left(|\mathbf{G}_a|^T|\mathbf{H}_a|\right)$, where $|\mathbf{A}|$ denotes the element-wise magnitude value of $\mathbf{A}$, with $\times$ replaced by a.

8.7 SPARSITY-ENHANCED MISMATCH MODEL-REVERSE DPSS (SEMMR)

In the SEMM, the ADR and eigenmode basis were utilized to model MIMO channels, but the former is susceptible to correlation between $\mathbf{H}$ and $\mathbf{G}$ and the latter is channel-dependent (Chang and Fung, 2014a). In addition, they both can offer substantial energy compaction capability, albeit under ideal circumstances. Hence, a good BEM requires the basis vectors to be invariant to channel response; does not suffer from the same strong correlation problem experienced by the ADR; and has good energy compaction, even under nonideal circumstances. Even though the DPSS is an ideal candidate to model doubly selective fading channels, as it possesses most of the above qualities of being a good basis, it still suffers from a strong correlation. Similar to the SEMM, the mismatch model proposed in Chang and Fung (2014b) is constructed by a dual-parameter concentric norm ball so that the mismatch energy corresponding to the PCs and NPCs can be judiciously bounded.

The channel model using the DPSS can be derived from (Equation 8.1) by letting $\mathbf{U}_V = \boldsymbol{\Phi}_f \in \mathbb{C}^{M\times M}$ and $\mathbf{U}_T = \boldsymbol{\Phi}_t \in \mathbb{C}^{N\times N}$, so that the doubly selective fading channel matrix can be written as $\mathbf{G}_d = \mathbf{\Phi}_f \mathbf{G}_{c,d} \mathbf{\Phi}_t^T \in \mathbb{C}^{M \times N}$, where $\mathbf{G}_{c,d}$ is the DPS coefficient matrix (so that $_\times$ in (Equation 8.1) becomes c) and $[\mathbf{G}_d]_{mn} = G(f,t)|_{f=m\Delta_f, t=nT_S}$, with $\Delta_f \triangleq \frac{1}{T_s}$ denoting the subcarrier spacing. $\mathbf{\Phi}_f$ and $\mathbf{\Phi}_t$ are unitary matrices whose column vectors are obtained by finding the eigenbasis of real-valued symmetric kernel matrices, i.e., $\mathbf{C}_\tau \phi_\ell^{(f)} = \lambda_\ell^{(\tau)} \phi_\ell^{(f)}$, and $\mathbf{C}_v \phi_q^{(t)} = \lambda_q^{(v)} \phi_q^{(t)}$, where $\varphi_\ell^{(f)} \in \mathbb{C}^M$ and $\lambda_\ell^{(\tau)}$ are the ℓth eigenvectors and the corresponding eigenvalue of $\mathbf{C}_\tau \in \mathbb{C}^{M \times M}$, respectively. Similarly, $\varphi_q^{(t)} \in \mathbb{C}^N$ and $\lambda_q^{(v)}$ are the qth eigenvectors and its corresponding eigenvalue of $\mathbf{C}_v \in \mathbb{C}^{N \times N}$ (Rossi and Muller, 2008). The elements of the kernel matrices are expressed as $[\mathbf{C}_\tau]_{m,m'} = \dfrac{\sin\left(2\pi(m'-m)\tilde{\tau}_{\max}\right)}{\pi(m'-m)}$, and $[\mathbf{C}_v]_{n,n'} = \dfrac{\sin\left(2\pi(n'-n)\tilde{v}_{\max}\right)}{\pi(n'-n)}$, with $m, m' \in \{1, \cdots, M\}$ and $n, n' \in \{1, \cdots, N\}$, where $\tilde{\tau}_{\max} = \tau_{\max}/T_S$ and $\tilde{v}_{\max} = v_{\max} T_S$ denote normalized maximum delay and Doppler spread, respectively. τ_{max} and v_{max} denote maximum delay and Doppler spread, respectively. T_S represents the symbol interval with cyclic prefix (CP), i.e., $T_S = T_{CP} + T_s$, with T_{CP} and T_s denoting the duration of the CP and useful OFDM symbol, respectively. It has been shown in Carrasco-Alvarez et al. (2012) that the DPSS is the best suboptimal basis for representing doubly selective fading wide-sense stationary uncorrelated scattering channels in the mean-squared sense, assuming the channel is $\tau-$ and $v-$ limited. $\mathbf{G}_d$ can be vectorized as $\mathbf{g}_d \triangleq \mathrm{vec}(\mathbf{G}_d) = (\mathbf{\Phi}_t \otimes \mathbf{\Phi}_f) \mathrm{vec}(\mathbf{G}_{c,d}) = (\mathbf{\Phi}_t \otimes \mathbf{\Phi}_f) \mathbf{g}_{c,d} \in \mathbb{C}^{MN}$. In summary, the bases and their associated coefficient vectors for various selective fading channels are denoted as

$$(\mathbf{\Phi}, \mathbf{g}_c) = \begin{cases} (\mathbf{\Phi}_t \otimes \mathbf{\Phi}_f, \mathbf{g}_{c,d}) & \text{doubly-selective} \\ (\mathbf{\Phi}_t, \mathbf{g}_{c,t}) & \text{time-selective} \\ (\mathbf{\Phi}_f, \mathbf{g}_{c,f}) & \text{frequency-selective} \end{cases} . \quad \text{Note that}$$

$\mathbf{\Phi} \in \mathbb{C}^{MN \times MN}$ and $\mathbf{g}_c \in \mathbb{C}^{MN}$, $\mathbf{\Phi} \in \mathbb{C}^{N \times N}$ and $\mathbf{g}_c \in \mathbb{C}^N$, and $\mathbf{\Phi} \in \mathbb{C}^{M \times M}$ and $\mathbf{g}_c \in \mathbb{C}^M$ for doubly, time-, and frequency-selective fading channels, respectively. Also note that the energy of $\mathbf{g}$ (which can equal $\mathbf{g}_d$, $\mathbf{g}_t$, or $\mathbf{g}_f$) is mostly concentrated on a few coefficients in $\mathbf{g}_c$, the amount of which reflects the number of PCs, or n_{pc}. Because the DPS coefficients of the channel are distributed nonuniformly, and most of the energy is concentrated on the greatest n_{pc} coefficients in the DPS domain, the energy of the mismatch can be bounded using the dual-parameter concentric NBMM called *Sparsity-Enhanced Mismatch Model—Reverse DPSS* (SEMMR), which is defined similarly as (Equation 8.2) by replacing $_\times$ with $_c$. Similar to the SEMM, the SEMMR can offer different levels of robustness between the PCs and NPCs, and thus be able to outperform the NBMM in the context of transceiver design, as shown in the sequel. It is assumed throughout that an LMMSE estimator is used to estimate $\mathbf{g}_{c,k}$ (implying the estimation is done in the DPS domain), the SNR during the estimation period is modestly high (this shall be known hereafter as estimation SNR, or ESNR), e.g., ESNR to 10 dB, and sufficiently large disparity exists between the PCs and the NPCs, so that a sufficiently large disparity will also exist between the PCs and the NPCs. The "reverse" part of the model is designed to reduce the correlation between $\mathbf{H}$ and $\mathbf{G}_k$ and will be made more apparent in Section 8.8.2.

8.8 PRECODER DESIGN USING THE SEMM AND SEMMR
8.8.1 SEMM PRECODER DESIGN

The SEMM precoder (and first-stage SEMMR precoder, except for the use of the effective channel as described in the sequel) can be designed by considering the interference constraint indicated in Section 8.5 (Chang and Fung, 2014a). That is (without the user index k),

$$
(\hat{\mathbf{g}} + \boldsymbol{\delta}_G)^{\mathrm{H}} \widetilde{\mathbf{Q}}_I (\hat{\mathbf{g}} + \boldsymbol{\delta}_G) \le I\text{th} \Rightarrow
\begin{aligned}
\mathbf{g}_{\times}^{\mathrm{H}} \widetilde{\mathbf{Q}}_{I,\times} \mathbf{g}_{\times} &= (\hat{\mathbf{g}}_{\times} + \boldsymbol{\delta}_{G,\times})^{\mathrm{H}} \widetilde{\mathbf{Q}}_{I,\times} (\hat{\mathbf{g}}_{\times} + \boldsymbol{\delta}_{G,\times}) \\
&= (\hat{\mathbf{g}}_{\times} + \boldsymbol{\delta}_{G\times\mathrm{pc}} + \boldsymbol{\delta}_{G\times\mathrm{npc}})^{\mathrm{H}} \widetilde{\mathbf{Q}}_{I,\times} (\hat{\mathbf{g}}_{\times} + \boldsymbol{\delta}_{G\times\mathrm{pc}} + \boldsymbol{\delta}_{G\times\mathrm{npc}}) \\
&\le I\text{th}, \ \forall \boldsymbol{\delta}_{G,\times} \in \mathcal{G}_{\mathrm{SEMM}} \text{ or } \mathcal{G}_{\mathrm{SEMMR}},
\end{aligned}
$$

$$(8.3)$$

where Ith denotes the instantaneous interference threshold. The inequality on the right is obtained by expressing $\mathbf{g}$ and $\boldsymbol{\delta}_G$ in the transform domain as $\mathbf{g}_{\times}$ and $\boldsymbol{\delta}_{G,\times}$, where
$$
\widetilde{\mathbf{Q}}_{I,\times} =
\begin{cases}
\widetilde{\mathbf{Q}}_{V,a} \triangleq (\mathbf{W}_T^{\mathrm{H}} \mathbf{Q} \mathbf{W}_T) \otimes \mathbf{I}_{n_V} & \text{for ADR} \\
\mathbf{Q}_{V,e} \triangleq (\mathbf{U}_T^{\mathrm{T}} \mathbf{Q} \mathbf{U}_T^*) \otimes \mathbf{I}_{n_V} & \text{for eigenmode}, \\
\widetilde{\mathbf{Q}}_c = \boldsymbol{\Phi}^{\mathrm{H}} \mathbf{Q} \boldsymbol{\Phi} & \text{for DPSS}
\end{cases}
\quad \text{with}
$$

$$
\boldsymbol{\Phi} =
\begin{cases}
\boldsymbol{\Phi}_t \otimes \boldsymbol{\Phi}_f & \text{doubly-selective} \\
\boldsymbol{\Phi}_t & \text{time-selective} \\
\boldsymbol{\Phi}_f & \text{frequency-selective}
\end{cases}
. \ \mathbf{W}_T \in n_T \times n_T \text{ is the DFT matrix used in the}
$$
ADR and $\mathbf{U}_T \in n_T \times n_T$ is the eigenvector matrix of the one-sided spatial correlation matrix used in the eigenmode.

Exploiting the large disparity between the values of the PCs and NPCs, (Equation 8.3) can be manipulated so that a new precoder design problem using the SEMM can be formulated as

SEMM precoder design formulation :

$$\max_{\mathbf{Q}, \gamma, \alpha, \beta} \gamma$$

$$
s.t. \ \begin{bmatrix} \widetilde{\mathbf{Q}}_R + \alpha \mathbf{B} & \widetilde{\mathbf{Q}}_R \hat{\mathbf{h}} \\ \hat{\mathbf{h}}^{\mathrm{H}} \widetilde{\mathbf{Q}}_R & \hat{\mathbf{h}}^H \widetilde{\mathbf{Q}}_R \hat{\mathbf{h}} - \alpha - \gamma \end{bmatrix} \succeq 0,
$$

$$
\begin{bmatrix} -\left[\left(\mathbf{i}_{pc} \mathbf{i}_{pc}^T \right) \circ \widetilde{\mathbf{Q}}_{V,\times} \right] + \beta \varepsilon_{G,\times,pc}^{-2} \mathbf{I}_{n_T n_V} & -\left[\left(\mathbf{i}_{pc} \mathbf{1}_{n_T n_V}^T \right) \circ \widetilde{\mathbf{Q}}_{V,\times} \right]^H \hat{\mathbf{g}}_{\times,k} \\ -\hat{\mathbf{g}}_{\times,k}^H \left[\left(\mathbf{1}_{n_T n_V} \mathbf{i}_{pc}^T \right) \circ \widetilde{\mathbf{Q}}_{V,\times} \right] & -\hat{\mathbf{g}}_{\times,k}^H \widetilde{\mathbf{Q}}_{V,\times} \hat{\mathbf{g}}_{\times,k} - \beta + I_{th}/4 \end{bmatrix} \succeq 0, \ \forall k,
$$

$$
\lambda_{\max}(\mathbf{Q}) \le p_{th} = \frac{I_{th}/3}{\varepsilon_{G,\times,pc}^2}, \ \left\| \widetilde{\mathbf{Q}}_{\times} \hat{\mathbf{g}}_{\times,k} \right\|_2 \le q_{th} = \frac{I_{th}/3}{2\varepsilon_{G,\times,npc}^2}
$$

$$
tr(\mathbf{Q}) \le P_{th}, \ \alpha, \beta \ge 0, \ \widetilde{\mathbf{Q}}_R \triangleq \mathbf{Q}^T \otimes \mathbf{I}_{n_R},
$$

$$
\widetilde{\mathbf{Q}}_{\times} =
\begin{cases}
(\mathbf{W}_T^H \mathbf{Q} \mathbf{W}_T)^T \otimes \mathbf{I}_{n_V}, \text{ for } \times = a, \\
(\mathbf{U}_T^T \mathbf{Q} \mathbf{U}_T^*)^T \otimes \mathbf{I}_{n_V}, \text{ for } \times = e.
\end{cases}
$$

$$(8.4)$$

Unfortunately, the coefficients in coefficient vectors $\mathbf{g}_c$ and $\mathbf{h}_c = \mathbf{\Phi}^H \mathbf{h}$ are strongly correlated, thus rendering the above problem formulation inappropriate for the first-stage SEMMR precoder. A second-stage transceiver will be used to bypass this problem, the design of which will be discussed in Section 8.8.2.

The power allocation property induced by the SEMM can be summarized by the power allocation-exploiting sparsity theorem-SEMM (PAST-SEMM), which can be deduced from (Equation 8.4), and it is stated in Chang and Fung (2014a) that the maximum amount of sparsity induced by the SEMM for precoder design to enhance SNR performance over the NBMM equals $(n_T - 1)n_V$. Even though the value of n_{pc} can be arbitrarily chosen, to attain optimal SNR performance, it should correspond to the true number of principal components in $\mathbf{G}$.

8.8.2 SECOND-STAGE SEMMR PRECODER AND DECODER DESIGN

The reason why $r_{H,G,c} \triangleq \mathrm{tr}\left(|\mathbf{g}_c|^T |\mathbf{h}_c|\right)$ is large is because the coefficients inside $\mathbf{g}_c$ and $\mathbf{h}_c$ are ordered in descending order; hence, it is no longer possible to attain performance gain via allocating more power to the sparse elements in $\mathbf{g}_c$ using only the first-stage precoder, because any allowable increase in power into the NPCs of $\mathbf{g}_c$ will be coupled directly into the NPCs of $\mathbf{h}$, rendering such power increase ineffective. This problem can, however, be amended by employing a second-stage precoder (and corresponding decoder) that linearly permutes (or reverses) the DPS coefficients of $\mathbf{g}$ such that the amount of correlation between $\mathbf{h}$ and $\mathbf{g}$ is minimized. To accomplish this, the functional block $\mathbf{W}$ at the transmitter is to reverse the ordering of DPS coefficients associated with $\mathbf{h}$ and $\mathbf{g}$. However, to guarantee that the coefficients in $\mathbf{h}_c$ and $\mathbf{g}_c$ are nearly orthogonal to each other, an additional decoder, $\mathbf{W}^{-1}$, is placed at the A-Rx, so the effect that $\mathbf{W}$ has on $\mathbf{h}$ will be canceled, thereby the coefficients in $\mathbf{h}_c$ and $\mathbf{g}_c$ will be near orthogonal in the DPS domain.

In the case of the doubly selective fading channel, $\mathbf{h}$ and $\mathbf{g}$ model doubly selective fading channels so that $\mathbf{g} = \mathbf{g}_d$. Note that the corresponding DPS coefficient matrix $\mathbf{G}_c = \mathbf{G}_{c,d}$ has its PCs concentrated on its upper left-hand corner, which is analogous to the descending order arrangement in $\mathbf{g}_{c,t}$ or $\mathbf{g}_{c,f}$. Because the goal of $\mathbf{W}$ is to move the DPS coefficients to the lower right-hand corner of $\mathbf{G}_c$, $\mathbf{W}$ is used to obtain the effective channel matrix $\mathbf{G}_{\mathrm{eff}} = \mathbf{W} \circ \mathbf{G}$, such that it should be equal to

$$\mathbf{G}_{\mathrm{rv}} = \mathbf{\Phi}_f \mathbf{P} \mathbf{G}_c \mathbf{P} \mathbf{\Phi}_t^T = \underbrace{\left(\mathbf{\Phi}_f \mathbf{P} \mathbf{\Phi}_f^H\right)}_{\triangleq \widetilde{\mathbf{R}}_{\Phi,f}} \mathbf{G} \underbrace{\left(\mathbf{\Phi}_t \mathbf{P} \mathbf{\Phi}_t^H\right)^T}_{\triangleq \widetilde{\mathbf{R}}_{\Phi,t}} = \widetilde{\mathbf{R}}_{\Phi,f} \mathbf{G} \widetilde{\mathbf{R}}_{\Phi,t}, \quad \text{where} \quad \mathbf{G} = \mathbf{G}_d \quad \text{and}$$

$$\mathbf{P}_{\mathrm{rv}} = \begin{bmatrix} & & 1 \\ & .^{.^{.}} & \\ 1 & & \end{bmatrix}$$ is the permutation matrix that reverses the ordering of the

elements of $\mathbf{G}_c$. Vectorizing $\mathbf{G}_{\mathrm{rv}}$ becomes $\mathbf{g}_{\mathrm{rv}} = \mathrm{vec}(\mathbf{G}_{\mathrm{rv}}) = \left(\widetilde{\mathbf{R}}_{\Phi,t} \otimes \widetilde{\mathbf{R}}_{\Phi,f}\right) \mathbf{g} = \widetilde{\mathbf{R}}_{\Phi,d} \mathbf{g}$, where $\mathbf{g} \triangleq \mathrm{vec}(\mathbf{G})$. Therefore, the optimal choice for $\mathbf{w} \triangleq \mathrm{diag}(\mathbf{W})$ becomes $\mathbf{w}^{\blacktriangle} = \left(\widetilde{\mathbf{R}}_{\Phi,d} \mathbf{g}\right) \oslash \mathbf{g}$, where $\oslash$ represents element-wise division. Unfortunately, $\mathbf{w}^{\blacktriangle}$ will be sensitive to not only imperfect CSI, but also to different $\mathbf{g}$'s in the case of multiple victims. To solve these problems, note that $[\mathbf{w}^{\blacktriangle}]_i = \sum_j \left[\widetilde{\mathbf{R}}_\Phi\right]_{ij}\left(\frac{[\mathbf{g}]_j}{[\mathbf{g}]_i}\right)$, where

$[\mathbf{g}]_i$ is the channel gain at the ith instant within one block duration. To desensitize the dependence of CSI, note that the variation of $([\mathbf{g}]_j/[\mathbf{g}]_i)$ is within one order of magnitude.[2] Hence, the ratio shall be approximated hereafter as $\left([\mathbf{g}]_j/[\mathbf{g}]_i\right) \approx 1,\ \forall i \neq j$. Using this approximation, the 2D channel-invariant second-stage precoder becomes $\overline{\mathbf{w}}_d \triangleq \text{vec}(\mathbf{W}) = \left(\widetilde{\mathbf{R}}_{\Phi,t} \otimes \widetilde{\mathbf{R}}_{\Phi,f}\right)\mathbf{1}_{MN} = \widetilde{\mathbf{R}}_{\Phi,d}\mathbf{1}_{MN}$, and the effective channel is $\mathbf{g}_{\text{eff},k} = \overline{\mathbf{w}}_d \circ \mathbf{g}_k$.

To summarize, the possible choices for the channel-invariant second-stage

precoder are $\overline{\mathbf{w}} = \begin{cases} \widetilde{\mathbf{R}}_{\Phi,d}\mathbf{1}_{MN} & \text{doubly-selective} \\ \widetilde{\mathbf{R}}_{\Phi,t}\mathbf{1}_N & \text{time-selective} \\ \widetilde{\mathbf{R}}_{\Phi,f}\mathbf{1}_M & \text{frequency-selective} \end{cases}$. The power allocation property

induced by the SEMMR can be summarized by the power allocation-exploiting sparsity theorem-SEMMR (PAST-SEMMR), It is stated in Chang and Fung (2014b) that the maximum amount of sparsity induced by the SEMMR that can be exploited for power allocation via proper transceiver design to enhance performance with a doubly selective fading channel equals $MN - 1$.

The SEMMR precoder design problem can be formulated as a semi-definite programming problem

First − stage SEMMR precoder design formulation :

$$\max_{\mathbf{Q},\gamma,\alpha,\beta} \gamma$$

$$s.t. \begin{bmatrix} \mathbf{Q}+\alpha\mathbf{B} & \mathbf{Q}\hat{\mathbf{h}} \\ \hat{\mathbf{h}}^H\mathbf{Q} & \hat{\mathbf{h}}^H\mathbf{Q}\hat{\mathbf{h}}-\alpha-\gamma \end{bmatrix} \succeq 0,$$

$$\begin{bmatrix} -\left(\mathbf{i}_{pc}\mathbf{i}_{pc}^T\right)\circ\widetilde{\mathbf{Q}}_c + \beta\varepsilon_{G,c,pc}^{-2}\mathbf{I}_{MN} & -\left[\left(\mathbf{i}_{pc}\mathbf{1}_{MN}^T\right)\circ\widetilde{\mathbf{Q}}_c\right]^H\hat{\mathbf{g}}_{\text{eff},c,k} \\ -\hat{\mathbf{g}}_{\text{eff},c,k}^H\left[\left(\mathbf{1}_{MN}\mathbf{i}_{pc}^T\right)\circ\widetilde{\mathbf{Q}}_c\right] & -\hat{\mathbf{g}}_{\text{eff},c,k}^H\widetilde{\mathbf{Q}}_c\hat{\mathbf{g}}_{\text{eff},c,k}-\beta+I_{th}/4 \end{bmatrix} \succeq 0, \forall k, \qquad (8.5)$$

$$\lambda_{\max}(\mathbf{Q}) \leq p_{th} = \frac{I_{th}/3}{\varepsilon_{G,c,pc}^2},\ \left\|\widetilde{\mathbf{Q}}_c\hat{\mathbf{g}}_{\text{eff},c,k}\right\|_2 \leq q_{th} = \frac{I_{th}/3}{2\varepsilon_{G,c,npc}^2},$$

$$tr(\mathbf{Q}) \leq P_{th}, \alpha, \beta \geq 0,$$

$$\widetilde{\mathbf{Q}}_c = \mathbf{\Phi}^H\mathbf{Q}\mathbf{\Phi}, \hat{\mathbf{g}}_{\text{eff},c,k} = \mathbf{\Phi}^H\hat{\mathbf{g}}_{\text{eff},k}$$

with $\hat{\mathbf{g}}_{\text{eff},k}$ equal to either $\overline{\mathbf{w}}_t \circ \hat{\mathbf{g}}_{k,t}$, $\overline{\mathbf{w}}_f \circ \hat{\mathbf{g}}_{k,f}$, or $\overline{\mathbf{w}}_d \circ \hat{\mathbf{g}}_{k,d}$, and $\hat{\mathbf{g}}_{k,\times}$ being the corresponding estimate, for $\times = t, f, d$. The problem formulation in (Equation 8.5) is different from (Equation 8.4) in that it uses the effective channel vector $\hat{\mathbf{g}}_{\text{eff},c,k}$ instead of $\hat{\mathbf{g}}_{c,k}$ to bypass the correlation problem.

---

[2]This has been confirmed by various simulations involving channels simulated under various conditions.

8.9 SIMULATION RESULTS

8.9.1 SEMM PRECODER

Simulation results for the SEMM precoder are generated using a half-rate 4×8 OSTBC, where $B=4$, $T=8$, $n_T=n_R=n_V=4$. Channels are simulated using the additive LOS model $\mathbf{H} = \sqrt{\frac{\kappa_H}{\kappa_H+1}}\mathbf{H}_{LOS} + \sqrt{\frac{1}{\kappa_H+1}}\mathbf{H}_{NLOS}$ and $\mathbf{G} = \sqrt{\frac{\kappa_G}{\kappa_G+1}}\mathbf{G}_{LOS} + \sqrt{\frac{1}{\kappa_G+1}}\mathbf{G}_{NLOS}$, where the definition of the LOS and NLOS (non-line-of-sight) matrices can be found in Paulraj et al. (2003). κ_H and κ_G denote the Rician factors of $\mathbf{H}$ and $\mathbf{G}$, respectively. Large-scale fading, such as path loss and shadowing effect of the channel, has been ignored. To make the model more realistic, the abstract ray-based frequency-flat MIMO channel model described in Tse and Viswanath (2005) is used to construct the NLOS channel matrices associated with $\mathbf{H}$ and $\mathbf{G}$. That is, the NLOS channel matrix associated with $\mathbf{H}$, and similarly with $\mathbf{G}$, is expressed as

$$\mathbf{H}_{NLOS} = \sum_{m=1}^{M} a_m \mathbf{a}(\theta_m^A)\mathbf{d}^T(\theta_m^D), \tag{8.6}$$

where a_m denotes the baseband complex channel gain, and $\mathbf{a}(\theta_m^A)$ and $\mathbf{d}^T(\theta_m^D)$ denote the spatial signature vectors corresponding to the angle-of-arrival (AoA) θ_m^A and angle-of-departure (AoD) θ_m^D of path m, respectively. AoA and AoD are assumed to be Gaussian distributed with two parameters: mean AoD/AoA and angular spread. The number of paths is assumed to be $M=120$. Herein antennas are aligned as a uniform linear array (ULA) with half-wavelength spacing. Note that the ray-based channel in (Equation 8.6) has been adopted as the channel model for the 3rd Generation Partnership Project (3GPP) standard (3GPP, 2003).

The transmit power threshold $P_{th}=10$ and instantaneous interference threshold $I_{th}=0.01$. The mean AoD and AoA for $\mathbf{H}$ are $90°$. For $\mathbf{G}$, the mean AoD for the A-Tx is $30°$ and the mean AoA at the victim is $0°$. Angular spread at the A-Rx and victim, denoted as $\sigma_{AS,R}$ and $\sigma_{AS,V}$, respectively, equals $30°$. The uncertainty bound associated with $\hat{\mathbf{H}}$ is $\epsilon_H=0.1$. The uncertainty bounds associated with $\hat{\mathbf{G}}_\times$ equal $\epsilon_{G\times pc}^2=0.5$ and $\epsilon_{G\times npc}^2=\rho_{npc}\epsilon_{G\times pc}^2$, where $\rho_{npc}=0.2$ (Chang and Fung, 2014a). This guarantees almost surely that $\delta_{G\times npc}$ will be bounded above by $\rho_{npc}\epsilon_{G\times pc}^2$. An ESNR of 10 dB was used to estimate $\mathbf{G}_\times$. To make fair comparison with the precoder design using the NBMM, the uncertainty bound associated with $\hat{\mathbf{G}}$ equals $\epsilon_G^2 = \epsilon_{G\times pc}^2 + \epsilon_{G\times npc}^2$. The cumulative distribution function (CDF) results below are obtained by averaging 100 channel realizations.

Figure 8.2 shows the CDF curves for receive signal power at the A-Rx, interference power at the victim for $\kappa_G=10$, and angular spread at the A-Tx $\sigma_{AS,T}=5°$. All SEMM precoders, including SEMM-angular (SEMMA) and (nonparametric) SEMME (labeled simply as SEMME), outperform the NBMM precoder by 3.4 dB on average in terms of receive signal power. Results for the pSEMME precoder are also included, where the one-sided correlation matrices are generated using a uniform PAS. Because the pSEMME is able to capture the same number of NPCs

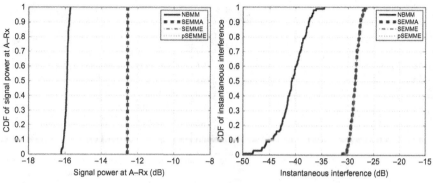

FIGURE 8.2

CDF curve of the received power at the A-Rx and instantaneous interference power toward the victim for $\kappa_H=0$, $\kappa_G=10$, $\sigma_{AS,T}=5°$, and $\sigma_{AS,R}=\sigma_{AS,V}=30°$. Mean AoD and AoA from the A-Tx to the A-Rx equals 90°. Mean AoD and AoA from the A-Tx to the victim equals 30° and 0°. SEMME uses nonparametric eigenmode. pSEMME uses parametric eigenmode.

with similar energy levels as the SEMME, this allows the pSEMME precoder to perform on par with the SEMME precoder. Furthermore, due to the presence of strong LOS, performance of the SEMMA precoder is comparable to those of the SEMME and pSEMME, and only one PC is selected in all three SEMMs to capture 90% of the total channel energy. The gain in receive signal power is carried over to performance gain in signal-enhancement ratio (SER), as shown in Chang and Fung (2014a). However, if the LOS component of **G** vanishes and $\sigma_{AS,T}$ increases from 5° to 15°, as shown in Chang and Fung (2014a), performance of the SEMMA precoder diminishes compared to the results in Figure 8.2. This is due to an increased number of PCs as a result of larger AS, which in turn decreases the number of NPCs that can be exploited by the precoder. Because the ADR is parameterized only by the number of antennas, it is more sensitive to changes in the angular properties of the channel than the eigenmode. Hence, this allows the SEMME and pSEMME precoders to outperform the NBMM precoder.

(a) CDF of the received power at the A-Rx. (b) CDF of the instantaneous interference power toward the victim.

8.9.2 SEMMR TRANSCEIVER

Consider the general case of doubly selective fading channels with three victims, with $N=M=4$. This is to prevent the simulation time from being prohibitive. The useful OFDM symbol duration is chosen as $T_s=50$ μs. The maximum delay spread is $\tau_{max}=10$ μs, with normalized Doppler shift $\tilde{v}_{max}=0.01$. The number of principal components is $n_{pc}=2$. Herein, continuous time-varying channels can be generated as $h(t)=\frac{1}{\sqrt{P}}\sum_{p=1}^{P}a_p e^{j(2\pi v_p t+\varphi_p)}$, where $\varphi_p \sim \text{Unif}[-\pi,\pi]$ denotes the random

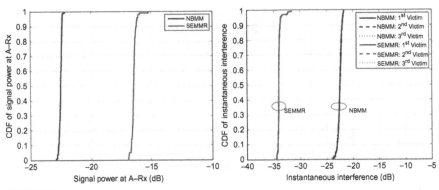

FIGURE 8.3

CDF curve of the received power at the A-Rx and instantaneous interference power toward the victims for doubly selective fading channels with three victims.

phase for path p. The number of paths is $P=20$. The system is operating at center frequency $f_c=2$ GHz. The angle of travel, AoT, θ_p, is Gaussian distributed, centered at $0°$ with standard deviation $30°$. $\mathbf{h}$ denotes the channel vector, with elements sampled from $h(t)$ with sampling period $T_S=0.1$ ms. The mismatch energy bounds are $\varepsilon_H=0.1$, $\varepsilon_{Gcpc}=0.3$, $\varepsilon_{Gcnpc}=0.09$. To make the comparison clear with the NBMM, $\varepsilon_G^2 = \varepsilon_{Gcpc}^2 + \varepsilon_{Gcnpc}^2$. Interference power for the NBMM precoder equals $\|\mathbf{GF}\|_F^2$, and equals $\|\mathbf{G\overline{W}F}\|_F^2$ for the two-stage SEMMR transceiver, where $\mathbf{\overline{W}} \triangleq \mathrm{Diag}(\mathbf{\overline{w}})$. Received signal power and interference toward multiple victims are shown in Figure 8.3. It can be seen from the figure that even with multiple victims, the signal power performance from the SEMMR transceiver is still more than 10 dB higher than the NBMM precoder, with less interference. By solving (Equation 8.5), the transmit power has been reallocated to the PCs of the communicating channel $\mathbf{H}$ to enhance signal power performance. Due to the reversing effect provided by $\mathbf{\overline{W}}$, the PCs of the communicating channel $\mathbf{H}$ coincide with the locations of the NPCs of the interfering channel $\mathbf{G}_k|_{k=3}$. This allows extra transmit power to be sent from the A-Tx to the A-Rx without causing noticeable interference to the victim, as the transmit power toward the victim is coupled into the NPCs of the interfering channel.

(a) CDF of the received power at the A-Rx. (b) CDF of the instantaneous interference power toward the victim.

8.10 CONCLUSION

Sparsity-enhanced mismatch models, SEMM and SEMMR, were proposed as a viable option for implementing terminals in cognitive small cells under different channel conditions. Such a strategy is useful for robust interference channel transmission when CSI in the communicating and interfering links is inaccurate. Results were presented in the context of intercell interference cancelation in which the A-Tx is allowed to transmit without causing noticeable interference at its victims. The design of the SEMM precoder and two-stage SEMMR transceiver involves bounding the instantaneous interference power in the interfering links. It has been shown in the PAST-SEMM and PAST-SEMMR that the performance gains of the SEMM precoder and SEMMR transceiver come from the allocation of the transmit power to the sparse elements in the interfering link, where this sparsity is brought forth by exploiting the energy compaction property of a BEM in the SEMM and SEMMR. Design of the SEMMR transceiver orthogonalizes the placement of the nonprincipal components of the interfering channels with respect to the principal components of the communicating channel, such that transmit power performance toward the receiver can be further optimized.

REFERENCES

3GPP, "Spatial channel model for MIMO simulations", TR 25.996 V6.1.0, Sep. 2003.

Arslan, H., 2007. Cognitive Radio, Software Defined Radio, and Adaptive Wireless Systems. Springer, Netherlands.

Carrasco-Alvarez, R., et al., 2012. Time-varying channel estimation using two-dimensional channel orthogonalization and superimposed training. IEEE Trans. Signal Process. 60 (8), 4439–4443.

Chang, C.-Y., Fung, C.C., 2012. Robust interference channel transmissionusing sparsity enhanced mismatch model. In: Proc. of Statistical Signal Processing Workshop (SSP). IEEE, Ann Arbor, Michigan, USA, pp. 836–839.

Chang, C.-Y., Fung, C.C., 2014a. Sparsity enhanced mismatch model for robust spatial intercell interference cancellation in heterogeneous networks. Accepted to the IEEE Trans. Comm.

Chang, C.-Y., Fung, C.C., 2014b. Sparsity enhanced mismatch model for heterogeneous networks with doubly-selective fading channels. Submitted to the IEEE Trans. Comm.

Du, H., et al., 2012. Joint transceiver beamforming in mimo cognitive radio network via second-order cone programming. IEEE Trans. Signal Process. 60 (2), 781–792.

Gardner, W., 1991. Exploitation of spectral redundancy in cyclostationary signals. IEEE Signal Proc. Mag. 8 (2), 14–36.

Gharavol, E.A., et al., 2010. Robust downlink beamforming in multiuser miso cognitive radio networks with imperfect channel-state information. IEEE Trans. Veh. Technol. 59 (6), 2852–2860.

Gharavol, E.A., et al., 2011. Robust linear tranceiver design in mimo ad hoc cognitive radio networks with imperfect channel state information. IEEE Trans. Wirelss Commun. 10 (5), 1448–1457.

Haykin, S., 2005. Cognitive radio: brain-empowered wireless communications. IEEE J. Sel. Area Commun. 23 (2), 201–220.

Huang, Y., et al., 2012. Robust multicast beamforming for spectrum sharing-based cognitive radios. IEEE Trans. Signal Process. 60 (1), 527–533.

Islam, M.H., et al., 2009. Robust precoding for orthogonal space-time block coded mimo cognitive radio networks. In: Proc. of the Workshop on Signal Processing Advances in Wireless Communications. Jun. 2009, pp. 86-90.

Ma, X., Giannakis, G.B., 2003. Maximum-diversity transmissions over doubly selective wireless channels. IEEE Trans. Inf. Theory 49 (7), 1832–1840.

Pascual-Iserte, A., et al., 2006. A robust maximin approach for mimo communications with imperfect channel state information based on convex optimization. IEEE Trans. Signal Process. 54 (1), 346–360.

Paulraj, A., et al., 2003. Introduction to Space-Time Wireless Communications. Cambridge University Press, UK.

Phan, K.T., et al., 2009. Spectrum sharing in wireless networks via qos-aware secondary multicast beamforming. IEEE Trans. Signal Process. 57 (6), 2323–2335.

Rossi, P.S., Muller, R.R., 2008. Slepian-based two-dimensional estimation of time-frequency variant mimo-ofdm channels. IEEE Signal Process. Lett. 15 (1), 21–24.

Slepian, D., 1978. Prolate spheroidal wave functions, fourier analysis, and uncertainty—v: the discrete case. Bell Syst. Tech. J. 57 (5), 1371–1430.

Tse, D., Viswanath, P., 2005. Fundamental of Wireless Communication. Cambridge University Press, New York.

Weichselberger, W., 2003. Spatial structure of multiple antenna radio channels: a signal processing viewpoint. Ph.D. Thesis, Vienna University of Technology.

Weichselberger, W., et al., 2006. A stochastic mimo channel model with joint correlation of both link ends. IEEE Trans. Wireless Commun. 5 (1), 90–100.

Zemen, T., Mecklenbrauker, C.F., 2005. Time-variant channel estimation using discrete prolate spheroidal sequences. IEEE Trans. Signal Process. 53 (9), 3597–3607.

Zhang, L., et al., 2009. Robust cognitive beamforming with partial channel state information. IEEE Trans. Wireless Commun. 8 (8), 4143–4153.

Zheng, G., et al., 2009. Robust cognitive beamforming with bounded channel uncertainties. IEEE Trans. Signal Process. 57 (12), 4871–4881.

Zheng, G., et al., 2010. Robust beamforming in cognitive radio. IEEE Trans. Wireless Commun. 9 (2), 570–576.

Ecologically Inspired Resource Distribution Techniques for Sustainable Communication Networks

Kumudu S. Munasinghe[1] and Abbas Jamalipour[2]

[1]*Faculty of Education, Science, Technology, and Mathematics, The University of Canberra, Canberra, Australia*
[2]*School of Electrical and Information Engineering, The University of Sydney, Sydney, NSW, Australia*

CHAPTER CONTENTS

9.1 INTRODUCTION

Future generations of communication networks, also known as next-generation networks (NGNs), have converged heterogeneous architectures (e.g., different access networks, traffic conditions, and user behavior). Hence, the core of an NGN can be seen as a converging point for heterogeneous classes of traffic. Therefore, it is absolutely essential to have appropriate controls for the NGN for optimizing resource utilization in an environment with multiservice session flows. The conventional method for guaranteeing fair allocation of resources is via admission control techniques. Regardless of the admission controlling scheme (e.g., measurement-based, transmission power-based, system load-based, throughput-based, and bandwidth-based algorithms (Tragos et al., 2008)), as the availability of resources diminishes with time, competition among the newly arriving flows become inevitable. Several studies have been carried out for observing these competitive outcomes for classical Internet traffic (i.e., homogeneous flows) (Voorhies et al., 2006; Guo and Matta, 2001). This initially encouraged us to revisit this problem from the perspective of a heterogeneous NGN. For the first time, we carried out an investigation on resource competition that takes place in an NGN via an ecologically inspired approach (Munasinghe and Jamalipour, 2011). The conclusion of this study was that the session class with the lowest resource requirements competitively dominates its presence over other classes, despite the presence of admission control mechanisms. This motivated us to explore a solution for equitable resource distribution, which does not disadvantage either session class. Hence, this chapter presents an ecologically inspired graphical theory for finding an equilibrium outcome, such that all competing classes in an NGN may coexist, despite competition among themselves. This will be an important step toward modeling the behavior and conditions for the stability of an NGN ecosystem, which to the best of our knowledge has not been explored as yet.

Next, we will apply the aforementioned ecologically inspired graphical theory for developing a mobility load-balancing optimization (MLBO) scheme for the NGN. In this case, the MLBO is designed for a long-term evolution (LTE) system as one of its self-organizing network (SON) functions. The objective of MLBO is to balance the load of an eNodeB to maintain its stability, so that the number of handoffs can be minimized. As per the 3rd Generation Partnership Project (3GPP) specification (3GPP, 2011a,b), an MLBO must be able to successfully balance not only its radio load, but also its transport network load (i.e., S1 interface). To the best of our knowledge, an MLBO algorithm capable of handling both aspects has not been proposed as yet. Therefore, as a very first step toward achieving this goal, our distributed MLBO algorithm will be capable of simultaneously balancing the load of the radio and transport interfaces. This ecologically inspired graphical theory will successfully identify an optimized resource usage region in order to achieve a balanced load outcome between both radio and transport interfaces. It also has necessary coordination with other SON algorithms such as mobility robustness/handover optimization (MR/HO) and energy savings to achieve better system stability.

The remainder of this chapter is organized as follows: The next section introduces the consumer resource interaction theory. After this, the theory's applicability to the NGN is presented. The next sections provide a simulation-based validation of the aforementioned graphical equilibrium theory and its proposed applicability to the NGN. The next sections describe how this theory could be applied for developing a novel MLBO framework for the LTE system, followed by results and conclusion.

9.2 CONSUMER-RESOURCE DYNAMICS

Resources are entities that contribute positively to population growth, and are consumed in the process. We have identified that the dynamics of abiotic resources such as energy, bandwidth, content, and so on adequately describe the characteristics of the resources in an NGN. Hence, as per the consumer resource interaction theory, the following four pieces of information are needed for predicting the equilibrium of a resource competition: the reproductive/growth response of the competing species against the resource, the mortality rate experienced by the same, the supply rate of the resource, and the consumption rate of each resource by each species. The two conditions of equilibrium to be satisfied are (1) the resource-dependent reproduction of each species must balance its mortality rate, and (2) a resource supply exactly balances total resource consumption for each resource. The relationship between the aforementioned four pieces for satisfying the conditions for equilibrium could be mathematically expressed as

$$\frac{dN_i}{dt} = \left[\mu_i(R_j) - m_i \right] N_i \quad \text{for } i = 1, \dots, n \tag{9.1}$$

$$\frac{dR_j}{dt} = S_j - \sum_{i=1}^{n} \mu_i(R_j) N_i Q_{ij} \quad \text{for } j = 1, \dots, m \tag{9.2}$$

As per Equation (9.1), the resource-dependent population growth dN_i/dt, is equal to μ_i, the growth rate at resource availability R_j minus the mortality rate m_i, where N_i is the population density for a species i. Further, Equation (9.2) represents the resource availability. Here, S_j is the maximum available level for a particular resource j that can ever be available in a closed habitat, and Q_{ij} is the resource quota (i.e., amount of resource j contained in a unit of population i). Also, because the types of resources considered in our discussion are abiotic in nature, there is no supply rate. Therefore, at any given time, such a resource is either in a bound or a free state. In order to enable a clear and concise validation of this graphical theory, we will consider a case of two resources. This gives special insight on how to classify resources based on their joint effects on the per capita growth rate of a population.

For two noninteractive and essential resources, the population growth requires consumption of each of the two resources, say, R_1 and R_2. Consider the first equilibrium arising from the growth of a single population on two such essential resources. By setting $dN_i/dt = 0$ in Equation (9.1), the growth isocline where a species' growth equals its mortality rate is obtained. This is called the zero net growth isocline (ZNGI) for that species. The population density will remain unchanged for a habitat when resource availabilities lie on the ZNGI, hence the satisfaction of the following condition:

$$\mu_i(R_1, R_2) = m_i. \tag{9.3}$$

As previously mentioned, the ZNGI is only half of the condition for establishing equilibrium. The second condition is where resource consumption rates balance resource supply rates. This balance in resource dynamics can be expressed from the following vector equation:

$$\vec{u} = \begin{bmatrix} S_1 - R_1^* \\ S_2 - R_2^* \end{bmatrix} = \begin{bmatrix} N_i^* Q_{i1} \\ N_i^* Q_{i2} \end{bmatrix} = \vec{c_i} \tag{9.4}$$

where $\vec{c_i}$ is the consumption vector and $\vec{u}$ is the supply vector, which indicated the trajectory that (R_1^*, R_2^*) would take if consumption ceased (Tilman, 1980). The simultaneous solution of Equations (9.3) and (9.4) for unknown R_1^*, R_2^*, and N_i^* defines

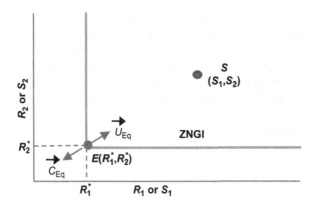

FIGURE 9.1

The equilibrium point, E, associated with a resource supply point, S, for two essential resource classes, R_1 and R_2.

the equilibrium for a single population (Leon and Tumpson, 1975). As per the graphical theory, out of all the points on the ZNGI, there is one point where Equation (9.4) will be satisfied—in other words, where $\vec{c}_i$ the consumption vector will be opposite in direction to $\vec{u}$ the resource supply vector. The slope of $\vec{c}_i$ the consumption vector is $-Q_{i2}/Q_{i1}$. The minus sign indicates that the consumption reduces the resource availability. The solution of Equation (9.4) also requires that $\vec{u}$ the supply vector has the same slope but points the opposite direction, in fact, directed toward the supply point. This is the resource equilibrium point, (R_1^*, R_2^*), as illustrated in Figure 9.1, directed toward $S(S_1, S_2)$ resource supply point. At this instance, N_i^* is the equilibrium density of species i and R^* is defined as the resource availability at which population growth rate equilibrates. Therefore, the graphical representation of the equilibrium of Equations (9.1) and (9.2) consists of finding the point lying on the isocline such that the supply vector of slope Q_{i2}/Q_{i1} also points toward the supply point. The final requirement is that the magnitudes of vectors $\vec{c}_i$ and $\vec{u}$ be equal.

Nevertheless, we wish to point out to the reader that there is much deeper meaning to the concept of R^*, which is governed by the R^* rule. Interested readers are referred to Grover (1997) for further information. Similarly, competition between two (or more) species for two essential resources could be represented by superimposing the graphical elements, as derived above. The condition for coexistence is where the ZNGIs of each competing population intersect (Phillips, 1973). This also implies that there is a set of resources available where all populations may coexist.

9.3 RESOURCE COMPETITION IN THE NGN

Similar adaptations of ecologically inspired solutions for the NGN are scarce in the literature. One such rare example is where an ecologically inspired approach is applied for developing an access selection algorithm and a price control algorithm for optimizing network revenue for a heterogeneous network (Chen et al., 2006). Furthermore, although not directly related, there are two noteworthy instances where ecologically inspired concepts have been applied to the field of information technology (IT). In one instance, a conceptual view of a sustainable IT ecosystem was proposed (Bash et al., 2008). In the other, a conceptual model toward homogenizing between two Internet enterprises through an ecologically inspired approach was presented (Yingzi et al., 2007).

In the proposed approach, the NGN can be considered as an ecosystem of heterogeneous networks over which multiservice application traffic flows. Hence, we argue that, similar to the competition between species for a limited supply of resources in an ecological system, sessions belonging to different classes may also compete for limited resources within a converged heterogeneous networking ecosystem. Therefore, if an NGN traffic/session class can be considered analogous to a species' group in an ecosystem, competition and distribution of resources could be modeled similar to an ecological environment. As illustrated in Figure 9.2, let's assume that one of the gateways at the backbone network that connects access networks to the Internet (i.e., the core of the NGN) becomes a bottleneck. As in the case of any bottleneck, this too has limited resource availability. The remainder of this discussion considers two basic resources, bandwidth and power, which are abiotic in nature and often affect the performance of a network bottleneck. Hence, let's assume that the bottleneck has a maximum bandwidth and power capacity of, say, S_b and S_p. For clarity and convenience, all available resources will be referred to as R_j, where $j = 1, \ldots, n$ (i.e., $j = 2$ in this case). As session flows corresponding to different service classes increase, resource levels at the gateway start to decrease.

If we consider these different service classes as various species groups indicated by $i = 1, \ldots, n$, the minimum required resource quota for each flow belonging to a service class can be denoted as Q_{ij}. Let the growth rate of flow arrivals attributing to service class (or species) i be denoted by the function f_i, where it is assumed that either $f_i > 0$ (i.e., the arrival request rate increases), or there exist enough requests for exhausting the gateway's resources. Note that existing flows have limited lifetimes due to sessions being hung up, and therefore frequently change the state of the abiotic resource from bound to free state and vice versa over a period of time. Such behavior is also shown by f_i. Therefore, the ecologically inspired dynamics for such a competition can be yielded by

$$\frac{dN_i}{dt} = [f_i - d_i(R_1, R_2)]N_i \quad \text{for } i = 1, \ldots, n \tag{9.5}$$

$$\frac{dR_j}{dt} = S_j - \sum_{i=1}^{n} f_i N_i Q_{ij} \quad \text{for } j = 1, 2 \tag{9.6}$$

where N_i is the population density of active flows attributed to class i, and function d_i is the drop rate for class i at resource availability R_j. It is obvious that flows of the service class i are dropped when the available resource $R_j < Q_{ij}$, as j falls below its

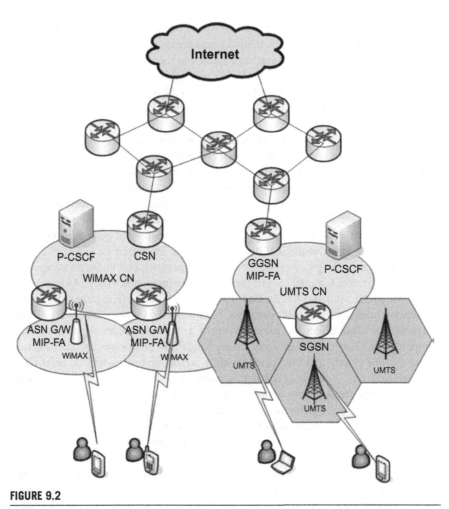

FIGURE 9.2

NGN core network architecture.

minimal quota. Similar to the scenario in Equations (9.1) and (9.2), in the case of an NGN, different classes have mutually negative effects on each other due to the consumption of limited resources. In other words, session classes in Equations (9.5) and (9.6) do not have any direct impact on each other, except through the competition for the diminishing resources. For an NGN where only one class exists, the population density (N_i) increases for $f_i > 0$. As the population grows, the available resource level declines. This level continues to decline as the number of flows increases, until the remaining resource R_j reaches R_j^*, which is considered the equilibrium (stable) point. At R_j^* the growth rate of the flows' arrival would exactly balance its drop rate (i.e., $[f_i = d_i(R_1, R_2)]$), and therefore the number of active sessions neither increases nor decreases.

This is similar to $dN_i/dt = 0$, in which case the population growth of a species stabilizes. This is called the ZNGI for that particular species group. As previously stated, the population density will remain unchanged for a habitat when resource availabilities lie on the ZNGI. It can be argued that R_j^* depends on the characteristics of a flow (e.g., duration, data rate, delay, resource consumption level, and so on) as well as its underlying transport protocol [Transmission Control Protocol (TCP) and User Datagram Protocol (UDP)]. For available resource levels smaller than R_j^*, flows are not admitted, as not enough resources are available for their Quality of Service (QoS) requirements. This trend continues until the resource level reaches an adequate level for accepting a new arrival.

Now for the case of two essential resource groups, the competition for i species groups is represented by superimposing the graphical elements, as derived above. Coexistence requires that the isoclines of the competing population intersect, as illustrated in Figure 9.3, implying that there is a set of resource availabilities for each population alone that can increase (regions 2 and 6), thus satisfying Phillips' coexistence conditions (Phillips, 1973). As per Phillips' conditions, one population ($i=1$) must be a superior competitor for one resource ($j=1$) and an inferior competitor for the other ($j=2$), and the second population ($i=2$) must be a superior competitor for the second resource ($j=2$) and an inferior competitor for the first ($j=1$).

Intersection of ZNGIs guarantees that there is an equilibrium point for both species. However, to see whether this point represents feasible and stable coexistence, consumption and supply vectors must be examined. Therefore, for a stable coexistence, the overall consumption vector $\vec{c}$, which is the resultant of single population vectors $\vec{c_1}$ and $\vec{c_2}$, and the supply vector must have the same magnitude and slope but point in opposite directions. As illustrated in Figure 9.3, for a habitat with two service

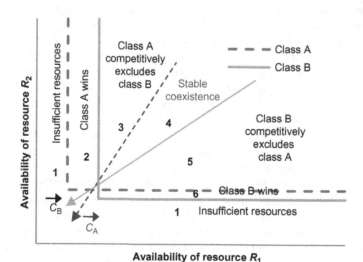

FIGURE 9.3

Competition between two service classes for two resources.

classes (say, A and B) and two resources (say, R_1 and R_2), resource ZNGIs and consumption vectors define six regions. Habitats with resource supply points falling in region 1 have insufficient resources for the survival of neither service class. Next, habitats with supply points falling in region 2 will only have sufficient resources for the survival of class A and insufficient resources for the survival of class B. Although it may seem that there are sufficient resources for the survival of both classes in region 3, class A reduces the resources down to the ZNGI of class B, thus competitively excluding class B.

A set of meaningful equilibrium solutions for both service classes will be possible in region 4 when a resource supply point lies between $\vec{c_1}$ and $\vec{c_2}$. This two-service class equilibrium point is locally stable because each species consumes proportionately more of the resource that limits its own growth (Phillips, 1973). This can be confirmed by noting that at the two-class equilibrium point, class A is limited by resource R_2 and class B is limited by resource R_1. Hence it is clear that stable coexistence is feasible only for a restricted set of supply points falling in region 4 of Figure 9.3. Opposite to what was observed for habitats in region 3, class B competitively excludes class A for habitats in region 5. Also, opposite to region 2, for habitats in region 6, available resources are only sufficient for the growth of class B. Then again, for supply points lying in region 1, neither species persists because the habitat is resource poor.

9.4 CONDITIONS FOR STABILITY AND COEXISTENCE

This section provides a simulation-based validation of the aforementioned graphical equilibrium theory of resources and its proposed applicability to the NGN. The aim of this simulation is to identify the level of resource supplies required for a stable coexistence for a given set of service classes. In this case, the two types of NGN core resources considered in this model are bandwidth and power. Similar to Figure 9.1, the basic concepts relating to coupling heterogeneous networks and vertical session handoff used for the simulation setup have been taken from Yingzi et al. (2007) (Munasinghe and Jamalipour, 2007). The MatLab-based simulation topology consists of four UMTS cells, six WLAN hotspots, five WiMAX cells, and a backbone mesh network consisting of nine routers (see Figure 9.1).

For the purpose of analyzing heterogeneous traffic classes at the NGN core network, a flow-level characterization has been used. A flow is defined as a unidirectional sequence of packets having the same identifier and transmitted with an interpacket interval smaller than a certain threshold. The WINNER project (WINNER, 2005) segregates these flows into 18 service classes based on their corresponding application types. Tragos et al. (2008) further summarizes this into eight representative classes by taking their QoS as well as resource requirements into consideration. Out of these eight classes, we have carefully chosen three classes to represent a fair characterization of low, medium, and high resource-consuming flows. As a result, simple telephony and messaging (64 kbps), multimedia telephony

(512 kbps), and HQ video streaming (1024 kbps) have been chosen to represent low, medium, and high resource-consuming classes of flows.

Next, flows belonging to these classes are generated via different networks according to a Poisson distribution, where the mean arrival rate grows over time. Durations of heterogeneous sessions are distributed between 30 and 120 s. This is based on the widely accepted assumption that human-initiated traffic is best modeled as a Poisson distribution (Roberts, 2004). Further, for simulation purposes, a measurement-based admission control algorithm is implemented at gateways (Tragos et al., 2008). Under this admission control algorithm, a flow is rejected if the estimated available resource is below a certain threshold (e.g., bandwidth and power in this case). The adopted routing mechanism for the mesh backbone is the Open Shortest Path First protocol.

Because resource competition at the NGN core mostly happens at gateway routers, as cited in Tucker et al. (2008), Cisco's recommendations are followed when estimating the required energy per transmitted bit at a bottleneck router. Hence, the average total per-bit energy consumption is assumed to be 10 nJ, which is a combination of processing energy, ingress/egress storage energy, transmit/receive energy, switch control energy, routing engine energy, and finally the energy overheads of the power supply inefficiency, including fans and blowers. Therefore, on average, a simple telephony session would consume 0.64 mW of energy from a gateway router. Similarly, an HQ video streaming would consume 5.12 and 20 mW of energy, respectively. Further, each of these routers has a resource availability of 100 Mbps bandwidth and 200 W of power.

We assume that at the NGN core network, streaming flows are handled by network nodes using a nonpreemptive priority queuing. Via application-specific prioritization, maximal responsiveness is ensured for streaming flows, which is otherwise not supported by its transport protocol UDP. Elastic-type flows have purposely been omitted because TCP plays a major role in manipulating the degree of fairness, which is beyond the scope of this discussion. Next, the arrival rates are increased for identifying the bottleneck nodes/routers in the backbone network.

As a result of heavy competition between arriving sessions, the resource availability at the bottleneck may rapidly diminish with time. The first step in this case is the determination of positions of ZNGIs of the flow classes. As previously mentioned, the ZNGIs are determined by the corresponding R^* values of the flow classes for essential resources (i.e., bandwidth and power). According to Section 9.2, R^* is defined as the resource availability at which population growth rate equilibrates (Grover, 1997). This is obtained by individually simulating bandwidth-limited and power-limited growth curves for each of the flow classes at the bottleneck node. Figure 9.4 illustrates the superimposed ZNGIs and the consumption vectors for the three flow classes.

As mentioned in Section 9.2, in order to achieve a locally stable equilibrium point, each competing class must consume relatively more of the resource that limits its growth at equilibrium (Tilman, 1980). This is also known as the coexistence condition for the considered classes. In other words, the ZNGIs of the flow classes must

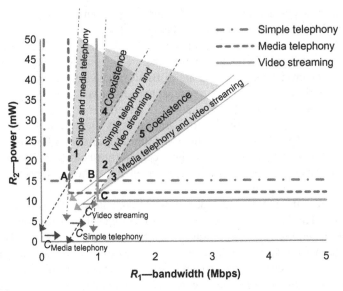

FIGURE 9.4

Competition between three service classes for two resources.

intersect, and their resource supply points must "focus" toward an equilibrium point for them to coexist. According to Figure 9.4, three points are noted (A, B, and C) where ZNGIs intersect each other. For example, at point A, media telephony is limited by R_1 (bandwidth) and simple telephony is limited by R_2 (power).

For example, the stable coexistence of simple and media telephony classes would be possible for a habitat with its resource supply point falling in region 1. In this case, the resource levels would eventually reduce down to A, at which point the growth of each flow class would stabilize. It further confirms with the equilibrium theory, because at point A, both of these classes consume proportionately more of the resource that limits their own growth (Bash et al., 2008). This can be seen by noting that at point A, the growth of the simple telephony class is limited by power, and the growth of the media telephony class is limited by bandwidth. In this region, the video-streaming class will be competitively excluded. This is illustrated in Figure 9.5a. Similarly, simple telephony and video-streaming classes coexist in region 2, and video telephony and video-streaming classes coexist in region 3. A further analysis into Figure 9.4 reveals two overlapped regions. Regions 4 and 5 correspond to the overlapping of regions 1 and 2, and 2 and 3, respectively. By applying the aforementioned argument, regions 4 and 5 can be identified as habitats for stable coexistence for all flow classes. Hence, fixing resource supply points within these regions could guarantee stable coexistence of all flow classes. Figure 9.5b illustrates how these three classes coexist when the resource supply point is fixed to region 4. A similar observation is noted when the resource supply point is fixed to region 5.

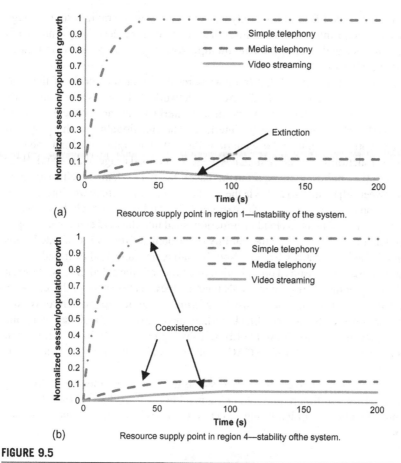

FIGURE 9.5

Dynamics of competition at different resource supply regions.

9.5 APPLICATION FOR LTE LOAD BALANCING

Next, we will apply the aforementioned ecologically inspired graphical theory for developing an MLBO scheme for an LTE system. The aim of MLB is to distribute user traffic across the system (i.e., between adjacent cells) in such a way that quality end user experience and higher system capacity is achieved. To this end, there have been two main approaches proposed thus far: distributed and centralized load balancing (LB). In the case of LTE, distributed LB approaches are better suited. For example, in the case of Lv et al. (2010), Zhang et al. (2011a,b), Kwan et al. (2010), Lobinger et al. (2011), and Zhang et al. (2011a,b), the algorithms run locally at eNodeBs and load information is exchanged via X2 interfaces. On the other hand, centralized LB approaches such as Suga et al. (2012) also exist, where the algorithm runs in a core network element. Nevertheless, a common deficiency in all of these algorithms is that the LB algorithm only takes into account the radio resource usage or simply the radio interface. Therefore, there is a need for designing a novel LB algorithm that is capable of optimizing multiple load aspects (e.g., radio and transport loads in this case).

The proposed distributed MLB framework resides at each eNodeB of the LTE Evolved - Universal Terrestrial Radio Access Network (E-UTRAN). The eNodeBs are interconnected with each other via the X2 interface and are connected to the LTE evolved packet core via the S1 interface. The functionality of the proposed MLB framework for an intra-LTE scenario is illustrated in Figure 9.6. First, the load information function senses and calculates the resource/load usage of the eNodeB for all interfaces (i.e., radio and transport in this case). Next, it exchanges cell-load information between adjacent eNodeBs via the X2 interface. This exchange of information happens periodically. However, under extreme load conditions, it could be an event-triggered operation. Then, this load information is fed into the load eco-inspired optimization function. This is where the stability of the actual eNodeB is evaluated. This algorithm defines three states for the eNodeB: stable, unstable, and overload.

If our proposed eco-inspired optimization function finds the eNodeB to be stable, it would then trigger the energy-saving SON function (beyond the scope of this chapter). If it finds the cell to be unstable, it would inform the call admission control (CAC) algorithm to selectively/partially block certain incoming session classes until the eNodeB regains its stability. Next, the LB hand-over process starts. Finally, if the cell is overloaded, it will first inform the CAC to fully block all sessions and initiate the LB hand-over process.

In each of these last two cases, the next step is to identify sessions for performing LB handoff. This will first require obtaining neighbor cell-load levels from the load information function and identifying target cells to perform load-balance handoff.

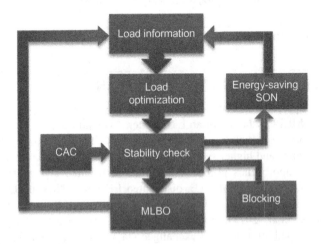

FIGURE 9.6

MLB framework for LTE system.

Once the source cell initiates handover to distribute some of its load, the target cell may perform CAC for the LB handovers. Then the handoff is performed. The key point to note is that a LB handoff must be clearly distinguishable from a normal hand-off for the target cell. Finally, the updating of handover and/or reselection configuration settings between the source and target cells takes place (Lv et al., 2010).

9.6 VALIDATION AND RESULTS

This section provides a simulation-based validation of the proposed MLB framework for an intra-LTE scenario. The LTE E-UTRAN setup consists of a 37 hexagonal-cell network connected via X2 interface. Because the proposed mechanism is distributed, each eNodeB will execute the MLB framework for balancing load between two of its interfaces (i.e., radio and transport) and then offloading to its neighboring cells. For the purpose of analyzing heterogeneous traffic classes at the eNodeB, a flow-level characterization has been used. A flow is defined as a unidirectional sequence of packets having the same identifier and transmitted with an interpacket interval smaller than a certain threshold. Based on the recommendations by the WINNER project (2005) and Tragos et al. (2008), we have carefully chosen three classes to represent a fair characterization of low, medium, and high resource-consuming flows. As a result, simple telephony and messaging (64 kbps), multimedia telephony (512 kbps), and HQ video streaming (1024 kbps) have been chosen to represent low, medium, and high resource-consuming classes of flows.

In the case of radio resource usage, the total (uplink and downlink) physical resource block (PRB) usage is considered (3GPP, 2011a,b). Therefore, the load is defined as the fraction of used PRBs in a cell. Constant bit rate is used for maintaining simplicity for this simulation. Based on the achievable throughput at 5 dB signal-to-interference noise ratio, the required number of PRBs for the above flow classes are calculated with the use of the look-up table in 3GPP (2009) for the best coding schemes (PRB bandwidth in LTE is 180 kHz). Hence, the required PRBs for 64, 512, and 1024 kbps flows are 1, 3, and 5, respectively. We consider each radio interface to have a 20 MHz channel bandwidth, and therefore 100 PRBs.

On the other hand, the transport interface is considered to have a fixed-link bandwidth of 10 Mbps connecting it to the backbone via the S2 link. Further, for simulation purposes, a measurement-based admission control algorithm is implemented at the transport interface (Voorhies et al., 2006). Under this admission control algorithm, a flow is rejected if the estimated available resource is below a certain threshold (e.g., bandwidth in this case). Next, a normalized user traffic pattern $A(t)$ is generated. This pattern is approximated by a simple sinusoidal profile, can be considered as a theoretical model, and is given by $A(t) = \frac{1}{2^b}\left[1 + \sin\left(\frac{\pi t}{12}\right)\right]^b + k$, where $A(t)$ is the instantaneous normalized traffic in Erlang, k is a constant, and $b \in \{1,3\}$, which determines the abruptness of the traffic profile (Marsan and Meo, 2009). A Poisson-distributed random process is added with both the patterns, modeling the random fluctuations of the total traffic. This is based on the widely accepted assumption that human-initiated traffic is best modeled as a Poisson distribution (Roberts, 2004).

As a result of heavy competition between arriving sessions at the eNodeB, the resource availability at the radio and transport interfaces may rapidly diminish with time. The first step of the optimization algorithm is the determination of positions of ZNGIs of the flow classes. As previously mentioned, the ZNGIs are determined by the corresponding R^* values of the flow classes for essential resources (i.e., bandwidth and power). R^* is defined as the resource availability at which population growth rate equilibrates (Grover, 1997). This is a predefined parameter for the eNodeB by individually simulating bandwidth-limited growth curves for each of the flow classes for each interface. Figure 9.7 illustrates the superimposed ZNGIs and the consumption vectors for the three flow classes.

As explained previously, in order to achieve a locally stable balanced-load condition, each competing class must consume relatively more of the resource that limits its growth at equilibrium (Phillips, 1973). According to Figure 9.7, under the given conditions, two stable regions can be identified for an eNodeB to perform. Hence, fixing resource supply points within these regions could guarantee stable coexistence of all flow classes. Further, Figure 9.7 shows three unstable regions. In these regions, one class may end up being competitively excluded by the others. In this case, the load-optimization algorithm will initiate selective blocking for overcoming this condition. Should the resource supply point fall on any other region, the optimization algorithm will trigger the overload case and hand off excess sessions until the resource supply points arrive at a stable region.

Figure 9.8 illustrates the number of unsatisfied users in the system against LB time for a 24-h period. Obviously, for the case of no MLB, a relatively higher number of unsatisfied users can be observed. Then, we compare our proposed model with a closely similar reference model (Zhang et al., 2011a,b) to evaluate

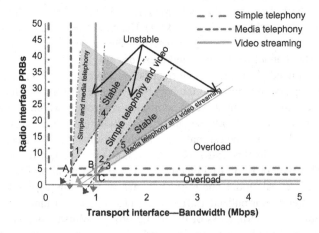

FIGURE 9.7

Decision making of the optimization algorithm.

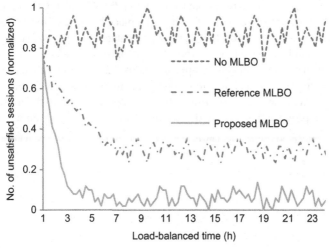

FIGURE 9.8

The number of unsatisfied users.

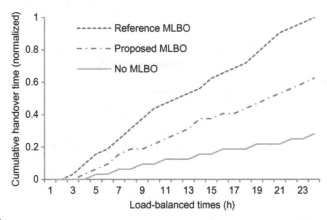

FIGURE 9.9

Cumulative handover times.

its performance. Our proposed model proves to be much more effective than Zhang et al's (2011a,b) in reducing the number of unsatisfied users. It also proves to be effectively stabilizing a load scenario before the system becomes overloaded. For example, when the system identifies an unstable condition, the eco-inspired optimization identifies specific flow classes that need to be blocked and handed off for the system to gain stability. This can also be noticed from Figure 9.8. Next, Figure 9.9 illustrates the cumulative hand-over time against the LB time. According to the results, it is clear how the proposed MLB framework introduces a relatively lower number of handoffs in comparison to the reference model in Zhang et al. (2011a,b). Obviously, when no MLB is applied, the handoffs only take place because of to user mobility, so they are minimal. Figure 9.10 illustrates the percentage of Load Balancing Handoffs (LBHOs) of total HOs. Due to the stability of the proposed optimization method, the total number of HOs is reduced by approximately 12-13% more than the reference method. Figure 9.11 illustrates how the load-distribution index varies against time. This reflects the degree of distribution among cells, where the most effective LB (i.e., the proposed) shows a distribution closest to one. Figure 9.12 shows

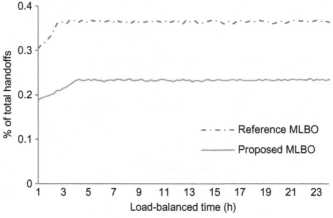

FIGURE 9.10

Percentage of LBHOs of total HOs.

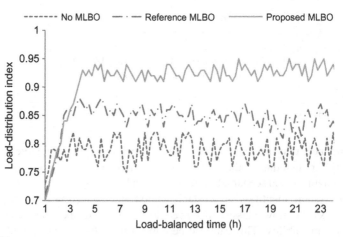

FIGURE 9.11

Load-distribution index.

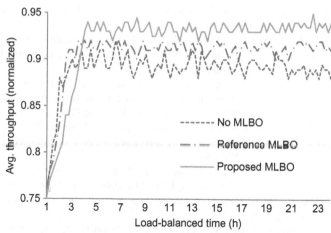

FIGURE 9.12

Average throughput.

the average of each cell's total throughput. After LB, some of the previous unsatisfied user's traffic can be successfully transmitted; hence the overall throughput increases, as illustrated.

9.7 CONCLUSIONS

This chapter develops an ecologically inspired graphical theory for equitable resource distribution in a multiresource, multiclass, heterogeneous NGN. With the proposed ecological graphical theory, optimal resource supply levels required for a stable coexistence could be accurately predicted for a heterogeneous network. Analytical proof and simulation results are provided on how a stable resource supply point could be determined for guaranteeing the coexistence of all flow classes in a closed system, thus achieving sustainability. The chapter further applies the aforementioned graphical theory to develop an MLB framework for LB in a multiresource, multiclass LTE E-UTRAN environment (a component of the NGN). The novelty of the proposed distributed MLB algorithm is that, for the first time, it is capable of simultaneously balancing the load of the radio and transport interfaces. With the help of the ecologically inspired load-optimizing algorithm, optimal load levels required for an eNodeB could be accurately predicted. Analytical proof and simulation results are provided on the effectiveness of our proposal against a conventional MLB, where the former outperforms the latter.

REFERENCES

3GPP, 2009. E-UTRA RF system scenarios (release 8), 3GPP TR 36.942 v8.2.0.
3GPP, 2011a. E-UTRA and E-UTRAN overall description, 3GPP TS 36.300 v10.3.0.

3GPP, 2011b. Self-configuring and self-optimizing networks (SON) use cases and solutions (release 9), 3GPP TR 136.902 v9.3.1.

Bash, C.E., Patel, C.D., Shah, A.J., Sharma, R.K., 2008. The sustainable information technology ecosystem. In: Proceedings of the 11th Intersociety Conference on Thermal and Thermomechanical Phenomena in Electronic Systems (ITherm 08).

Chen, J., et al., 2006. An ecology-based adaptive network control scheme for radio resource management in heterogeneous wireless networks. In: Proceedings of Bio-Inspired Models of Network, Information and Computing Systems, Madonna Di Capiglio, Italy.

Grover, J.P., 1997. Resource Competition. Chapman and Hall, London, UK.

Guo, L., Matta, I., 2001. The war between mice and elephants. In: Proceedings of Conference on Network Protocols, pp. 180–188.

Kwan, R., et al., 2010. On mobility load balancing for LTE systems. In: Proceedings of IEEE 72nd Vehicular Technology Conference Fall (VTC 2010-Fall), Ottawa, Canada.

Leon, J., Tumpson, D., 1975. Competition between two species for two complementary or substitutable resources. J. Theor. Biol. 50, 185–201.

Lobinger, A., Stefanski, S., Jansen, T., Balan, I., 2011. Coordinating handover parameter optimization and load balancing in LTE self-optimizing networks. In: Proceedings of IEEE Vehicular Technology Conference Spring (VTC 2011-Spring), Budapest, Hungry.

Lv, W., et al., 2010. Distributed mobility load balancing with RRM in LTE. In: Proceedings of 3rd IEEE International Conference on Broadband Network and Multimedia Technology (IC-BNMT), Beijing, China.

Marsan, M.A., Meo, M., 2009. Energy efficient management of two cellular access networks. In: Proceedings of Green Metrics Workshop, Washington, USA, pp. 1–5.

Munasinghe, K.S., Jamalipour, A., 2007. A unified mobility and session management platform for next generation mobile networks. In: Proceedings of IEEE Global Communications Conference, Washington, DC, USA.

Munasinghe, K.S., Jamalipour, A., 2011. Resource competition at the NGN core network: an ecologically inspired analysis. In: Proceedings of International Conference on Telecommunications (ICT2011), Cyprus.

Phillips, O.M., 1973. The equilibrium and stability of simple marine biological systems. I. Primary nutrient consumers. Am. Nat. 107, 73–93.

Roberts, J., 2004. Internet traffic, QoS and pricing. Proc. IEEE 92 (9), 1389–1399.

Suga, J., Kojima, Y., Okuda, M., 2012. Centralized mobility load balancing scheme in LTE systems. In: Proceedings of 8th International Conference on Wireless Communication Systems, Limassol, Cyprus.

Tilman, D., 1980. Resources: a graphical mechanistic approach to competition and predation. Am. Nat. 116, 362–393.

Tragos, E.Z., Tsiropoulos, G., Karetsos, G.T., Kyriazakos, S.A., 2008. Admission control for QoS support in heterogeneous 4G wireless networks. IEEE Netw. 22 (3), 32–37.

Tucker, R.S., et al., 2008. Energy consumption in IP networks. In: Proceedings of the European Conference and Exhibition on Optical Communications, Brussels.

Voorhies, S., Lee, H., Klappenecker, A., 2006. Fair service for mice in the presence of elephants. Inf. Process. Lett. 99 (3), 96–101.

WINNER Deliv. D4.4, 2005. Impact of cooperation schemes between RANs, IST-2003-507581 WINNER.

Yingzi, X., Zhenyu, L., Yupei, Z., Jie, L., 2007. Lotka-Volterra model: new perspectives in the study of homogeneity among Internet enterprises. In: Proceedings of International Conference on Wireless Communications (WCNC2007), Networking & Mobile Computing.

Zhang, L., et al., 2011a. A two-layer mobility load balancing in LTE self-organization networks. In: Proceedings of 13th International Conference on IEEE Communication Technology (ICCT), Jinan, China.

Zhang, M., Li, W., Jia, S., Zhang, L., Liu, Y., 2011b. A lightly-loaded cell initiated load balancing in LTE self-optimizing networks. In: Proceedings of 6th International Conference on Communications and Networking in China (CHINACOM), Harbin, China.

Multiobjective Optimization in Optical Networks 10

Benjamín Barán and Diego P. Pinto-Roa

Polytechnic, National University of Asuncion - UNA, Asuncion, Paraguay

CHAPTER CONTENTS

10.1 INTRODUCTION

Internet traffic growth motivates a lot of efforts to focus on increasing communication network capability to satisfy this impressive growth. In this context, optical networks have been established as the main technology for huge volumes of data traffic, subject to adequate levels of QoS (quality of service).

Optical fiber is the main element of optical networks and possesses high immunity to electromagnetic noise. With the arrival of wavelength-division multiplexing (WDM) technology, the electronic bottleneck problem was mainly solved (Somani, 2005). WDM divides the bandwidth of optical fiber into several optical channels, called wavelengths. Each wavelength is handled by adequate optoelectronic devices. Another basic component of an optical network is the optical cross-connect (OXC), which is installed in network optical nodes. OXCs give scalability to WDM networks connecting input wavelengths to output optical fibers (Somani, 2005).

10.1.1 COMMON OPTICAL NETWORK PROBLEMS IN A MULTIOBJECTIVE CONTEXT

To put an optical network to work, three fundamental areas can be observed: (1) *network design* (Mukherjee et al., 1996), (2) *network management* (Somani, 2005), and (3) *network protection* (Zhou and Subramanian, 2000). These will be briefly discussed here.

1. *Optical network design.* The main objective of a network design is to define the internal structure of optical nodes and the characteristics of optical links in order to satisfy traffic requirements without incurring a high investment. Depending on the technology used, network design can implicate several subproblems:
   - Wavelength-converter allocation (WCA) (Xiao and Leung, 1999)
   - Wavelength-splitter allocation (Rouskas, 2003)
   - Optical amplifier allocation (Ramamurthy et al., 1997)
   - Add-drop multiplexer allocation (Somani, 2005)
   - Slot-time interchanger (Zhu et al., 2005; Somani, 2005)
   - Number of optical fiber sizing (Jeong and Ayanoglu, 1996)
   - Number of wavelengths per optical fiber sizing (Somani, 2005; Zhu et al., 2005; Harada et al., 1999)
   - Number of wavelengths per waveband (Lee et al., 1993, 2004; Gerstel et al., 2000)

2. *Optical network management.* WDM network management aims to optimize what resources should do in order to connect source-destination nodes using a lightpath. Typically, the assigned resources are
- Optical links that conform a path (Barry and Humblet, 1996; Karasan and Ayanoglu, 1998; Mokhtar and Azizoglu, 1998)
- Optical fibers in each optical link (Somani, 2005; Yates, 1997)
- Wavelengths in each optical fiber (Barry and Humblet, 1996; Karasan and Ayanoglu, 1998, Yates, 1997)
- Time slots at each wavelength channel (Somani, 2005; Zhu et al., 2005)

  The optical routing problem has as its objective the calculation of optimal lightpaths for a request set.
3. *Optical network protection.* WDM network protection is a natural extension of network management, but it considers contingencies in case of network failure. Basically, optical network protection should give alternative paths when the principal path is broken by failure. The literature considers three generalized types of failure (Somani, 2005):
- Links
- Nodes
- Shared risk link group

  These failures can be caused by optical fibers, transmitters, receivers, amplifiers, splitters, and/or converters. As an example, a cable gets damaged about one time each 228 years per kilometer (4.39 cuts/year/1000 km) (To and Neusy, 1994).

Every mentioned problem has been treated as a mono-objective optimization problem subject to multiple constraints. Typically, constraints are defined by network architecture and the type of traffic to be handled. However, when resources are limited and traffic demand increases very fast, the performance of a network may not be acceptable. In fact, the minimization of network resources negatively impacts the QoS for the users (Arteta et al., 2007; Pinto-Roa et al., 2009, 2011). Consequently, the design, management, and protection of optical networks should benefit from handling resources with a multiobjective optimization approach, considering several objectives at the same time. For example, cost and performance of a network used to be contradictory objective functions; that is, to improve one objective, the other should be penalized.

With the above justifications, this work chooses one representative subproblem of each area to demonstrate the usefulness of using multiobjective optimization when studying optical networks. The three subproblem examples are

- Management: *routing and wavelength-assignment* (RWA) problem (Arteta et al., 2007)
- Design: WCA problem (Maciel et al., 2009)
- Protection: *cycles of preconfigured protection* (p-cycles) (Colman et al., 2008a,b, c, 2009)

With these considerations, this chapter is organized as follows: The next section will establish the notation and the concepts of multiobjective optimization, while in

Sections 10.3–10.5, the RWA, WAC, and p-cycle problems will be studied in a pure multiobjective context. Finally, conclusions and future research will be summarized in Section 10.6.

10.2 MULTIOBJECTIVE OPTIMIZATION

In order to understand formulations of an optical network in a multiobjective optimization context, some basic and common necessary notation is presented. In the following sections, extra notation will be introduced for each specific subproblem, as needed.

$\lvert . \rvert$	Cardinality of a set
x	Decision vector
X	Decision space
y	Objective vector
Y	Objective space
X_{fea}	Feasible decision space
Y_{fea}	Feasible objective space
X_{true}	Pareto-optimal set
Y_{true}	Pareto-optimal front
Y_{known}	Pareto front
A	Set of algorithms
B	Set of performance metrics
Π	Number of independence run of algorithms
ϑ	Stop criterion used in algorithms
SP	Set of average performance of algorithms

10.2.1 MULTIOBJECTIVE OPTIMIZATION FORMULATION

A general multiobjective optimization problem (MOP) (Coello et al., 2007) includes a set of n decision variables, k objective functions, and m restrictions. Objective functions and restrictions are functions of decision variables. This can be expressed as

$$\text{Optimize } y = f(x) = (f_1(x), f_2(x), \ldots, f_k(x)) \tag{10.1}$$

subject to

$$e(x) = (e_1(x), e_2(x), \ldots, e_m(x)) \geq 0, \tag{10.2}$$

where $x = (x_1, x_2, \ldots, x_n) \in X$ is the decision vector, and $y = (y_1, y_2, \ldots, y_k) \in Y$ is the objective vector. Depending on the kind of problem, "optimize" could mean minimize or maximize. The set of restrictions $e(x) \geq 0$ determines the set of feasible solutions $X_{fea} \subseteq X$ and its corresponding set of objective vectors $Y_{fea} \subseteq Y$.

A multiobjective problem consists of finding the decision vector x that optimizes $f(x)$. In general, there is no unique "best" solution but a set of compromise solutions, none of which can be considered better than the others when all k objectives are considered at the same time. This derives from the fact that there can be conflicting objectives, that is, trade-offs between different objectives.

If a solution $u \in X_{fea}$ is better than another solution $v \in X_{fea}$ in at least one of the k objectives and is not worst in any other objective, it is said that u dominates v, denoted as $u \succ v$. If neither v dominates u nor u dominates v, u, and v are noncomparable, denoted as $u \bullet v$. Alternatively, for the rest of this work, $u \triangleright v$ will denote that u dominates v, or $f(v) = f(u)$. At the same time, a decision vector $x \in X_{fea}$ is nondominated with respect to a set $Q \subseteq X_{fea}$ if and only if $x \triangleright q$, $\forall q \in Q$. If x is nondominated with respect to the whole set X_{fea}, it is called an optimal Pareto solution; therefore, the *Pareto-optimal set* X_{true} may be formally defined as

$$X_{true} = \{x \in X_{fea} : \neg \exists x' \succ x, \ \forall x' \in X_{fea}\}. \tag{10.3}$$

The corresponding set of objective vectors $Y_{true} = f(X_{true})$ constitutes the *optimal Pareto front*.

10.2.2 MULTIOBJECTIVE PERFORMANCE METRICS

To compare and evaluate the performance of multiobjective algorithms during different experiments, this work considers seven performance figures proposed in Coello et al. (2007) and described below.

Overall nondominated vector generation (OG) is the number of Y_{known} solutions generated by a MOP algorithm, as shown in Equation (10.4). A high value of OG is generally preferred, but it doesn't necessarily imply a good solution. OG indicates cardinality.

$$OG = |Y_{known}| \tag{10.4}$$

Overall nondominated vector-generation ratio (OR) is the ratio of the number of solutions calculated by a MOP algorithm to the real number of solutions in the Pareto-optimal front Y_{true}, as shown in Equation (10.11). A higher value of OR is generally preferred.

$$OR = \frac{|Y_{known}|}{|Y_{true}|} \tag{10.5}$$

Number of nondominated solutions (NS) indicates the number (quantity) of nondominated solutions of Y_{known} inserted in Y_{true}, as shown in Equation (10.6). Logically, a high value is preferred and it indicates the efficiency of a MOP algorithm.

$$NS = Y_{known} \cap Y_{true} \tag{10.6}$$

Error ratio (ER) is the proportion of Y_{known} which is not part of Y_{true}, as shown in Equation (10.13). If ER $= 0$, all solutions of Y_{known} are in Y_{true}; in the worst case, ER $= 1$.

$$ER = \frac{|Y_{known} - (Y_{known} \cap Y_{true})|}{|Y_{known}|} \tag{10.7}$$

Generation distance (GD) represents how far Y_{known} is from Y_{true}, as shown in Equation (10.14). If GD$=0$, it means that every solution in Y_{known} is also in Y_{true}; therefore, the smaller, the better.

$$GD = \frac{1}{|Y_{true}|} \sum_{S \in Y_{true}} \min[d(S,S') : S' \in Y_{known}], \tag{10.8}$$

where $d(S, S')$ denotes Euclidian distance.

Overall nondominated vector generation σ (OG$_\sigma$) gives an idea of the proportion of Y_{known} solutions generated by a MOP algorithm considering a given distance of σ; therefore, the larger the OG$_\sigma$, the better. In this work, the value used for σ has been established at 10% of the distance between the point with better evaluation in the first objective and the point with better evaluation in the last objective function.

$$OG_\sigma = \frac{1}{|Y_{known}| - 1} \sum_{S \in Y_{known}} |\{q \in Y : d(p,q) > \sigma\}| \tag{10.9}$$

Extension (E) of a Pareto front Y_{known} is defined as

$$E = \sqrt{\sum_{i=1}^{k} \max \{d(S_i, S_i') : S_i, S_i' \in Y_{known}\}} \tag{10.10}$$

Contributed solutions (CS) to the Pareto front Y_{true} indicates the part of Y_{true} that was found by a MOP algorithm that finds Y_{known}.

$$CS = \frac{|Y_{known} \cap Y_{true}|}{|Y_{true}|} \tag{10.11}$$

A high value of CS indicates that most solutions of Y_{true} were found in Y_{known}.

10.2.3 EXPERIMENTAL METHODOLOGY

In some metrics, such as contributed solutions (CS), generation distance (GD), error ratio (ER), number of nondominated solutions (NS), and overall nondominated vector generation (OG), it is necessary to know the Pareto-optimal set X_{true}, which is not easy in many practical problems. In fact, this may be almost impossible for some problems. In this context, an approximation X_{app} may be calculated as a set of Pareto-suboptimal solutions that may be supposed near to X_{true}. As already stated, let A be the set of evolutionary algorithms (EAs) to be evaluated, Π the number of independent executions of each algorithm, and some stop criterion ϑ for the considered set of algorithms. The scheme used in this work to estimate X_{app} is presented in lines 1-15 of Algorithm 1 that shows the procedure *Performance-of-Algorithms*.

ALGORITHM 1

Performance-of-Algorithms
Input: A, B, Π, ϑ
Output: *SP*
  1: $X \leftarrow \varnothing$
  2: **for** $i \in \{1, 2, \ldots, |A|\}$ **do** /*|A| is the number of algorithms of set A
  3: $X_i \leftarrow \varnothing$
  4:   **for** $j \in \{1, 2, \ldots, \Pi\}$ **do** /* Π is the number of independent executions
  5:   Execute algorithm $A_i(\vartheta)$ to obtain a set of non-dominated solution X_{ij}
  6: $X_i \leftarrow X_i \cup X_{ij}$
  7:   **end for**
  8: $X \leftarrow X \cup X_i$
  9: **end for b**
10: $X_{app} \leftarrow \varnothing$
11: **for** $k \in \{1, 2, \ldots, |X|\}$ **do** /*|X| is the number of solutions of set X
12:   **if** $x_k \nprec X$ **then** /*according Pareto dominance concept
13: $Z_{app} \leftarrow Z_{app} \cup \{x_k\}$
14:   **end if**
15: **end for**
16: **for** $i \in \{1, 2, \ldots, |A|\}$ **do**
17:   **for** $b \in \{1, 2, \ldots, |B|\}$ **do**
18: $p_{ib} \leftarrow 0$ /* p_{ib} is the b-th performance of i-th algorithm
19:   **for** $j \in \{1, 2, \ldots, \Pi\}$ **do**
20: $p_{ib} \leftarrow p_{ib} +$ Pareto-Comparison (Z_{ij}, Z_{app}, B_b) /*calculate b-th performance of i-th algorithm
21:   **end for**
22: $SP_{ib} \leftarrow {}^{p_{ib}}/_{\Pi}$ /* SP_{ib} is the average the b-th performance of i-th algorithm
23:   **end for**
24: **end for**
25: **return** *SP* /**SP* is the performance vector

The main idea of Performance-of-Algorithms is to approximate X_{true}, considering all solutions found using every algorithm of the set to later estimate performance metrics. In fact, the set of solutions X for all algorithms is calculated considering all executions, in lines 1-9. Then, in lines 10-15, dominated solutions are eliminated by Pareto dominance and X_{app} is obtained. After obtaining $X_{\text{app}} \approx X_{\text{true}}$, each set X_{ij} is compared with X_{app} in order to calculate performance metrics for each algorithm, considering presented metrics. In lines 16-24, the vector of average performance SP, whose entry (i, b) is the average performance B_b of algorithm A_i, is finally calculated.

10.2.4 ALGORITHMS TO SOLVE MOPs

10.2.4.1 Types of optimization problems and WDM networks

A fundamental aspect in the process of developing efficient algorithms for optimization problems is to define the nature of the problem treated, which depends on the types of variables, objective functions, and constraints (Rao, 2009). In this context, the optimization problems can be classified in several ways.

- Classification based on the existence of constraints: constrained or unconstrained problem
- Classification based on the nature of the design variables: static or dynamic problem
- Classification based on the physical structure of the problem: optimal control or nonoptimal control problem
- Classification based on the nature of the equations involved: nonlinear, geometric, quadratic, or linear problem (to name just a few)
- Classification based on the permissible values of the design variables: integer, real-valued, or mixed problem
- Classification based on the deterministic nature of the variables: stochastic or deterministic problem
- Classification based on the separability of the functions: separable or nonseparable objective function
- Classification based on the number of objective functions: multiobjective (several objectives) or mono-objective (only one objective) problem

Considering the above classifications, the optical network problems addressed in this work may be characterized as shown in Table 10.1.

It should be said that the literature reports two main types of optimization techniques (Rao, 2009; Hillier et al., 2010): (a) *classical* and (b) *modern* optimization techniques. Classical approaches focus on calculating the optimal solution. However, in many situations, problems are so complex that the calculation of the optimal

Table 10.1 Characteristics of considered optical network problems

Problem	Constraint	Variables	Equations	Objective Function	Reference
RWA	Constrained	Integer, static, deterministic	Nonlinear, separable	Multiobjective	Arteta et al. (2007)
WCA	Constrained	Integer, static, deterministic	Linear, separable	Multiobjective	Maciel et al. (2009)
p-Cycle	Constrained	Integer, static, deterministic	Nonlinear, separable	Multiobjective	Colman et al. (2008a,b,c)

solution is too hard, taking too many resources. In this context, modern approaches called metaheuristic optimization achieve the calculation of suboptimal solutions, but in a reasonable computation time. Particularly, the RWA, WCA, and p-cycle are very complex problems (Somani, 2005) that benefit from the use of appropriated metaheuristic resolutions, as will be shown shortly.

Metaheuristic techniques are iterative procedures that have been developed in recent years (Hillier et al., 2010). There is a wide variety of metaheuristic alternatives that can be used (Brownlee, 2011). Within this range of options, EAs and ant colony optimization (ACO) are highlighted as the most used to solve problems in the area of optical networks, and therefore will be introduced next.

10.2.4.2 Evolutionary algorithms

EAs are inspired by the evolutionary process of nature. At each evolutionary step, new individuals (solutions) are generated by a combination of current individuals, known as population. The combination process is manipulated by selection, crossover, and mutation operators as operators to build descendants. It is clear that individuals with high fitness value will have more chances to have descendants (similar new solutions), called children. Usually, at each evolutionary step, the best individual remains (elitism). At the end of an evolutionary process, good solutions are expected as a consequence of the evolutionary process, as in nature.

Unlike traditionally mono-objective EAs, a multiobjective evolutionary algorithm (MOEA) tries to compute the whole Pareto set. In this new context, at each evolutionary step, the best set of nondominated solutions obtained so far is saved (elitism). Basically, MOEAs differ from each other by methods of updating the set of nondominated solutions. The most outstanding MOEAs in the literature are (Coello et al., 2007): the nondominated sorting genetic algorithm (NSGA), the NSGA-II, the strength Pareto evolutionary algorithm (SPEA), and SPEA2. These algorithms have been successfully applied to several engineering problems, and are by far de facto benchmark algorithms.

The basic procedure for an MOEA is presented in Algorithm 2. The first population is chosen randomly. At each iteration, a new population is created by applying evolutionary operators to the current population and the best current set of nondominated solutions.

ALGORITHM 2

MOEA

Input: Evolutionary parameters
Output: X_{app}
1: $i \leftarrow 0$
2: $X_{app} \leftarrow \varnothing$
3: Generate X_i /* generate first population

Continued

```
 4: X_app ← Get_Set_Non_Dominated_Solution(X_i, X_app)
 5: while stop condition is false do
 6:        X_{i+1} ← Build_New_Population(X_i, X_app, Evolutionary parameters)
 7:        X_app ← Get_Set_Non_Dominated_Solution(X_{i+1}, X_app)
 8:        i ← i + 1
 9: end while
10: return X_app    /* an approximation Pareto set
```

10.2.4.3 Ant colony optimization

ACO is a metaheuristic approach inspired by the foraging behavior of ant colonies (Dorigo and Di Caro, 1999). In particular, ACO accumulates knowledge obtained by each ant of a colony in a pheromone matrix τ. Typically, at each iteration, an ant builds a complete solution using the pheromone information as well as local information η called visibility. This calculated solution is used to reinforce pheromone levels according to the quality of the solution itself. Similar to EAs, at each iteration, ACO keeps the best solution (elitism). Pheromone evaporation is applied after each completed iteration in order to avoid stagnation in local optima.

Like MOEAs, multiobjective ACOs (MOACOs) preserve the best set of nondominated solutions found so far. The number of colonies, reinforcement, and evaporation pheromone techniques, as well as the specific methodology for updating the set of nondominated solutions, identify the type of MOACO. The literature has reported several MOACO approaches (Arteta et al., 2007), as highlighted in Table 10.2, to be used in later sections of this work.

Algorithm 3 presents a basic procedure for MOACOs. Initially, a pheromone matrix is initialized to a determined constant level. Unlike in MOEAs, each solution of a MOACO is built progressively by one ant using information from the pheromone's matrix. The information from the nondominated solutions is reflected in the pheromone matrix at the end of each iteration.

Table 10.2 MOACO of state of the art

#	MOACO	Abbreviation	Reference
1	Multiple-objective ant Q algorithm	MOAQ	Mariano and Morales (1999)
2	Bicriterion ant	BIANT	Iredi et al. (2001)
3	Pareto ant colony optimization	PACO	Doerner et al. (2002)
4	Multiobjective ant colony system	MOACS	Schaerer and Barán (2003)
5	Multiobjective max-min ant system	M3AS	Pinto-Roa and Barán (2005)
6	COMPETants	COMP	Doerner et al. (2003)
7	Multiobjective omicron ACO	MOA	Gardel et al. (2006)
8	Multiobjective ant system	MAS	Paciello et al. (2006)

ALGORITHM 3

MOACO
Input: Swarm parameters
Output: X_{app}
  1: $i \leftarrow 0$
  2: $\tau \leftarrow \tau_o$
  3: $X_{app} \leftarrow \emptyset$
  4: Generate X_i /* generate first population
  5: $X_{app} \leftarrow$ Get_Set_Non_Dominated_Solution(X_i, X_{app})
  6: **while** stop condition is false **do**
  7:       $X_{i+1} \leftarrow$ Build_New_Population(τ, Swarm parameters)
  8:       $X_{app} \leftarrow$ Get_Set_Non_Dominated_Solution(X_{i+1}, X_{app})
  9:       $\tau \leftarrow$ Update_Pheromones(τ, X_{app}, swarm parameters)
  10:      $i \leftarrow i + 1$
  11: **end while**
  12: **return** X_{app} /* an approximation Pareto set

10.3 RWA PROBLEM

Optical networks may increase their traffic capability exploiting the bandwidth of optical fiber using WDM technology. In addition to the classic routing problem, a wavelength assignment is also needed for a complete lightpath; this new problem is called routing and wavelength assignment (RWA). Basically, for each pair of nodes' source-destination, RWA strategies try to calculate the right lightpaths. The RWA problem is known to be a nondeterministic polynomial-time complete (NP-complete) problem (Mukherjee, 2006), and therefore, heuristic solutions are fundamental. Note that RWA is the central problem of a WDM network, given that the design, management, and protection problems are all intrinsically associated to it.

10.3.1 TRADITIONAL RWA

Traditionally, RWA has been broadly studied in the literature as a mono-objective optimization problem (Mukherjee, 2006; Rouskas and Perros, 2002) with the same approaches as mixed-integer programming formulation (Ramaswami et al., 1996; Dutta and Rouskas, 2000; Leonardi et al., 2000). However, exact methods are not scalable; therefore, heuristic (Ramaswami et al., 1996; Dutta and Rouskas, 2000; Leonardi et al., 2000, 2005) and metaheuristic techniques (Mukherjee et al., 1996; Le et al., 2005; Saha and Sengupta, 2005) have been also proposed.

A mono-objective RWA seeks to optimize a cost function subject to constraints. The cost can be a function of used resources (by lightpath), while constraints can be imposed by the physical layer or the QoS (Mukherjee et al., 1996).

10.3.2 MULTIOBJECTIVE RWA FORMULATION

In this work, a WDM network is modeled as a direct graph $G = (V, E, \Lambda)$, where V is the set of nodes, E is the set of links between nodes, and Λ is the set of available wavelengths for each optical link in E. Let

e	Optical link $e = (i, j) \in E$ from node $i \in V$ to node $j \in V$
λ	Wavelength in the set Λ
u	Unicast request $u = (s, d)$ with a source node $s \in V$ and a destination node $d \in V$
U	Set of unicast requests, $U = \{u_1, \ldots, u_{|U|}\}$
λ_e^u	Wavelength λ assigned to the unicast request u at link e
ll	Light-link $ll = (e, \lambda)$ conformed by wavelength λ at link e
lp_u	Lightpath for request u; $lp_u = \{ll_1, ll_2, \ldots, ll_{|lp_u|}\}$, where if $ll_k = (e = (i, j), \lambda)$ then $ll_{k+1} = (e' = (j, l), \lambda')$
L	Solution of the RWA problem considering the set of U requests; $L = \{lp_1, lp_2, \ldots, lp_{|U|}\}$

Notice that L is the decision variable x presented in the previous section. Using the above definitions, the RWA problem may be stated as a MOP searching for the best solution L that simultaneously minimizes the following objective functions.

Total hop count:

$$\text{TH} = \sum_{lp_u \in L} |lp_u| \qquad (10.12)$$

Total number of wavelength conversions:

$$\text{TC} = \sum_{lp_u \in L} C_{ip} \qquad (10.13)$$

subject to wavelength-conflict constraint

$$\lambda_e^u \neq \lambda_e^{u'}, \quad \forall u, u' \in U, \qquad (10.14)$$

where

- $|lp_u|$ is the hop count of lp_u
- $C_{ip_u} = \sum_{k=2}^{|lp_u|} X_{ll_k}$ is the number of wavelength conversions of lp_u
- X_{ll_k} is a binary variable so that $X_{ll_k} = 1$ if $\lambda_{e_k}^u \neq \lambda_{e_{k-1}}^u$, $X_{ll_k} = 0$ in otherwise

10.3.3 ACO FOR RWA

ACO makes use of simple ants and a pheromone matrix $\tau = \{\tau_{ij}\}$ for iteratively constructing RWA candidate solutions. The initial values are $\tau_{ij} = \tau_0 \forall (i, j) \in E$, where $\tau_0 > 0$. The relative influence between the visibility and the pheromone levels is controlled using parameters α and β. While an ant is visiting node i, N_i represents the set of neighbor nodes that are not yet visited while staying at node i. The probability of choosing a node j, while being at node i, is defined by the following equation:

$$p_{ij} = \begin{cases} Z_{ij} & \text{if } j \in N_i, \\ 0 & \text{otherwise,} \end{cases} \qquad (10.15)$$

where

$$Z_{ij} = \frac{\tau_{ij}^{\alpha} \cdot \eta_{ij}^{\beta}}{\sum\limits_{g \in N_i} \tau_{ig}^{\alpha} \cdot \eta_{iig}^{\beta}}. \qquad (10.16)$$

In particular, for the considered RWA problem, this work uses two visibilities. The first visibility reflects the desirability of choosing a link with the same wavelength as the previous link, if available. Therefore, the conversion visibility ψ_{ij} used is defined as

$$\psi_{ij} = \begin{cases} 1 & \text{if } (i,j) \text{ does not change wavelength,} \\ \Delta & \text{if a conversion at node } i \text{ is needed,} \end{cases} \qquad (10.17)$$

where $\Delta \ll 1$ (in this work, $\Delta = 0.01$ is used).

The second visibility is the hop-count visibility δ_{ij}; that is, it tries to minimize the hop count from a feasible neighbor i to a destination node j. This visibility has been calculated using the well-known Dijkstra shortest-path algorithm (Choi et al., 2000). This way, Equation (10.16) is extended to

$$Z_{ij} = \frac{\tau_{ij}^{\alpha} \cdot \psi_{ij}^{r \cdot \beta} \cdot \delta_{ij}^{(1-r)\beta}}{\sum\limits_{g \in N_i} \tau_{ig}^{\alpha} \cdot \psi_{ig}^{r\beta} \cdot \delta_{ig}^{(1-r)\beta}}, \qquad (10.18)$$

where r and $(1-r)$ are the relative influence of the two proposed visibilities. It is also possible to implement other approaches that use more than one pheromone matrix (Iredi et al., 2001; Doerner et al., 2002; Gardel et al., 2006).

Note that the hop-count visibility matrix δ_{ij} is an input data of the problem. In order to avoid premature convergence toward a local optimum, pheromone evaporation is applied for all links (i, j) in the pheromone matrix according to $\tau_{ij} = (1 - \rho)\tau_{ij}$, where parameter $\rho \in (0, 1]$ determines the evaporation rate (Dorigo and Di Caro, 1999). Considering an elitist strategy, the best solutions found so far L_{best} update τ according to $\tau_{ij} = \tau_{ij} + \Delta\tau$, where typically $\Delta\tau = 1/f(L_{best})$ if $(i, j) \in L_{best}$ and $\Delta\tau_{ij} = 0$ otherwise. ACO is especially appealing when constructing solutions are needed; therefore, it is interesting to study its application to the RWA problem.

10.3.4 MOACO FOR RWA

Alternative implementations of MOACO algorithms derive from traditional ACO. These implementations are based mainly on the number of colonies, the way to manage the pheromone information, and the heuristic functions (Lopez-Ibañez et al., 2004). The combinations of these characteristics are used to build different MOACO algorithms.

The first alternative is to have multiple colonies, where a group of all the ants is divided into disjoint sets of smaller sizes, and each set is considered a colony. Each colony specializes in a region of the Pareto front, and it uses its own pheromone information. In a cooperative approach, solutions can be exchanged between colonies in such a way that an ant from one colony can update the pheromone matrix of other colonies, trying to achieve a synergetic effect.

For pheromone information and visibilities, two alternatives exist. The first uses one pheromone matrix with the information related to all the objectives. The second alternative uses multiple pheromone matrices, where each matrix contains the pheromone information with respect to only one objective of the problem, or a specialized colony. The transition rule can be based on a single objective, or it can be based on the aggregation of the different pheromone matrices, assigning a weight to each objective. The visibility has similar alternatives; that is, for the same problem, several visibilities can be defined and combined.

It is worth mentioning that in a mono-objective context, each ant that finds a new solution that is better than the best-found solution so far, updates the pheromone matrix τ. Analogously, in a multiobjective context, the ants that build a good solution must also update Y_{known}, the set of best nondominated solutions found by a running algorithm if and only if the new solution is nondominated with respect to the whole set Y_{known} found so far. At the same time, dominated solutions of the Pareto set under construction must be eliminated. Algorithm 4 presents a general MOACO approach considered for this work. The different MOACO algorithms implemented in Arteta et al. (2007) are summarized in Table 10.2.

ALGORITHM 4

RWA_MOACO
Input: $\alpha, \beta, U, \rho, \tau_0, m, G = (V, E, \Lambda)$;
Output: Y_{known};
  1: $t \leftarrow 0$
  2: $\tau \leftarrow$ Inicialize_Pheromones(τ_0);
  3: **while** stop criterion is not verified **do**
  4:        $t \leftarrow t + 1$ /* generations
  5:        **for** ant =1 **to** m /* m is the number of ants
  6:           $L \leftarrow$ Ant_RWA($\alpha, \beta, U, G, \tau$);
  7:           Evaluate_Solution(L);
  8:           $Y_{known} \leftarrow$ Update_Y_{known}(L);
  9:        **end for**
10:        $\tau \leftarrow$ Update_Pheromones(Y_{known});
11: **end do**
12: **return** Y_{known};

All presented MOACO algorithms are independent, general-purpose methods of optimization. To particularize them to the RWA problem, each artificial ant should build the set of lightpaths that constitutes a solution L. In this context, an artificial ant for the RWA is generically called Ant-RWA. Each Ant-RWA builds a solution traveling the wavelength graph (WG) (Zang et al., 2002; Li and Wang, 2004; Xu et al., 2005). The WG is obtained by the mapping of the original graph $G = (V, E, \Lambda)$, transforming the RWA problem into a simple routing problem by considering $|\Lambda|$ parallel links between nodes i and j. Algorithm 5 depicts the job of an Ant-RWA.

Basically, an Ant-RWA builds, in WG, a path for each unicast request u, to satisfy the demand U, if it is possible. Once the problem is regularly solved in WG, the lightpaths belonging to the solution L' are found, and finally, the L' solution is mapped over the original graph G to have the desired solution in the original network topology.

ALGORITHM 5

Ant-RWA

Input: $\alpha, \beta, \rho, U, G = (V, E, \Lambda)$;
Output: L;
  1: $L' \leftarrow \varnothing$;
  2: $WG \leftarrow G$ /* mapping of G according to (Arteta, 2007)
  3: **for** each $u = (s, d) \in U$ **do**
  4:      $R \leftarrow \varnothing$;
  5:      $R \leftarrow R \cup \{ s \}$;
  6:      **do**
  7:          Select node i of R and build set N_i ;
  8:          **if** $N_i == \varnothing$ **then**
  9:              $R \leftarrow R - \{ i \}$;   /* erase node without feasible neighbor
 10:          **else**
 11:              Assign probability p_{ij} to each node of N_i ;
 12:              Select node j of N_i ;
 13:              $lp_u \leftarrow lp_u \cup \{ (i, j) \}$; $R \leftarrow R \cup \{ j \}$;
 14:              OnLineEvaporation (ρ);   /*for MOACS, PACO, MOA or MAS
 15:          **end if**
 16:      **while** $R \neq \varnothing$ or $j \neq d$ **do**
 17:      **if** $R == \varnothing$ **then**
 18:          Request u not satisfy, **return** Error;
 19:      **else**
 20:          Prune Tree lp_u;   /* eliminate not used links
 21:          $L' \leftarrow L' \cup \{ lp_u \}$;
 22:      **end if**
 23: **end for**
 24: $L \leftarrow$ Mapping(L', WG, G);
 25: **return** L;

The procedure Ant-RWA can be used by any of the proposed MOACOs to build solutions. The differences among them are mainly given by the way they select a connection using different pheromone matrices and visibilities. For the MOACS, PACO, and MAS metaheuristics, a pseudorandom rule (Schaerer and Barán, 2003) is used, while the other algorithms choose node j with probability p_{ij} given by (10.15). Another difference is the online evaporation that is carried out by MOACS, PACO, MOA, and MAS. This evaporation is implemented with the purpose of improving the search with the newest solutions, while slowly forgetting old solutions.

10.3.5 CLASSICAL HEURISTICS

Classical heuristic (CH) approaches (Choi et al., 2000; Zang et al., 2000; Arteta et al., 2007) make the RWA problem tractable, dividing it into two subproblems: routing on one hand and wavelength assignment on the other. Each subproblem can be solved separately, using a different heuristic.

The traditional routing problem is solved by well-known techniques, based on shortest-path algorithms such as

- *Dijkstra shortest path* (SP): This algorithm finds the shortest route from a given source to a destination in a graph. The route is a path whose cost is the least possible one.
- *K-shortest path* (K-SP): K-shortest-path algorithms find more than one route for each source and destination pair. K-alternative paths provide flexibility in route selection.

For the wavelength-assignment problem, a number of heuristics have been proposed in the literature; the ones used in this work are

- *Random* (RR): To establish a connection, this wavelength-assignment algorithm randomly chooses one of the free available wavelengths on all links.
- *First-fit* (FF): This algorithm assumes that the wavelengths are arbitrarily ordered. The first-fit algorithm checks the status of the wavelengths sequentially and chooses the first available wavelength to establish a connection.
- *Most-used* (MU): The free wavelengths that are used on the greatest number of links in the network are chosen first to establish a connection.
- *Least-used* (LU): The free wavelength that is used on the least number of links in the network is chosen to establish a connection.

These algorithms were implemented to compare them with the MOACO algorithms (see Algorithm 4). Basically, the CHs are applied for each unicast request u, until the demand U (set of requests) is satisfied, when possible.

First, Algorithm 6 calculates the route for a request, applying the SP or the K-SP algorithms; then, for the path obtained, it selects the wavelengths for each optical link, applying one of the wavelength-assignment heuristics explained above, in order to build a complete lightpath.

Table 10.3 Set of algorithms for the classical heuristics

#	Approach	Routing Algorithm		Wavelength-Assignment Algorithm			
		SP	K-SP	RR	FF	MU	LU
1	KSPFF		✓		✓		
2	KSPLU		✓				✓
3	KSPMU		✓			✓	
4	KSPRR		✓	✓			
5	SPFF	✓			✓		
6	SPLU	✓					✓
7	SPMU	✓				✓	
8	SPRR	✓		✓			

This way, eight possible combinations of CHs are implemented in this work, as shown in Table 10.3, to experimentally compare them to the proposed MOACO algorithms.

ALGORITHM 6

Classical Heuristics
Input: $G = (V,E,\Lambda),U, K$;   /* K: Number of shortest path
Output: Y_{known};
  1: Initialize Parameters;
  2: **while** stop criterion is not verified **do**
  3:     Randomize order of set U;
  4:     **for** each $u \in U$ **do**
  5:         Path ← CalculateShortestPath(u, K);
  6:         lp_u ← AssignWavelength(Path);   /*FF, RR, LU or MU
  7:         Evaluate_Solution(lp_u);
  8:         $L \leftarrow L \cup \{ lp_u \}$;
  9:     **end for**
10: **while**
11: Y_{known} ← Update_$Y_{known}(L)$;
12: **return** Y_{known};

10.3.6 SIMULATIONS

Simulations were carried out using the NTT network topology (Arteta et al., 2007). The algorithms have been implemented on a 1910 MHz AMD Athlon computer with 256 MB of RAM with a GCC compiler. For these experiments, the results of the proposed MOACOs have been compared to the eight CH combinations shown in Table 10.3.

For each demand group U, the vector of performance metric P was calculated using the procedure given in Algorithm 1, where the set of algorithms A is made by MOACO and CH algorithms, given in Tables 10.1 and 10.2, respectively; the number of independence executions is $\Pi = 10$, the stop criterion is $\vartheta = 100$ iterations, and the set of performance metric $B = \{GD, OG_\sigma, E, ER\}$. Furthermore, the number of blocked requests (NB) was calculated in average for the Π runs. The MOACO parameters are: 40 ants, $\beta/\alpha = 4$ relative importance, evaporation percent $\rho = 0.10$, a pseudorandom probability of $q_0 = 0.95$ (for MOACS), and for the K-shortest path ($K = 3$).

10.3.7 EXPERIMENTAL RESULTS

Table 10.4 shows four sets of requests that were used for the experiments, while Table 10.5 presents experimental results normalized for all groups. In this table, $R = GD + IG_\sigma + E + ER$, while NB indicates the number of blocked requests. The dark square indicates the best performance for each metric.

Table 10.6 presents general averages of the comparison metrics already defined, considering all performed experiments. MOACOs presented a better performance on average than the CHs in metrics GD, OG_σ, and E. However, considering the error metric ER, the CHs presented a slightly better performance on average, where the 3SPLU got the best average performance.

It is interesting to note that all MOACOs outperform all CHs on average, indicating an encouraging potential for future studies. Finally, it should be mentioned that considering all implemented algorithms, on average, MOAQ is the best MOACO according to the presented experimental results.

Table 10.4 Unicast groups used for the tests

Test Group	$|U|$	U
Group 1	10	(18,49) (15,1) (11,42) (8,33) (32,10) (37,13) (19,17) (28,31) (40,24) (11,41)
Group 2	20	(47,30) (47,48) (23,17) (21,4) (28,42) (54,11) (9,49) (51,12) (11,47) (17,8) (20,50) (3,25) (11,48) (46,30) (18,53) (17,36) (48,9) (18,1) (28,49) (3,29)
Group 3	30	(47,30) (47,48) (23,17) (21,4) (28,42) (54,11) (9,49) (51,12) (11,47) (17,8) (20,50) (3,25) (11,48) (46,30) (18,53) (17,36) (48,9) (18,1) (28,49) (3,29)
Group 4	40	(30,28) (45,31) (8,44) (15,31) (49,1) (39,30) (50,38) (39,21) (43,37) (19,23) (13,43) (14,25) (53,0) (53,31) (26,33) (30,44) (19,30) (50,44) (25,0) (7,26) (3,12) (45,19) (35,36) (49,34) (8,47) (20,37) (10,9) (53,40) (23,8) (3,21) (1,52) (16,19) (12,17) (6,46) (30,17) (30,29) (29,2) (23,1) (48,29) (34,43)

Table 10.5 Experimental Results

A	Group 1						Group 2						Group 3						Group 4					
	NB	GD	OG_σ	E	ER	R	NB	GD	OG_σ	E	ER	R	NB	GD	OG_σ	E	ER	R	NB	GD	OG_σ	E	ER	R
BJANT	0	0.99	0.10	0.05	0.90	0.51	0	0.82	0.95	0.71	0.02	0.62	0	0.76	0.98	0.62	0.00	0.53	0	0.64	1.00	1.00	0.00	0.66
COMP	0	0.85	0.80	0.54	0.30	0.62	0	0.74	0.98	0.76	0.00	0.62	0	0.70	0.87	1.00	0.00	0.64	0	0.46	0.82	0.91	0.00	0.55
MOAQ	0	0.86	1.00	1.00	0.30	0.79	0	0.69	0.90	1.00	0.00	0.65	0	0.57	0.73	0.89	0.00	0.55	0	0.37	0.70	0.92	0.00	0.50
MOACS	0	0.98	0.20	0.11	0.80	0.52	0	0.78	0.81	0.80	0.00	0.60	0	0.76	0.91	0.85	0.00	0.63	0	0.61	0.93	0.89	0.00	0.61
M3AS	0	1.00	0.00	0.00	1.00	0.50	0	0.82	0.86	0.77	0.00	0.61	0	0.75	0.80	0.60	0.00	0.54	0	0.52	0.88	0.88	0.00	0.57
MAS	0	0.98	0.20	0.10	0.70	0.49	0	0.79	1.00	0.86	0.00	0.66	0	0.60	0.81	0.84	0.02	0.57	0	0.60	0.93	0.97	0.00	0.62
PACO	0	0.99	0.10	0.05	0.90	0.51	0	0.77	0.80	0.77	0.00	0.59	0	0.77	0.95	0.82	0.00	0.63	0	0.55	0.82	0.84	0.00	0.55
MOA	0	1.00	0.00	0.00	1.00	0.50	0	0.82	0.94	0.88	0.00	0.66	0	0.79	1.00	0.84	0.00	0.66	0	0.60	0.87	0.88	0.00	0.59
3SPFF	0	0.00	0.00	0.00	0.00	0.00	0	0.00	0.03	0.06	0.00	0.02	0	0.00	0.10	0.15	0.00	0.06	0	0.30	0.41	0.30	0.00	0.25
3SPLU	0	1.00	0.00	0.00	1.00	0.50	0	0.35	0.08	0.07	0.65	0.29	0	0.43	0.07	0.17	0.20	0.22	0	0.86	0.38	0.26	0.13	0.41
3SPMU	0	0.03	0.00	0.00	0.00	0.01	0	0.08	0.11	0.09	0.00	0.07	0	0.05	0.13	0.15	0.00	0.08	0	0.00	0.32	0.24	0.00	0.14
3SPRR	0	1.00	0.00	0.00	1.00	0.50	0	0.36	0.14	0.10	0.10	0.17	0	0.39	0.12	0.20	0.00	0.18	0	0.81	0.31	0.24	0.08	0.36
SPFF	0	0.00	0.00	0.00	0.00	0.00	2	-	-	-	-	-	2	-	-	-	-	-	4	-	-	-	-	-
SPLU	0	1.00	0.00	0.00	1.00	0.50	2	-	-	-	-	-	2	-	-	-	-	-	4	-	-	-	-	-
SPMU	0	0.03	0.00	0.00	0.00	0.01	2	-	-	-	-	-	2	-	-	-	-	-	4	-	-	-	-	-
SPRR	0	1.00	0.00	0.00	1.00	0.50	2	-	-	-	-	-	2	-	-	-	-	-	4	-	-	-	-	-

Table 10.6 General averages of comparison metrics

A	Performance Metrics					
	NB	GD	OG$_\sigma$	E	ER	T
BIANT	0.00	0.80	0.76	0.60	0.23	0.60
COMP	0.00	0.69	0.87	0.80	0.08	0.61
MOAQ	0.00	0.62	0.83	0.95	0.08	0.62
MOACS	0.00	0.78	0.71	0.66	0.20	0.59
M3AS	0.00	0.77	0.64	0.56	0.25	0.55
MAS	0.00	0.74	0.74	0.69	0.18	0.59
PACO	0.00	0.77	0.67	0.62	0.23	0.57
MOA	0.00	0.80	0.70	0.65	0.25	0.60
3SPFF	0.00	0.07	0.13	0.13	0.00	0.08
3SPLU	0.00	0.66	0.13	0.12	0.49	0.35
3SPMU	0.00	0.04	0.14	0.12	0.00	0.07
3SPRR	0.00	0.64	0.14	0.13	0.30	0.30
SPFF	2.00	0.00	0.00	0.00	0.00	0.00
SPLU	2.00	0.25	0.00	0.00	0.25	0.13
SPMU	2.00	0.01	0.00	0.00	0.00	0.00
SPRR	2.00	0.25	0.00	0.00	0.25	0.13

10.4 WCA PROBLEM

The design and expansion of WDM networks are critical problems in optical tele-communications, in order to allow ever-growing traffic. In that sense, one main objective is to minimize request blocking with minimum investment and management costs (Somani, 2005). A blocking problem appears when the system is unable to assign a lightpath to a request; therefore, this request is lost. This problem is related to several issues, such as (1) efficiency of the routing algorithms (Chu et al., 2002), (2) number of available optical fibers at links (Somani, 2005), and (3) the wavelength-conversion capability of optical routers (Kovacevic and Acampora, 1996). When wavelength conversion is not possible, just one wavelength can be used in a given lightpath. This is known as the wavelength-continuity constraint problem (Somani, 2005). To overcome the blocking generated by this constraint, it is necessary to add wavelength converters in the optical routers. A wavelength converter is a device that changes a signal with wavelength (λ) into another wavelength (λ'). The WCA problem calculates at which optical nodes wavelength converters should be located, and in what number. This WCA problem is known to be an NP-hard problem when dealing with irregular network topologies (Xiao and Leung, 1999).

The WCA problem was first treated as a single-objective optimization problem. More specifically, there are two optimization approaches reported in the literature:

- *Minimum blocking*: Given a dynamic traffic pattern, it decides where to locate a fixed number of wavelength converters to minimize the blocking probability (Roy and Naskar, 2008).
- *Minimum cost*: Given a dynamic traffic pattern and an upper bound for the blocking probability, it tries to locate a minimum number of wavelength converters to satisfy a given blocking probability upper bound (Fang and Low, 2006).

However, several authors (Roy and Naskar, 2008; Jeong and Seo, 2005; Foo et al., 2005) have already detected a conflict between the minimization of the blocking probability and the minimization of the number of wavelength converters; that is, in order to get solutions with minimal blocking probability, it is necessary to locate a great number of wavelength converters, and vice versa. This conflict implies that the WCA problem should be treated in a multiobjective context, considering both objectives at the same time. Therefore, we analyze the WCA problem as a MOP that may be solved using a MOEA (Coello et al., 2007). This algorithm simultaneously optimizes the blocking probability and the number of wavelength converters. This study considers a dynamic traffic scenario.

10.4.1 RELATED WORK

Several works notice the need to design WCA algorithms that exploit the characteristics of RWA algorithms. Typically, these approaches propose analytic blocking measures under strict constraints. One open question in this research line is what should be the first design: RWA algorithms in function of WCA algorithms, or vice versa (Chu et al., 2002).

Recently, some works have proposed optical WCA (OWCA) approaches that adapt to any RWA algorithm (Hei et al., 2004; Roy and Naskar, 2008; Xiao and Leung, 1999). They use traffic direct or indirect simulations as a fundamental tool. On one hand, a direct-simulation approach evaluates each solution, applying an RWA algorithm under a dynamic traffic pattern (Hei et al., 2004). On the other hand, an indirect-simulation approach solves problems in two stages: simulation stage and optimization stage (Roy and Naskar, 2008; Xiao and Leung, 1999). Given a traffic pattern and all resources available in a WDM optical network, the simulation stage is carried out to obtain statistical information about the amount of time the wavelength converters are used at each node. With the previous information, an indirect-simulation approach begins the optimization stage.

Algorithms based on an indirect-simulation approach are simpler and faster than those based on a direct-simulation approach, but they depend on statistical information (Roy and Naskar, 2008). However, a direct simulation gives full information about the performance of the proposed solution (Hei et al., 2004). Other works reported heuristic approaches, which allocate a great number of wavelength converters at random (Lee et al., 1993; Subramaniam et al., 1996). Their objective was to analyze the advantage of using wavelength conversion in WDM optical networks (Roy and Naskar, 2008).

In a recent work (Roy and Naskar, 2008), an indirect-simulation approach based on a differential evolutionary algorithm (DEA) (Storn and Price, 1997) is proposed as an optimizer, and the use of a utilization matrix (UM) as the storage of statistical information. They considered a fixed number of wavelength converters to be allocated. Their work is an improvement over the one of Xiao and Leung (Xiao and Leung, 1999), which was the first to propose the indirect-simulation approach using a heuristic as the main optimizer.

Essentially, each row of a UM corresponds to an index of one WDM optical network node, and each column of UM indicates the number of used wavelength converters. UM_{ij} represents the percentage of time taken up by wavelength converters j at node i. UM is obtained from the simulation stage and represents the overall usage time after the pattern traffic has been tested (Xiao and Leung, 1999).

In the optimization stage, the main objective of DEA is to find how many wavelength converters to allocate at each node in such a way that the sum of total utilization (SU) is maximized. A solution with high SU should give low blocking probability (Roy and Naskar, 2008; Xiao and Leung, 1999).

10.4.2 CLASSICAL PROBLEM FORMULATION

Before presenting the WCA formulation, a few extra symbols are presented:

F_v	Number of input/output optical fibers at node $v \in V$
m_v	Number of converters at node $v \in V$
M	Solution of a WCA problem; $M = \{m_1, m_2, \ldots, m_{|V|}\}$
u_b	Variable of blocking; if u was blocked then $u_b = 1$ otherwise $u_b = 0$
k_v	Max number of wavelength converters endured by node $v \in V$
k_{max}	Max value of set $\{k_1, k_2, \ldots, k_{|V|}\}$
ξ_j^i	Binary variable; if $i \leq j$ then $\xi_j^i = 1$ otherwise $\xi_j^i = 0$

Roy (2008) proposed to solve the WCA problem as a single-objective optimization problem under the indirect-simulation approach. Statistical information is given in UM that was obtained in the previous simulation stage. Given a topology $G = (V, E, \Lambda)$, the UM, and a fixed number of wavelength converters C^*, a DEA tries to calculate a solution M that maximizes the SU in the optimization stage according to

$$SU = \sum_{v \in V} \sum_{j=1}^{k_{max}} \left(UM_{v,j} \cdot \xi_{m_v}^j \right) \tag{10.19}$$

subject to

$$m_v \leq k_v = \Lambda \cdot F_v; \quad \forall v \in V \tag{10.20}$$

$$C^* = \sum_{v \in V} m_v, \tag{10.21}$$

where $UM_{v,j}$ denotes the percentages of time that j wavelength converters are being utilized simultaneously at node $v \in V$. Constraint (10.20) assures that wavelength converters installed in network nodes do not go beyond the maximum number of optical switch input, while constraint (10.21) indicates that the total number of wavelength converters is exactly C*.

10.4.3 MULTIOBJECTIVE FORMULATION

This work analyzes the solution of the WCA problem as a MOP under a direct-simulation approach. Given a topology G, an RWA algorithm, and a set of dynamic requests U, the goal is to calculate a solution M that simultaneously minimizes

Number of wavelength converters

$$C = \sum_{v \in V} m_v \tag{10.22}$$

Blocking number

$$BN = \sum_{u \in U} u_b \tag{10.23}$$

subject to (10.20), where $u_b \in \{0, 1\}$ is obtained by simulation (Algorithm 8) of the RWA, as explained in the next subsection.

10.4.4 TRAFFIC MODELS AND SIMULATION ALGORITHM

A dynamic unicast request U is composed of Γ time slots and $|U|$ requests. Γ is selected according to a maximum time of the simulation (MTS) and the request's arrival period of time (PT) estimated. That is, Γ can be estimated by $\Gamma = \text{MTS/PT}$. The set U is made according to Algorithm 7, where w_{min} and w_{max} indicate minimum and maximum bandwidth, respectively.

ALGORITHM 7

Dynamic Traffic Generator
Input: Γ, $|U|$, V, w_{min}, w_{max}
Output: U
1: $U \leftarrow \varnothing$
2: **for** $i = 1$ to $i = |U|$ **do**
3:     $s \leftarrow$ random-set(V)
4:     $d \leftarrow$ random-set(V-$\{s\}$)
5:     $t_o \leftarrow$ random-numerical($1, \Gamma$)
6:     $t_f \leftarrow$ random-numerical(t_o, Γ)
7:     $w \leftarrow$ random-numerical(w_{min}, w_{max})
8:     $U \leftarrow U \cup \{(s, d, t_o, t_f, w)\}$
9: **end for**
10: **return** U

The direct-simulation algorithm is carried out as detailed in Algorithm 8. Given a dynamic unicast request U, a topology G, an RWA algorithm, and a solution M, Algorithm 8 calculates the blocking number (BN) of solutions M; Equation (10.23).

ALGORITHM 8

Direct-Simulation Algorithm

Input: U, G, RWA, M, Γ

Output: BN

  1: BLN$\leftarrow$0
  2: **for** $t = 1$ **to** $t = \Gamma$ **do**
  3:    **for each** $u \in$ U **do**
  4:       **if** $u(t_o) == t$ **then**
  5:          Compute lp_u
  6:          **if** $lp_u ==$ Null **then**
  7:             BLN $\leftarrow$ BLN + 1
  8:          **else** reserve lp_u resources in G
  9:       **else if** $u(t_f) == t$ **then**
10:           free lp_u resources
11:    **end for**
12: **end for**
13: **return** BN

10.4.5 EA FOR WCA

This work proposes to use the SPEA (Zitzler and Thiele, 1999; Zitzler et al., 2000) to solve the WCA problem. SPEA is a second-generation MOEA, successfully used to solve several engineering problems (Coello et al., 2007). SPEA is based on three populations (set of possible solutions). The first one is the current population (PA). This is replaced by individuals (solutions) from the evolutionary population (PX). An external population (PE) keeps the best individuals calculated during the evolutionary cycles. The chromosome (digital solution) representation and the evolutionary operators utilized are explained below.

Chromosome. Each individual M is represented by a chromosome, which is composed of $|V|$ genes. Each gene $v \in \{1, 2, \ldots, |V|\}$ represents the number of wavelength converters ($m_v \in \{0, 1, \ldots, k_v\}$) to be installed at node $v \in V$. It should be noticed that every single chromosome must respect restriction (10.20).

Evaluation. The number of wavelength converters C is the sum of the values m_v given by genes; Equation (10.22). Given an RWA algorithm, a topology G, and a dynamic request U, the blocking number BN of each individual M is assessed applying a direct-simulation approach. The implemented simulation is given in Algorithm 8.

Adjustment. After assessing the blocking number BN, the unused wavelength converters are removed. For example, let us assume that $m_v = 4$ and the simulation registered only 3 wavelength conversions in node $v \in V$. Then, the value of m_v is modified to $m_v = 3$. This process tries to accelerate convergence.

Fitness. Each individual receives a quality value (or *fitness*), which is calculated according to the strength Pareto approach proposed in SPEA (Zitzler and Thiele, 1999). The fitness of one chromosome is calculated on the basis of its number of wavelength converters C and blocking number BN. Basically, each individual $M \in$ PF receives a value of strength proportional to the number of individuals that it dominates. The fitness of each individual in a Pareto front (PF) is equal to its strength. Each individual $M' \notin$ PF gets a fitness inversely proportional to the sum of the individuals' strength that belong to the PF, and dominates the PF (Zitzler and Thiele, 1999).

Selection. A binary tournament approach is adopted because of its simplicity and robustness, reported in Goldberg (1989).

Crossover. According to the structure of the chromosome, a one-point crossover operator is considered. The probability of applying crossover p_c is selected according to Coello et al. (2007).

Mutation. The mutation operator is implemented in two stages. First, an individual is selected with a probability of mutation p_m. Then, each gene is modified with a probability p_g. The values of p_m and p_g are input data, selected randomly with uniform distribution.

10.4.6 EXPERIMENTAL RESULTS

In this study, we assume the following conditions:

- The bandwidth of requests w is equal to the bandwidth of a complete wavelength.
- The number of optical fibers at each link is one.
- All optical fibers have 10 wavelengths ($\Lambda = 10$, $\forall e \in E$).
- For a given G topology, we consider the maximum supported load capacity as $LM = |V| \times \Psi \times \Lambda$, where Ψ is the link average of WDM nodes.
- The shortest-path (SP) routing and first-fit (FF) wavelength-assignment algorithms are used to build the UM by simulation (Roy and Naskar, 2008).
- The shortest-path aware (SPA) routing and first-fit (FF) wavelength-assignment algorithms are used to evaluate the solutions in the direct simulations.

Experimental tests have as their main objective a fair comparison between SPEA and DEA approaches. Note that SPEA is a multiobjective approach, while DEA is a single-objective approach. DEA takes as input data the number of wavelength converters C^* to be allocated. The strategy to achieve a fair comparison consists of executing DEA for each value of $C^* \in \{C_{min}, C_{min}+1, \ldots, C_{max}\}$. C_{min} and C_{max} are obtained from SPEA runs. This way, it can get a set of trade-off solutions. The details of our experimental tests are given below.

1. *Dynamic scenario*. The dynamic request U is generated using $\Gamma = 1000$, with a uniform distribution for simplicity.

2. *Statistical use.* UM is calculated applying a simulation considering a dynamic request U, a topology G, an RWA algorithm, and all WDM resources available.
3. SPEA and DEA are compared using the experimental methodology based on Algorithm 1, where the set of algorithm $A = \{\text{SPEA, DEA}\}$, the number of independence runs $\Pi = 10$, the stop criterion $\vartheta = 1000$ iterations, and the set of performance measures $B = \{\text{CS, GD, ER}\}$.
4. In this experiment, it is necessary to modify Algorithm 1 a little bit when DEA is considered, because it is a single-objective algorithm and we want to calculate a set of nondominated solutions. After the SPEA algorithm is run, the values $C_{\min}$ and $C_{\max}$ are obtained from the set X_1. Line 5 of Algorithm 1, when DEA is evaluated, is stated as follows:

DEA is run $(C_{\max} - C_{\min} + 1)$ times. For each run, DEA receives as input data the stop criterion ϑ and the number of wavelength converters C^* to allocate. In this context, the set of nondominated solutions is given by $Z_{2j} = \overset{C_{\max}}{\underset{C^* = C_{\min}}{\cup}} A_2(\vartheta, C^*)$, where $A_2 = \text{DEA}$.

10.4.6.1 Numerical results

The experimental tests presented below were performed on National Science Foundation (NSF) network topology (Maciel et al., 2009), composed of 14 nodes and 42 links. The evolutionary parameters are the following: $p_c = 1$, $p_m = 0.3$, $p_g = 0.4$, and 100 individuals of the population were adopted experimentally.

The steps presented in Section 10.4.6 were applied to the NSF network under the following traffic scenarios: low load $|U|_{\text{low}} = 30\%\text{LM}$, half load $|U|_{\text{half}} = 60\%\text{LM}$, high load $|U|_{\text{high}} = 90\%\text{LM}$, and saturation load $|U|_{\text{sat}} = 100\%\text{LM}$. Experimental results under different traffic scenarios are presented in Figure 10.1.

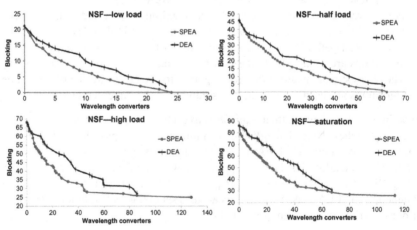

FIGURE 10.1

Pareto front for low, medium, high, and saturated traffic load.

Table 10.7 Experimental results for NSF network

Alg	Traffic Scenarios					
	Low Load			**Half Load**		
	CS	GD	ER	CS	GD	ER
SPEA	1	0	0	1	0	0
DEA	0	3.1	1	0	5.18	1
	High Load			**Saturated Load**		
SPEA	0.97	0	0	1	0	0
DEA	0.03	7.42	0.95	0	12.49	1

A summary of all experimental results considering CS, GD, and ER measures are given in Table 10.7. In almost all scenarios, it can be noticed that $Y_{app} \cong Y^*_{spea}$; this is the reason why the best Pareto front X_{app} is not included in Figure 10.1. The ER and GD of Y_{spea} are null values, while Y_{dea} obtains very high values. Clearly, all experimental results report that Y_{spea} is better than Y_{dea}. We can notice in Table 10.7 that SPEA has contributed better solutions than DEA has in all experimental tests. The Pareto fronts (Y_{spea}) generated by SPEA dominate all solutions found by DEA (Y_{dea}). The SPEA solutions dominate almost all DEA solutions in all scenarios, as shown in Table 10.7; just one good solution calculated by DEA was not discovered by SPEA. The solutions found by SPEA (Y_{spea}) have null average distance, and the solutions found by DEA (Y_{spea}) have a larger one. At the same time, the error obtained by DEA is also greater than the error obtained by SPEA. This numerically demonstrates that $Y_{app} \cong Y^*_{spea}$, as we exposed above (see Table 10.7).

Notice that when traffic increases, DEA decrements the quality of its solution set. This suggests that DEA generates relative good solutions only for low traffic. SPEA obtains good performance considering every traffic load, with the disadvantage of using more resources than DEA; this is because SPEA uses the direct-simulation approach. Finally, to emphasize a better appraisal of the experimental results, a consolidated time (in seconds) spent per solution (s/sol) is presented in Table 10.8, where TRT is a full time of experimental test, TSA is the total number of trade-off solutions calculated by the algorithm in all scenarios, and TSA_{best} is the total number of

Table 10.8 Total run time

Alg	TRT (s)	Total Solutions in PF_best			Total Solutions	
		TSA_best	TRT/TSA_best (s/sol)		TPF	TRT/TPF (s/sol)
SPEA	11,991	285 sol	42		285 sol	42
DEA	23,663	1 sol	23,663		89 sol	265

trade-off solutions obtained by an algorithm that belongs to the best Pareto front in all scenarios.

We can observe that DEA running time is longer than SPEA's, given that DEA is forced to calculate every single solution in the range between C_{min} and C_{max} calculated by SPEA (indirect simulations are faster than direct simulations).

10.5 p-CYCLE PROTECTION

Considering the importance of fault tolerance in WDM networks, Grover and Stamatelakis (1998) proposed a new method inspired by strategies with speedy recovery networks ringed synchronous optical networking/synchronous digital hierarchy (SONET/SDH), applied to mesh networks, called preconfigured protection cycle (p-cycle).

p-Cycle is a link-based protection method, with precalculated and dedicated protection designed to protect the working capacity of each link, reserving up to 50% of the total capacity of it. Essentially, the method provides protection of one alternative path in an on-link cycle ("*on-link*"), and two alternatives when a link is surrounded by a p-cycle ("*straddling-link*") (Liu and Ruan, 2004). An example can be seen in Figure 10.2, where before any failure occurs in protected links, the network already has an alternative way in case of a problem.

The problem of selecting optimal p-cycles or optimal cycles is studied in the literature as an NP-complete problem (Johnson, 1975). Earlier works have used a combination of heuristic algorithms, splitting the problem into two stages: first generating a set of candidates' cycles (Grover and Stamatelakis, 1998; Stamatelakis and Grover, 2000), and then selecting a subset of cycles or optimal solutions to solve the problem (Grover and Stamatelakis, 1998; Liu and Ruan, 2004; Stamatelakis and Grover, 2000). To generate candidate cycles, Stamatelakis and Grover (2000) suggested some heuristics for a topology of N links, getting NN candidate cycles. The complexity of the algorithm to generate cycles, proposed in Liu and Ruan (2006), was previously studied in Johnson (1975), proposing an

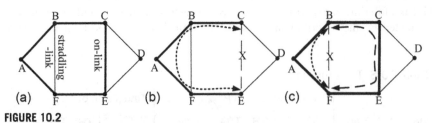

FIGURE 10.2

Preconfigured protection cycle. (a) An example of p-cycle. (b) Protection after failure of an "on-link." (c) Protection before failure of a "straddling-link," where p-cycle allows the use of two alternative routes.

algorithm called straddling-link algorithm (SLA), which generates only N candidate cycles. In addition, three algorithms were proposed—ADD, EXPAND, and GROW—to achieve a substantial improvement of candidates' quality cycles.

Topological degree (*topological score*—TS), efficiency (*a priori efficiency*), and redundancy (*network redundancy*—NR) were proposed in Somani (2005), Liu and Ruan (2004), and Farkas et al. (2005) to improve the selection of optimal cycles and as elements of assessment and screening. An integer linear programming (ILP) formulation was proposed in Grover and Stamatelakis (1998) to calculate a solution, but it is impractical for large and complex networks because of the high complexity of the method.

To solve the abovementioned problem, a heuristic algorithm called capacitated iterative design algorithm (CIDA) was proposed in Farkas et al. (2005). It achieved good performance when combined with cycle-generators SLA+GROW, as proposed in Farkas et al. (2005).

The protection of dynamic random traffic with p-cycles was proposed in Somani (2005) and Liu and Ruan (2006), and routing process protection was proposed in Schupke et al. (2002) and Liu and Ruan (2006). The high redundancy of the set of p-cycles selected by the CIDA approach was identified in Drid et al. (2007) and Coello et al. (2007). To reduce the redundancy, Drid et al. (2007) proposed a new heuristic approach, which selects a set of short p-cycles with fewer redundancies than the set obtained by CIDA.

The circumference of p-cycle protection directly affects the quality of restoration routes. Onguetou and Grover (2008) proposed a combination of ILP+GA (genetic algorithm) to optimize restoration routes over p-cycle protection in order to reduce the longitude of the restoration routes, but only using a single-objective context. Later, Colman et al. (2008a,b,c) proposed a single-objective GA to maximize the protection efficiency of a set of selected p-cycles, obtaining better results than CIDA did.

For the first time, the optimal selection of p-cycle protection problems was solved in a pure multiobjective context, minimizing the total cost and maximizing total protection, implementing two state-of-the-art MOEAs, and reaching a better solution than CIDA and the proposed GA did (Colman et al., 2008a,b,c). Additionally, Colman et al. (2008a,b,c) demonstrated that the SPEA approach gets better results than the tested approaches, using the nondominated relation and one performance figure to compare two implemented MOEAs.

In a later work of the same authors (Colman et al. (2008a,b,c)), the GAs and MOEAs minimize the redundancy and p-cycle circumference. This work proposes an extension of related articles (Colman et al., 2008a,b,c), including three new optimization objectives to control p-cycle circumference and redundancy of protection, using two state-of-the-art MOEAs implemented in Colman et al. (2008a,b,c) and using five performance figures for a better comparison. Figure 10.3 and Table 10.8 resume the problem addressed in this work.

Figure 10.3d and Table 10.9 (gray color indicates superiority) show only two restoration routes for link 1-3, with an average of 3.5 links of restoration. Figure 10.3e

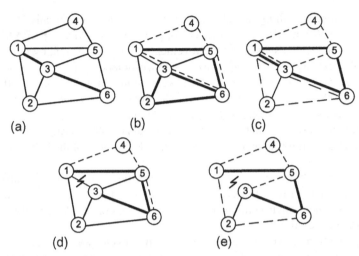

FIGURE 10.3

Restoration process with p-cycle. (a) An example route between nodes 1 and 6. (b) Set P1 two big p-cycles $p_1=\{(1\text{-}4), (4\text{-}5), (5\text{-}6), (6\text{-}3), (3\text{-}1)\}$, $p_2=\{(1\text{-}5), (5\text{-}6), (6\text{-}3), (3\text{-}1)\}$. (c) Set P2 three small p-cycles $p_3=\{(1\text{-}5), (5\text{-}6), (6\text{-}3), (3\text{-}1)\}$, $p_4=\{(1\text{-}4), (4\text{-}5), (5\text{-}3), (3\text{-}1)\}$, $p_5=\{(1\text{-}3), (3,6), (6\text{-}2), (2\text{-}1)\}$. (d) Failure of the example route in link 1-3, restoration using a set A. (e) Restoration using set B.

Table 10.9 Sets P1 and P2 redundancy and circumference values

Set	Routes Restoration Link 1-3	Average Longitude Routes	Redundancy	Average Circumference
P1	R1 = (1-4-5-6) R2 = (1-5-6)	3.5	0.66	5
P2	R1 = (1-4-5-3) R2 = (1-5-6) R3 = (1-2-6)	2.66	0.75	4

shows three restoration routes for the same failure, provided by set P2, giving better restoration than set P1. In terms of redundancy (Equation (10.26)), set P1 is better, but in average p-cycle circumference (Equation (10.27)), set P2 is superior. Considering the standard deviation of p-cycle circumference, both sets are almost the same. This demonstrates that the three additional objectives are in contradiction. It is critical to improve p-cycle selection, and this confirms the requirement to solve this protection problem in a multiobjective context, complementing protection and cost (Colman et al., 2008a,b,c).

10.5.1 PROBLEM FORMULATION

The following extra nomenclature is needed in this section:

p_k	Set of links that form a cycle that passes through $\|p_k\|$ nodes, $p_k = \{e_1, e_2, \ldots, e_{\|p_k\|}\}$
P	Set of cycles protecting G; $p = \{p_1, p_2, \ldots, p_{\|p\|}\}$
$X_k(e)$	Variable indicating the type of protection provided to e by cycling p_k. If e is a *straddling-link* in p_k then $X_k(e) = 2$, if e is an *on-link* in p_k then $X_k(e) = 1$, otherwise $X_k(e) = 0$
$Z_k(e)$	Variable indicating whether p_k provides protection to e_j. If e_j is an *on-link* or a *straddling-link* in p_k, $Z_k(e) = 1$, otherwise $Z_k(e) = 0$

Considering the above definitions, the proposed problem consists of the following: given a graph G, calculate a set of p-cycles $S \subset P$ that simultaneously minimize

Total cost—TC:

$$TC = \sum_{p_k \in S} |p_k| \tag{10.24}$$

Total protection—TP:

$$TP = \sum_{p_k \in S} TS(p_k) \tag{10.25}$$

Total redundancy—TR:

$$TR = \sum_{p_k \in S} NR(p_k) \tag{10.26}$$

Average p-cycle circumference—$\overline{PC}$:

$$\overline{PC} = \frac{\sum_{p_k \in S} |p_k|}{|S|} \tag{10.27}$$

Standard deviation p-cycle circumference—PC_σ:

$$PC_\sigma = \frac{1}{|S|} \sum_{p_k \in S} \left(\overline{PC} - |p_k| \right) \tag{10.28}$$

where the cyclical protection or TS and NR are defined in Somani (2005), Schupke et al. (2002), and Doucette et al. (2003) as

$$TS(p_k) = \sum_{e \in E} X_k(e) \tag{10.29}$$

$$NR(p_k) = \frac{|p_k|}{2|E| - |p_k|} \tag{10.30}$$

subject to the restriction of full protection of G given by

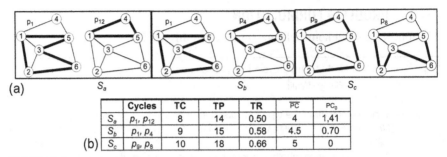

(a)

	Cycles	TC	TP	TR	$\overline{PC}$	PC_0
S_a	p_1, p_{12}	8	14	0.50	4	1,41
S_b	p_1, p_4	9	15	0.58	4.5	0.70
S_c	p_9, p_8	10	18	0.66	5	0

(b)

FIGURE 10.4

Example of a protection based on p-cycles. (a) Three subsets $S \subset P$ as possible protection schemes. (b) Calculations of the solutions. $S_a \not\succ S_b, S_c; S_b \not\succ S_a, S_c; S_c \not\succ S_a, S_b$.

$$\sum_{e_j \in E} \sum_{p_k \in S} Z_k(e_j) \geq |E| \tag{10.31}$$

As an example, consider the network topology presented in Figure 10.3. Figure 10.4a and b show three possible solutions: S_a, S_b, and S_c.

The first solution is $S_a = \{p_1, p_{12}\}$, the second one is $S_b = \{p_1, p_4\}$, and the last solution is $S_c = \{p_8, p_9\}$. Note that the solutions are not comparable (none of them dominates the others). S_a has the lowest cost, protection, redundancy, and p-cycle circumference, but a high standard deviation of p-cycle circumference. On the other hand, S_c has the largest cost and the best protection, high redundancy, and p-cycle circumference, but the lowest standard deviation of p-cycle circumference. At the same time, S_b is a trade-off solution. Note that for this example, the calculated Pareto set is $Y = \{S_a, S_b, S_c\}$.

10.5.2 GENERATING CANDIDATE CYCLES

Calculating the set of all candidates' cycles P_{full} in high-complexity network topologies is completely impractical (Doucette et al., 2003). Several publications propose careful construction of a set of candidates' cycles $P \subset P_{full}$ (Liu and Ruan, 2004; Zhao et al., 2006). SLA and GROW are the most referenced and studied algorithms to generate candidates' cycles (Colman et al., 2008a,b,c; Doucette et al., 2003; Farkas et al., 2005). Both approaches calculate a set of efficient cycles' candidates adaptable to topologies of different sizes, nodes, and number of links (Doucette et al., 2003; Farkas et al., 2005; Zhao et al., 2006).

Basically, the SLA approach makes two calls to the Dijkstra shortest-path algorithm to build a cycle for each link, generating a maximum number of cycles, equivalent to the number of links in a given topology. This way, a small, scalable, and efficient set of candidate cycles is built (Zhang and Yang, 2002). The version of the GROW algorithm considered in this work is the exhaustive version proposed in Doucette et al. (2003). GROW builds new cycles from a set of cycles calculated in advance by another algorithm.

Therefore, SLA is used to generate a set of cycles P1, while GROW generates another set of cycles P2 with new cycles of high quality, which verifies the restrictions shown in Equation (10.31). Experimental results in Colman et al. (2008a,b,c) indicate that several runs of GROW over the same set of candidate cycles increment the quality and the quantity of the candidates' cycles. For this work, we run GROW two times over the set of candidate cycles generated by SLA, using the same candidate set used in Colman et al. (2008a,b,c).

10.5.3 MULTIOBJECTIVE EVOLUTIONARY ALGORITHMS

The NSGA-II (Deb and Goel, 2002) and SPEA2 (Zitzler and Thiele, 1999) MOEA were implemented in this work, taking as data input the topology $G = (V, E, \Lambda)$ and a set of candidate cycles P, calculating a Pareto front $Y = \{S_1, S_2, \ldots, S_{|Y|}\}$, where $S_i \subset P$ is a cyclical protection. The following steps are the most important functions on the implemented approaches (Colman et al., 2008a,b,c).

Build a protection table: Let $P = \{p_1, p_2, \ldots, p_{|P|}\}$ be a set of candidate cycles. The protection table is an ordered storage of $p_k \in P$ to simplify the representation of a chromosome. Each chromosome represents a solution $S \subset P$, which is divided into two columns: (a) *CantCycle,* number of cycles of S; and (b) *IDs p-cycles* indexed to a table of protection cycles. Note that the representation of a chromosome dynamically depends on CantCycle. Figure 10.5 shows the relationship between the table of protection cycles and the structure of a chromosome, inspired by the one proposed in Crichigno and Barán (2004).

Generate evolutionary population: Each individual is built randomly. Initially, a positive integer of CantCycle is randomly assigned, CantCycle $\in \{1, 2, \ldots, |P|\}$. CantCycle defines the number of gens to be used. Each gen_i ($i = 1, \ldots,$ CantCycle) is randomly assigned with an ID (identification unique index) pointing to a position

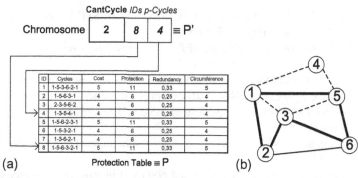

(a)   Protection Table $\equiv$ P   (b)

FIGURE 10.5

Relation among chromosome, protection, and solution. (a) *CantCycle* (=2) indicates the number of cycles in chromosome *S*. The following numbers indicate a position in the protection table $p_k \in S$. (b) The cycle selected by the chromosome.

of the protection table. To maintain consistency, the following restriction is imposed on genes: $gen_i \neq gen_j$, when $i \neq j$.

Ensure protection: The solution proposed by a chromosome must verify the restriction of total protection of the links, according to Equation (10.30). Thus, if necessary, a new subroutine adds cycles to a chromosome until it complies with this restriction.

Evolutionary operators: This work implements a binary tournament as a method for selecting the parents (Goldberg, 1989). The crossing operator takes two parents and creates a new son. The operator has two stages: first crossing CantCycle, and then crossing the ID of the parents' genes. To cross CantCycle, a blend crossover operator is used (Goldberg, 1989). Once CantCycle is defined for a child, genes from both parents are copied directly to the given child. The rest of the genes are randomly chosen from one of the parents. Finally, a mutation operation may totally replace some individuals with new randomly generated individuals, considering a probability of mutation as $p_m = 0.2$.

10.5.4 EXPERIMENTAL RESULTS

Experiments were conducted considering the following environment. Algorithms were implemented in Mathworks MatLab 2008 for the Microsoft Windows Vista platform. We employed an IBM ThinkPad computer with Intel Dual Core 2 1.8 GHz, 4 GB RAM, on a Microsoft Windows Vista Professional operating system. For this study, 12 topologies of different sizes and densities were considered, which are presented in Table 10.10. The experiments were carried out according to Algorithm 1 of Section 10.2.4, considering $A = \{$NSGA-II, SPEA2$\}$ as the set of algorithms, the number of executions $\Pi = 50$, and each execution was over $\vartheta = 10$ min. Furthermore, the set of considered performance measures was $B = \{$OG, OR, E, ER, GD$\}$. Table 10.11 shows the performance evaluation of both MOEAs (gray color indicates superiority).

Based on the results shown in Table 10.11, the follow behavior is observed:

(a) In NSF, PAN, ELHN, GOBN, and ECN topologies, SPEA2 outperformed NSGA-II in all performance figures.
(b) In ULHN and ARPA topologies, SPEA2 obtained better results than NSGA-II with only one exception: OG value.
(c) Considering CNBN topology, SPEA2 obtained better results in most performance figures over NSGA-II, except for E value.
(d) In NTT topology, SPEA2 outperformed NSGA-II in most performance figures, except ER value.
(e) In IRIS topology, NSGA-II obtained better results in most performance figures, except in ER.
(f) In overall values, SPEA2 outperformed NSGA-II in most performance figures, except in OG, but this is not relevant because the difference is very small.

The behavior observed confirms the superiority of SPEA2 over NSGA-II, in this experiment. Additionally, other experimental results using SPEA2 confirm and

Table 10.10 Tested topologies

ID	Network Topology	Nodes	Links	$|P_3|$	Reference
NSF	USA National Science Foundation Backbone	14	25	756	Somani (2005)
BCBY	Bell Core Backbone Yerse LATA	15	27	780	Grover and Stamatelakis (2000)
ULHN	USA Long Haul	28	45	780	Liu and Ruan (2006)
ARPA	USA ARPA Backbone II	21	25	30	Shen and Grover (2003)
BCN	Bell Core Backbone	25	28	404	Shen and Grover (2003)
PAN	PAN European COST 239	11	26	366	Shen and Grover (2003)
ELHN	European Long Haul— France Telecomm	43	71	1683	Liu and Ruan (2006)
GOBN	German Optical Backbone	17	27	212	Schupke (2005)
ECN	ECNet European Backbone	18	39	420	Huang and Copeland (2001)
CNBN	China National Backbone	66	120	4257	Zhang and Yang (2002)
NTT	Nippon Telephone Telegraph Backbone	49	66	545	Crichigno and Barán (2004)
IRIS	RedIRIS Spanish Research	19	32	183	Colman et al. (2008a,b,c)

Table 10.11 Overall performance figure values

Average of Performance	SPEA2	NSGA-II
OG	611.75	626.916667
OR	0.53044278	0.48524387
E	437	414
ER	0.27384629	0.35472663
GD	32.8565984	61.9436591

improve the results obtained in Colman et al. (2008a,b,c), reducing redundancy and controlling the p-cycle circumference, threatening the problem identified in Drid et al. (2007) and Onguetou and Grover (2008).

10.6 CONCLUSIONS

This work has presented bio-inspired EAs to solve optical WDM network problems from the viewpoint of a pure multiobjective optimization. The three considered problems, RWA, WCA, and p-cycle, are clear examples of corresponding management, design, and protection areas, respectively. In this sense, the main characteristic of this multiobjective approach is that a nondominated solution set is calculated within a reasonable time, considering the complexity of the problems. This way, the decision maker can know a complete set of trade-off solutions, better understanding the problem and improving the spectrum of possibilities to help his cost-benefit analysis for a better-informed final decision.

Another fundamental aspect to consider is that the design of network architecture, management schemas, and strategies for protection are all interrelated when a real optimal operation is searched for (Pinto-Roa et al., 2011). In this context, in the future, these problems should be considered simultaneously to calculate more optimal and robust solutions.

Finally, it should be mentioned that in WDM networks, there is a vast number of complex, mostly interrelated problems. In this regard, there is a clear need to step up efforts in solving different optical network problems in a joint fashion to obtain optimal solutions, even considering different objective functions in pure multiobjective contexts, as the ones presented in this chapter.

REFERENCES

Arteta, A., Barán, B., Pinto-Roa, D.P., 2007. Routing and wavelength assignment over WDM optical networks. A comparison between MOSCOs and classical approaches. In: II IFIP/ACM Latin America Networking Conference 2007 LANC'07, San José, Costa Rica.

Barry, R.B., Humblet, P.A., 1996. Models of blocking probability in all-optical networks with and without wavelength changers. IEEE J. Select. Areas Commun. 14 (5), 858–867.

Brownlee, J., 2011. Clever Algorithms: Nature-Inspired Programming Recipes. ISBN 978-4467-8506-5.

Choi, J.S., Golmie, N., Lapeyrere, F., Mouveaux, F., Su, D., 2000. A functional classification of routing and wavelength assignment schemes in DWDM networks: static case. Proceeding of VII International Conference on Optical Communications and Networks, January.

Chu, X., Li, B., Liu, J., Li, L., 2002. Wavelength converter placement under a dynamic RWA algorithm in wavelength-routed all-optical networks. In: IEEE 2002 International Conference on Communications, Circuits and Systems and West Sino Expositions, pp. 865–870. IEEE Conference Proceedings, Chengdu, China.

Coello, C.A.C., Van Veldhuizen, D.A., Lamont, G.B., 2007. Evolutionary Algorithms for Solving Multi-objective Problems. Kluwer Academic, New York, 242.

Colman, C.E., Pinto-Roa, D.P., Barán, B., 2008a. Optimal selection of p-cycles in WDM networks, an approach based in genetic algorithm. In: Proceeding, Latin American Conference of Informatics 2008 (CLEI 2008), Santa Fe, Argentina.

Colman, C.E., Pinto-Roa, D.P., Barán, B., 2008b. Optimal selection of p-cycles in WDM networks, an approach based on MOEA. In: Proceeding, IEEE Colombian Communication Congress 2008 (IEEE COLCOM 2008), Popayan, Colombia.

Colman, C.E., Pinto-Roa, D.P., Barán, B., 2008c. Solving multi-objective p-cycle protection problem in WDM optical networks in an evolutionary algorithm approach. In: Proceeding, IEEE 7th International Information and Telecommunication Technologies Symposium 2008 (IEEE 7TH I2TS 2008), Iguazu Falls, Brazil.

Colman, C.E., Pinto-Roa, D.P., Barán, B., 2009. Optimizing p-cycles selection with MOEAs approach to protect WDM optical networks. In: 24th IFIP TC7 Conference on System Modeling and Optimization.

Crichigno, J., Barán, B., 2004. Multiobjective multicast routing algorithm for traffic engineering. In: IEEE International Conference on Computers and Communication Networks – ICCCN 2004. Chicago, IL, USA, pp. 301–306.

Deb, K., Goel, T., 2002. A fast and elitist multi-objective genetic algorithm: NSGA-II. IEEE Trans. Evol. Comput. 6 (2), 182–197.

Doerner, K., Gutjahr, W., Hartl, R., Strauss, C., 2002. Pareto ant colony optimization: s meta-heuristic approach to multiobjective portfolio selection. In: Proceedings of the 4th Meta-heuristics International Conference. Porto, pp. 243–248.

Doerner, K., Hartl, R., Reimann, M., 2003. Are COMPETants more competent for problem solving? The case of a multiple objective transportation problem. CEJOR 11 (2), 115–141.

Dorigo, M., Di Caro, G., 1999. The ant colony optimization meta-heuristic. New Ideas in Optimization. McGraw Hill, London, UK.

Doucette, J., He, D., Grover, W.D., Yang, O., 2003. Algorithmic approaches for efficient enumeration of candidate p-cycles and capacitated p-cycle network design. In: Proceedings of the Fourth International Workshop on the Design of Reliable Communication Networks (DRCN 2003), Banff, Alberta, Canada, pp. 212–220.

Drid, H., Cousin, B., Molnar, M., 2007. A heuristic solution to protect communications in WDM optical networks using p-cycles. In: Proceeding Workshop on Traffic Engineering, Protection and Restoration for Future Generation Internet, Oslo, Norway.

Dutta, R., Rouskas, G.N., 2000. A survey of virtual topology design algorithms for wavelength routed optical networks. Opt. Network Mag. 1 (1), 73–89.

Fang, C., Low, C.P., 2006. Optimal wavelength converter placement with guaranteed wavelength usage. In: Networking 2006, pp. 1050–1061. LNCS 3976.

Farkas, A., Szigeti, J., Cinkler, T., 2005. P-cycle based protection scheme for multi-domain networks. In: 5th International Workshop on Design of Reliable Communication Networks (DRCN), Naples, Italy.

Foo, Y.C., Chien, S.F., Low, A.L.Y., Teo, C.F., Lee, Y., 2005. New strategy for optimizing wavelength converter placement. Opt. Soc. Am. (OSA), Opt. Exp. 13 (2), 545–551.

Gardel, P., Baran, B., Estigarribia, H., Fernandez, U., Duarte, S., 2006. Multiobjective reactive power compensation with an ant colony optimization algorithm. In: AC and DC Power Transmission. ACDC 2006. The 8th IEE International Conference on, 28–31 March 2006, pp. 276–280.

Gerstel, O., Ramaswami, R., Wang, W., 2000. Making use of a two stage multiplexing scheme in a WDM network. In: Optical Fiber Communication Conference, vol. 3, pp. 44–46.

Goldberg, D.E., 1989. Genetic Algorithms in Search, Optimization, and Machine Learning. Addison-Wesley Publishing Company.

Grover, W.D., Stamatelakis, D., 1998. Cycle-oriented distributed pre-configuration: ring-like speed with mesh-like capacity for self-planning network restoration. In: Proceedings of the IEEE International Conference on Communications (ICC) 1998, Atlanta, GA, USA, pp. 537–543.

Grover, W.D., Stamatelakis, D., 2000. Bridging the ring-mesh dichotomy with p-cycles. In: Proceedings of the Second International Workshop on the Design of Reliable Communication Networks (DRCN), Munchen, Germany, pp. 92–104.

Harada, K., Shimizu, K., Kudou, T., Ozeki, T., 1999. Hierarchical optical path cross-connect systems for large scale WDM networks. In: Optical Fiber Communication Conference. Optical Society of America, p. WM55.

Hei, X., Zhang, J., Bensaou, B., Cheung, C.C., 2004. Wavelength converter placement in least-load-routing-based optical network using genetic algorithms. J. Opt. Netw. 3 (5), 363–378.

Hillier, F.S., Lieberman, G.J., Hillier, F., Lieberman, G., 2010. Introduction to Operations Research. McGraw-Hill, New York, NY.

Huang, H., Copeland, J.A., 2001. Hamiltonian cycle protection: a novel approach to mesh WDM optical network protection. In: Proceedings of the IEEE Workshop on High Performance Switching and Routing (HPSR), Dallas, TX, USA, pp. 31–35.

Iredi, S., Merkle, D., Middendorf, M., 2001. Bi-criterion optimization with multicolony ant algorithms. In: Proceedings of the First International Conference on Evolutionary Multi-criterion Optimization (EMO'01). In: Lecture Notes in Computer Science, vol. 1993, pp. 359–372.

Jeong, J., Ayanoglu, E., 1996. Comparison of wavelength interchanging and wavelength-selective cross-connects in multi wavelength all-optical networks. In: The Conference on Computer Communications, Fifteenth Annual Joint Conference of the IEEE Computer and Communications Societies, Networking the Next Generation, San Francisco, CA, USA, pp. 156–163.

Jeong, H.Y., Seo, S.W., 2005. A binary (0–1) linear program formulation for the placement of limited-range wavelength converters in wavelength-routed WDM networks. J. Lightwave Technol. 23 (10), 3076–3091.

Johnson, D.B., 1975. Finding all the elementary circuits of a directed graph. SIAM J. Comput. 4 (1), 77–84.

Karasan, E., Ayanoglu, E., 1998. Effects of wavelength routing and selection algorithms on wavelength conversion gain in WDM optical networks. IEEE/ACM Trans. Networking 6 (2), 186–196.

Kovacevic, M., Acampora, A., 1996. Benefits of wavelength translation in all-optical clear-channel networks. IEEE J. Sel. Areas Commun. 14 (5), 868–880.

Le, V.T., Ngo, S.H., Jiang, X., Horiguchi, S., Guo, M., 2005. A genetic algorithm for dynamic routing and wavelength assignment in WDM networks. In: Cao, J., Yang, L.T., Guo, M., Lau, F. (Eds.), Parallel and Distributed Processing and Applications. In: Lecture Notes in Computer Science, vol. 3358. Springer, Berlin, Heidelberg, pp. 893–902.

Lee, M., Yu, J., Kim, Y., Kang, C., Park, J., 1993. Design of hierarchical cross connects WDM networks employing a two-stage multiplexing scheme of waveband and wavelength. IEEE J. Select. Areas Commun. 20 (1), 166–171.

Lee, S.S.W., Yuang, M.C., Tien, P.L., 2004. A Lagrangean relaxation approach to routing and wavelength assignment for multi granularity optical WDM networks. In: IEEE Global Telecommunications Conference, Proceedings IEEE CLOBECOM, pp. 1936–1942.

Leonardi, E., Mellia, M., Marsan, M.A., 2000. Algorithms for the logical topology design in WDM all-optical networks. Opt. Network Mag. 1 (1), 35–46.

Leonardi, E., Mellia, M., Marsan, M.A., 2005. A genetic algorithm for dynamic routing and wavelength assignment in WDM networks. In: Cao, J., Yang, L.T., Guo, M., Lau, F. (Eds.), Parallel and Distributed Processing and Applications. In: Lecture Notes in Computer Science, vol. 3358. Springer, Berlin, Heidelberg, pp. 893–902.

Li, T., Wang, B., 2004. Cost effective shared path protection for WDM optical mesh networks with partial wavelength conversion. Photon Netw. Commun. 8 (3), 251–266.

Liu, C., Ruan, L., 2004. Finding good candidate cycles for efficient p-cycle network design. In: Proceedings of the 13th International Conference on Computer Communications and Networks (ICCCN 2004), Chicago, Illinois, EUA, pp. 321–326.

Liu, C., Ruan, L., 2006. P-cycle design in survivable WDM networks with shared risk link groups (SRLGs) (extended version). Photon Netw. Commun. 11 (3), 301–311.

Lopez-Ibañez, M., Paquete, L., Stützle, T., 2004. On the design of ACO for the biobjective quadratic assignment problem. In: Dorigo, et al., (Ed.), Proceedings of the Fourth International Workshop on Ant Colony Optimization (ANTS 2004). In: Lecture Notes in Computer Science, Springer Verlag.

Maciel, R., Sobrino, M., Pinto-Roa, D.P., Barán, B., Brizuela, C.A., 2009. Optimal wavelength converter allocation: a new approach based MOEA. In: Proceedings of the 5th International Latin American Networking Conference LANC.

Mariano, C., Morales, E., 1999. A multiple objective ant-q algorithm for the design of water distribution irrigation networks. Technical report HC-9904, Instituto Mexicano de Tecnología del Agua, México.

Mokhtar, A., Azizoglu, M., 1998. Adaptive wavelength routing in all-optical networks. IEEE/ACM Trans. Networking 6 (2), 197–206.

Mukherjee, B., 2006. Optical WDM Networks. Springer, New York.

Mukherjee, B., Banerjee, D., Ramamurthy, S., 1996. Some principles for designing a wide-area WDM optical network. IEEE/ACM Trans. Networking 4 (5), 684–706.

Onguetou, D.P., Grover, W.D., 2008. Approaches to p-cycle network design with controlled optical path lengths in the restored network state. J. Opt. Netw. 7 (7), 673–691.

Paciello, J., Martínez, H., Lezcano, C., Barán, B., 2006. Multiobjective optimization algorithms based on ant colony. In: XXXII Latin-American Conference on Informatics 2006 – CLEI2006. Santiago de Chile, Chile, August.

Pinto-Roa, D.P., Barán, B., 2005. Solving multiobjective multicast routing problem with a new ant colony optimization approach. In: IFIP/ACM Latin America Networking Conference 2005 – LANC'05. Cali, Colombia.

Pinto-Roa, D.P., Barán, B., Brizuela, C.A., 2009. Wavelength converter allocation in optical network. An evolutionary multiobjective algorithm approach. In: Ninth International Conference on Intelligent Systems Design and Applications, ISDA 2009, Pisa, Italy, pp. 411–419.

Pinto-Roa, D.P., Barán, B., Brizuela, C.A., 2011. Routing and wavelength converter allocation in WDM networks: a multi-objective evolutionary optimization approach. Photon Netw. Commun. 22, 23–45. http://dx.doi.org/10.1007/s11107-011-0304-4.

Ramamurthy, B., Iness, J., Mukherjee, B., 1997. Minimizing the number of optical amplifiers needed to support a multi wavelength optical LAM/MAN. In: The Conference on Computer Communications, Sixteenth Annual Joint Conference of the IEEE Computer and Communications Societies, Driving the Information Revolution, Kobe, Japan. Proceedings IEEE INFOCOM, pp. 261–268.

Ramaswami, R., Kumar, N., Sivarajan, K.N., 1996. Design of logical topologies for wavelength-routed optical networks. IEEE J. Select. Areas Commun. 14 (5), 840–851.

Rao, S.S., 2009. Engineering Optimization: Theory and Practice. John Wiley & Sons, Hoboken, NJ.

Rouskas, G.N., 2003. Optical layer multicast: rationale, building blocks, and challenges. IEEE Netw. 17 (1), 60–65.

Rouskas, G.N., Perros, H.G., 2002. A tutorial on optical networks. In: Advanced Lectures on Networking, Springer, Berlin, Heidelberg, pp. 155–193.

Roy, K., Naskar, M.K., 2008. Genetic evolutionary algorithms for optimal allocation of wavelength converters in WDM optical networks. Photonic Netw. Commun. 16 (1), 31–42.

Saha, M., Sengupta, I., 2005. A genetic algorithm based approach for static virtual topology design in optical networks. In: INDICON, Annual IEEE, pp. 392–395.

Schaerer, M., Barán, B., 2003. A multiobjective ant colony system for vehicle routing problems with time windows. In: Proceedings of the Twenty First IASTED International Conference on Applied Informatics, Insbruck, Austria, pp. 97–102.

Schupke, D.A., 2005. Automatic protection switching for p-cycles in WDM networks. OSN 2 (1), 35–48.

Schupke, D.A., Gruber, C.G., Autenrieth, A., 2002. Optimal configuration of p-cycles in WDM networks. In: Proceedings of the IEEE International Conference on Communications (ICC) 2002, New York City, USA, pp. 2761–2765.

Shen, G., Grover, W.D., 2003. Extending the p-cycle concept to path segment protection for span and node failure recovery. IEEE J. Select. Areas Commun. 21 (8), 1306–1319.

Somani, A.K., 2005. Survivability and Traffic Grooming in WDM Optical Networks. Cambridge University Press, New York.

Stamatelakis, D., Grover, W.D., 2000. Network restorability design using pre-configured trees, cycles, and mixtures of pattern types. Technical report TR-1999-05, TRLabs, Edmonton, AB, Canada.

Storn, R., Price, K., 1997. Differential evolution a simple and efficient heuristic for global optimization over continuous spaces. J. Glob. Optim. 11 (4), 341–359.

Subramaniam, S., Azizoglu, M., Somani, A.K., 1996. All-optical networks with sparse wavelength conversion. IEEE/ACM Trans. Netw. 4 (4), 544–557.

To, M., Neusy, P., 1994. Unavailability analysis of long-haul networks. IEEE J. Select. Areas Commun. 12 (1), 100–109.

Xiao, G., Leung, Y., 1999. Algorithms for allocating wavelength converters in all-optical networks. IEEE/ACM Trans. Networking 7 (4), 545–557.

Xu, S., Li, L., Wang, S., 2005. A genetic algorithm for dynamic routing and wavelength assignment in WDM networks. In: Cao, J., Yang, L.T., Guo, M., Lau, F. (Eds.), Parallel and Distributed Processing and Applications. In: Lecture Notes in Computer Science, vol. 3358. Springer, Berlin, Heidelberg, pp. 893–902.

Yates, J. 1997. Performance analysis of dynamically-reconfigurable wavelength-division multiplexed networks. PhD Thesis. The University of Melbourne, Australia.

Zang, H., Jue, J., Mukherjee, B., 2000. A review of routing and wavelength assignment approaches for wavelength-routed optical WDM networks. Opt. Network Mag. 1 (1), 47–60.

Zang, H., Huang, R., Pan, J., 2002. Methodologies on designing a hybrid shared-mesh protected WDM network with sparse wavelength conversion and regeneration. In: Proceedings of APOC 2002, Shanghai, China, pp. 188–196.

Zhang, H., Yang, O., 2002. Finding protection cycles in DWDM networks. In: Proceedings of the IEEE International Conference on Communications (ICC), New York City, USA, pp. 2756–2760.

Zhao, T., Yu, H., Li, L., 2006. A novel algorithm for node-encircling and link candidate p-cycles design in WDM mesh network. J. Chin. Inst. Eng. 29 (7), 1227–1233.

Zhou, D., Subramanian, S., 2000. Survivability in optical networks. IEEE Netw. 14 (6), 16–23.

Zhu, K., Zhu, H., Mukherjee, B., 2005. In: Traffic Grooming in Optical WDM Mesh Networks. In: Optical Networks, Springer, New York.

Zitzler, E., Thiele, L., 1999. Multi-objective evolutionary algorithms: a comparative case study and the strength Pareto approach. IEEE Trans. Evol. Comput. 3 (4), 257–271.

Zitzler, E., Deb, K., Thiele, L., 2000. Comparison of multiobjective evolutionary algorithms: empirical result. Evol. Comput. 8 (2), 173–195.

Cell-Coverage-Area Optimization Based on Particle Swarm Optimization (PSO) for Green Macro Long-Term Evolution (LTE) Cellular Networks

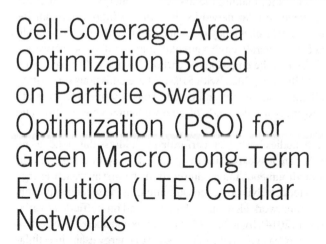

Mohammed H. Alsharif, Rosdiadee Nordin, and Mahamod Ismail

Department of Electrical, Electronics and Systems Engineering, Faculty of Engineering and Built Environment, Universiti Kebangsaan Malaysia, Bangi, Selangor, Malaysia

CHAPTER CONTENTS

11.1 INTRODUCTION

The number of evolved Node Bs (eNBs) is expected to reach 2.43 million by 2018, to achieve a population coverage target of 1.3 billion long-term evolution (LTE) subscribers (Ericsson, 2013). According to Alsharif et al. (2013), eNBs are considered the primary source of energy consumption in cellular networks and account for 57%

of the total energy used. Therefore, making the eNBs more energy efficient is the primary focus in creating green cellular networks. It is increasingly important for cellular operators to achieve energy efficiency (EE) in cellular networks to maintain profitability and to reduce the overall environmental impact of these networks. According to Suarez et al. (2012), the emissions of carbon dioxide (CO_2) have been predicted to increase to 349 $MtCO_2$ by 2020, with 51% of these emissions originating from the mobile sector, if this issue is not addressed. Thus, mobile operators are under immense pressure to meet the demands of environmental conservation and of cost reduction.

Recently, many research studies have been performed to address this issue, using *greener* cellular networks that are less expensive to operate. Two approaches can generally be used to reduce the environmental impact and costs. In the first approach, energy efficient hardware is used to reduce eNB power consumption. In the second approach, the intelligent management of network elements, which is based on traffic-load variations, is adopted (Alsharif et al., 2014). Switching eNBs on/off is considered an effective method of increasing EE for two reasons: (i) the main source of energy usage in cellular networks is the operation of eNB equipment, and (ii) the infrastructures of cellular networks are designed to support daytime traffic. Traffic loads during the day differ from those at night. Hence, the amount of energy that is wasted from the inefficient use of resources cannot be neglected, particularly for low loads (Niu et al., 2010).

The principle of this switch-off approach is as follows: if the traffic is low in a given area, some cells will be switched off, and the radio coverage and service will be provided by the remaining active cells, which may lead to a lack of coverage because the eNB maximum power is limited. In addition, the received signal power rapidly decreases as the transmit-receive distance increases. In this study, the cell-switching on/off approach has been adopted to achieve significant reductions in the energy consumption of the network, while achieving the maximum coverage possible. To this end, a particle swarm optimization (PSO) algorithm has been adopted to maximize coverage under constraints of the transmission power of the eNB (P_{tx}), the total antenna gain (G), the bandwidth (BW), the signal-to-interference-plus-noise ratio (SINR), and shadow fading (σ).

The remainder of this chapter is organized as follows. Related works are listed in Section 11.2. Section 11.3 includes the mechanism of the cell-switching scheme that is proposed in the present study. The system model and the cell-coverage-area optimization problem are described in Section 11.4. Section 11.5 presents a brief introduction to the considered PSO algorithm. The simulation, the corresponding results, and a discussion are presented in Section 11.6. Finally, the conclusion is given in Section 11.7.

11.2 RELATED WORKS

Several studies have investigated the switch-off approach. In Chiaraviglio et al. (2008, 2009) and Marsan et al. (2009), different approaches for switching off a specific number of base stations (BSs) in Universal Mobile Telecommunications System (UMTS) cellular networks during low-traffic periods were presented. Chiaraviglio et al. (2008) switched off a randomly chosen number of BSs, and the energy reduction was computed by simulating UMTS cellular networks. The same authors presented an improvement of their previous work. They proposed a dynamic network planning scheme for switching BSs on and off and considered a uniform and a hierarchical scenario (Chiaraviglio et al., 2009). In another study, Marsan et al. (2009) demonstrated how to optimize energy savings by assuming that any fraction of cells can be switched off based on a deterministic traffic-variation pattern over time. In addition, two approaches that achieve energy savings were proposed in Zhou et al. (2009): (i) a greedy centralized algorithm, where each BS is examined based on its traffic load to determine if the BS will be switched off; and (ii) a decentralized algorithm, where each BS locally estimates its traffic load and independently decides if it is going to be switched off. Gong et al. (2010) proposed a dynamic switch on/off algorithm based on blocking probabilities. The BSs are switched off based on the traffic-variation with respect to a blocking probability constraint. Xiang et al. (2011) studied the optimal number of active BSs that will be deployed based on the trade-off between fixed power and dynamic power. Lorincz et al. (2012) presented a novel optimization model that can be used for energy-saving purposes at the level of a UMTS cellular-access network. Bousia et al. (2012) proposed a switch-off decision-making scheme based on the average distance between BSs and user equipment (UE), where the BS that is at the maximum average distance will be switched off.

In the literature, there are a number of examples that consider dynamic cell-size adjustments to reduce energy consumption. Among them, Niu et al. (2010) introduced the cell-zooming concept, which adaptively adjusts the size of the cells based on the current traffic load, to obtain energy savings. In their work, they used a cell-zooming server, which is a virtual entity in the network controls, to manage the cell-zooming procedure. The cell-zooming server collects information, such as the traffic load, channel conditions, and user requirements, and subsequently determines if there are opportunities for cell-zooming. They also proposed centralized and distributed versions of user association algorithms for cell-zooming. Another study that considered variable cell sizes for energy savings is Bhaumik et al. (2010). In this work, Bhaumik et al. considered two types of BSs: subsidiary BSs with a low transmission power, and umbrella BSs with a high transmission power. They proposed a self-operating network that adaptively turns subsidiary and umbrella BSs on and off based on the current traffic demands. Similarly, Kokkinogenis and Koutitas (2012) assumed a cellular network consisting of micro- and macro-BSs, where micro-BSs can be switched on/off, while macro-BSs can iteratively adjust their transmission power until the required quality of service (QoS) is achieved. They proposed static centralized, dynamic distributed, and hybrid topology management schemes to reduce the overall energy consumption of the network while satisfying certain QoS requirements. Figure 11.1 provides a summary of related works that have investigated the possibility of reducing energy consumption via the switch-off approach.

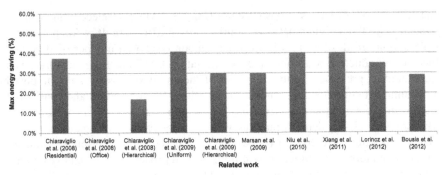

FIGURE 11.1

Summary of previous studies that have investigated the possibility of energy savings.

11.3 MECHANISM OF PROPOSED CELL-SWITCHING SCHEME

In this section, we present an algorithm that is proposed to achieve reduced energy consumption in cellular networks by reducing the number and size of active macro-cells based on traffic conditions. Figure 11.2 shows the traffic load that is considered in this simulation.

If the traffic load in a given area is low, several cells will be switched off, and the other cells will provide radio coverage and service for the entire region. The question that naturally arises is which cells should be switched off, and which cells should remain active? In this work, the decision to determine which cells remain active

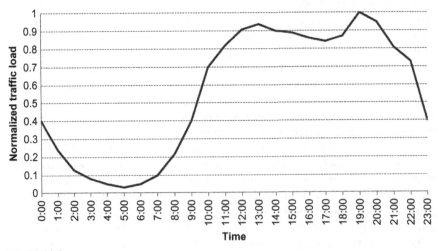

FIGURE 11.2

The daily traffic-load pattern of the eNB (Chen et al., 2011).

depends on two considerations. The first consideration is the ease with which radio coverage can be provided to neighboring cells to guarantee service. The second consideration is that the largest possible number of neighboring cells should be switched off to obtain the greatest decrease in energy requirements. The cells that satisfy these conditions are located in the middle of a cluster (and are called *master cells* in this study, indicated in Figure 11.3 as gray cells) and can easily provide coverage to neighboring cells that will be switched off later. Herein, two cases are discussed:

1. The normal case of a high-traffic load ($0.4 < \lambda \leq 1$): The network consists of 29 identical cells, which provide coverage to a 42.38 km^2 area, where each cell has an optimum radius of 750 m. The radius is chosen based on two considerations: (a) the small radius of the cell or the small cell-size results in a radio transmitter (eNB) and receiver (UEs) that are closer to each other, which mitigates phenomena that affect the properties of the received signal, such as propagation path loss, multipath fading, and shadow fading, which are due to the higher SINR and the spectral efficiency. (b) The cost of the infrastructure of a wireless system, which is linearly proportional to the number of eNBs; therefore, the macrocells yield the lowest cost for many scenarios because fewer eNBs are used to provide coverage for a large area, indicating that coverage (i.e., the cell range) is an important parameter in designing wireless systems (Johansson et al., 2004).

   In this case, all cells in the network (29 cells) are active, and the BSs operate with full functionality to provide the full coverage needed to guarantee radio service, as

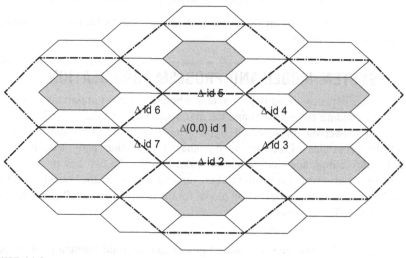

FIGURE 11.3

Cellular network structure. (The solid black cells represent a normal case with $R_{org} = 750$ m, and the dash-dot black cells represent low-traffic loads with $R = 2R_{org} = 1.5$ km.)

shown in Figure 11.3 for the cells framed in solid black. However, this category of traffic continues for only 13 h (10 a.m. to 11 p.m.), as shown in Figure 11.2. After this period, the mobile traffic becomes less than 0.4, which represents a low-traffic-load case. This is the focus of this study: we seek to achieve a balance between reducing energy consumption in the network and maximizing coverage to guarantee radio service and to achieve energy savings.

2. A low-traffic load ($0 < \lambda \le 0.4$): It is well known that power consumption grows proportionally with the number of cells. In this case, the number of cells is reduced by 75.86% (22 cells will be switched off) by increasing the area coverage of the cells that are located in the middle of the cluster, that is, the gray *master cells* in Figure 11.3, which provide coverage for the whole area. The following paragraph discusses the mechanism of the cell-switching process.

During operation, the cells are monitoring the traffic load and are able to switch off when the traffic load drops below a certain threshold and stays below the threshold for a certain period of time. At this time, the master cells inform their six neighboring eNBs in the same cluster to prepare to switch off by sending a multicast control signal via an X2 signaling interface. Upon receipt of the switch-off control signal, neighboring cells immediately begin to gradually decrease their transmission power, and their resident UEs will inter-handover to a master cell based on a stronger UE-eNB path. As the master cells monitor the traffic load when the traffic load increases, they will send wake-up control messages via X2 signaling to the neighboring cells to signal these cells to return to operation. The neighboring cells will begin gradually increasing their transmission power, and the resident UEs in the network will inter-handover to these cells based on a stronger UE-eNB path.

The following section will highlight the mathematical model for the cell-coverage-area optimization problem.

11.4 SYSTEM MODEL AND PROBLEM FORMULATION

The coverage depends on many parameters, whereby the most important parameters are the surrounding environment and the maximum radius of the cell, which have a significant impact on the received signal. However, three phenomena primarily affect the properties of the received signal: propagation path loss, multipath (small-scale) fading, and shadow (large-scale) fading. The two key properties are modeled as a zero-mean Gaussian random variable with a variance on a logarithmic scale. Therefore, a basic propagation model for the received power (P_{rx}) can be written as follows (Debus, 2006):

$$P_{rx}(\text{dBm}) = P_{tx} + G - L_{Hata} - \sigma \tag{11.1}$$

where P_{tx} and G denote the transmitted power and the total antenna gain, respectively; L_{Hata} represents the Hata path-loss model; and σ is the shadow-fading margin.

The Hata path-loss model is expressed as a function that includes the frequency (f), eNB antenna height (h_b), UE antenna height (h_m), and radius of the cell (R). The basic formula for the Hata path loss is (Debus, 2006)

$$L_{\text{Hata}} \, (\text{dB}) = 69.55 + 26.16 \log_{10}(f_{\text{MHz}}) - 13.82 \log_{10}(h_b) - a(h_m)$$
$$+ [44.9 - 6.55 \log_{10}(h_b)] \times \log_{10}(R_{\text{km}}) - K_{\text{suburban}} \tag{11.2}$$

$$a(h_m) = [1.1 \log_{10}(f_{\text{MHz}}) - 0.7] \times h_m - [1.56 \log_{10}(f_{\text{MHz}}) - 0.8] \tag{11.3}$$

$$K_{\text{suburban}} = 4.78 \, [\log_{10}(f_{\text{MHz}})]^2 - 18.33 \log_{10}(f_{\text{MHz}}) + 40.94 \tag{11.4}$$

Each radio receiver can only detect and decode signals with strengths greater than the receiver sensitivity power (P_{min}), which is another parameter affecting the availability of the coverage, especially at the cell boundary. P_{min} can be expressed as follows (Sesia et al., 2011):

$$P_{\text{min}} \, (\text{dBm}) = N_o \text{BW} + N_f + \text{SINR} + \text{IM} \tag{11.5}$$

where $N_o\text{BW}$ represents the thermal noise level for a specified noise bandwidth, N_f is the noise figure for the receiver, SINR is the signal-to-interference ratio, and IM is the implementation margin.

The cell-coverage area in a cellular system is defined as the percentage of the area within a cell that has received a signal at a power above a given minimum P_{min}. A cell requires some minimum received SINR for acceptable performance; the SINR requirement translates to a minimum P_{min} throughout the cell. The transmission power at the BS is designed for an average received power at the cell boundary of P_{min}. However, random shadowing and path loss will cause some locations within the cell to have a received power below P_{min}. According to Goldsmith (2005), the closed form for the cell coverage can be expressed as follows:

$$C \, (\%) = Q(a) + \exp\left(\frac{2 - 2ab}{b^2}\right) Q\left(\frac{2 - ab}{b}\right) \tag{11.6}$$

$$a = \left(\frac{P_{\text{min}} - P_{\text{rx}}(r)}{\sigma_\varphi}\right), \quad b = \left(\frac{10\alpha \log_{10}(\exp)}{\sigma_\varphi}\right) \tag{11.7}$$

where σ_φ is the standard deviation of the shadow fading and α is a path-loss exponent. In Equation (11.6), the cell-coverage area is expressed as a function $C = f(a, b) = f(P_{\text{min}}, P_{\text{rx}}, \alpha, \sigma_\varphi)$, where the minimum received power is expressed as a function $P_{\text{min}} = f(N_o, \text{BW}, N_f, \text{SINR}, \text{IM})$ and where the received power is $P_{\text{rx}} = f(P_{\text{tx}}, G, L, \sigma)$. In this study, a PSO has been adopted to maximize coverage under the constraints of P_{tx}, G, BW, SINR, and σ. The problem formulation is described as follows:

$$(p:) \underset{P_{\text{tx}}, G, \text{BW}, \text{SINR}, \sigma}{\text{maximize}} \left[Q(a) + \exp\left(\frac{2 - 2ab}{b^2}\right) Q\left(\frac{2 - ab}{b}\right) \right] \tag{11.8}$$

subject to the following constraints:

$$0 < P_{\text{tx}} \leq P_{\text{tx}}^{\text{max}} \tag{11.9}$$

$$G_{min} \leq G \leq G_{max} \qquad (11.10)$$

$$BW_{min} \leq BW \leq BW_{max} \qquad (11.11)$$

$$SINR_{min} \leq SINR \leq SINR_{max} \qquad (11.12)$$

$$\sigma_{min} \leq \sigma \leq \sigma_{max} \qquad (11.13)$$

It is clear that the problem posed by Equation (11.8) is a nonlinear optimization problem. A PSO algorithm is used to solve the optimization problem.

11.5 PSO ALGORITHM

A PSO is a population-based algorithm used for determining optimal solutions. This algorithm has several advantages, such as lower computational costs, better performance, and fewer adjustable parameters compared to other global optimization algorithms (Kennedy and Eberhart, 1995), which is the motivation behind the adoption of this algorithm to solve the nonlinear problem posed by Equation (11.8).

A PSO is initialized using a group of randomly positioned particles, and it subsequently searches for an optimal point; the position and velocity of each particle in the swarm, with N decision parameters in the optimization problem, are defined as $X_i = (x_{i1}, x_{i2}, \ldots, x_{in})$ and $V_i = (v_{i1}, v_{i2}, \ldots, v_{in})$, respectively. The best previous position of each particle is defined as $P_i = (p_{i1}, p_{i2}, \ldots, p_{in})$, and the global best position of all particles is represented by $P_g = (p_{g1}, p_{g2}, \ldots, p_{gn})$. Therefore, the velocity and position of each particle are updated as follows:

$$v_{new} = w \times v_{old} + c_1 \times r_1 (p_{in} - x_{in}) + c_2 \times r_2 (p_{gn} - x_{in}) \qquad (11.14)$$

where w is the inertia weight; r_1 and r_2 are random numbers, which are usually chosen between [0, 1]; c_1 is the self-recognition component coefficient, which is a positive constant; c_2 is the social component coefficient, which is a positive constant; and the choice of the values $c_1 = c_2 = 2$ is generally referred to as the learning factor. The following weighting function is usually utilized in Equation (11.14):

$$w = \frac{w_{max} - [(w_{max} - w_{min}) \times iter]}{iter_{max}} \qquad (11.15)$$

where w_{max} is the initial weight, usually chosen as a large value but less than 1; w_{min} is the final weight; iter is the current iteration number; and $iter_{max}$ is the maximum iteration number. A large w enables a global search, whereas a small w enables a local search. Linearly decreasing the inertia weight from a relatively large value to a small value through the course of the PSO run gives the best PSO performance comparisons with fixed inertia weight settings.

From Equation (11.14), a particle decides where to move next, considering its own experience, that is, the memory of its best past position, and the experience of the most successful particle in the swarm. The new position is then determined using the previous position and the new velocity and can be written as

$$x_{in\_new} = x_{in\_old} + v_{new} \tag{11.16}$$

Figure 11.4 illustrates the particle position update process.

Based on these expressions, the pseudocode of the PSO algorithm is as in Figure 11.5.

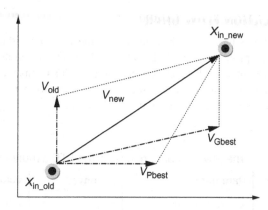

FIGURE 11.4

Updating particle position.

Algorithm of the proposed scheme
1: Initial population of particles with random positions (X_i) and velocities (V_i)
2: Evaluate the initial fitness values $f(X_i)$ of each particle
3: Store the best initial fitness value, Pbest (P_i), and Gbest (P_g).
4: **while** i< iteration **do**
5:     r_1 = rand (); r_2 = rand (); $c_1 = c_2 = 2$; $w_{min} = 0.4$; $w_{max} = 0.9$;
6:     calculate w according to Eq. (15),
7:     update V_i according to Eq. (14),
8:     update X_i according to Eq. (16),
9:     compare the position of each particle under the constrains
10:       **if** (X_i> max. constrains) **then**
11:         X_i = max. constrains
12:       **end if**
10:       **if** (X_i < min. constrains) **then**
11:         X_i = min. constrains
15:       **end if**
16:     Evaluate a new fitness values $f(X_i)$ for each particle
17:     Compare each particle's fitness evaluation with the current particles to obtain the individual best position
18:     Compare fitness evaluation with the population's overall previous best to obtain the global best position
19: **end while**

FIGURE 11.5

Pseudocode of the considered PSO algorithm.

11.6 SIMULATION RESULTS AND DISCUSSION

11.6.1 SIMULATION SETUP

The network consists of a set of macrocells characterized by the same coverage radius. Table 11.1 summarizes the simulation parameters.

11.6.2 SIMULATION FLOW CHART

Figure 11.6 illustrates the operational process of the PSO algorithm that has been applied to solve the problem posed in Equation (11.8). The initial positions are randomly selected, where at each iterative step, the particles exchange information to update their movements toward the global best position to satisfy the maximum coverage under the constraints of P_{tx}, G, BW, SINR, and σ.

Table 11.1 List of the physical parameters of the LTE simulation

Item	Parameter	Acronym	Value	Unit
Network parameters	Carrier frequency	f_c	2.6	GHz
	LTE bandwidth	BW	1.4-20	MHz
	Max. cell radius	R	1.5	km
	Max. number of cells	cell	29	#
Base station parameters	Max. BS transmission power	P_{tx}^{max}	46	dBm
	eNodeB antenna height	h_b	10	m
	Tx antenna gain	G	5-10	dB
	Number of antennas	N_{ant}	2	#
	Max. total power consumption	$P_{max\_tot}^{BS}$	965	W
Mobile station parameters	Thermal noise density	N_o	174	dBm/Hz
	Noise figure	N_f	9	dB
	Implementation margin	IM	3	dB
	UE antenna height	h_m	1.5	m
Propagation losses	Morphology	Suburban		
	Propagation model	Hata path-loss model		
	SINR	$SINR_{min}$	−5.1	dB
		$SINR_{max}$	18.6	
	Shadow-fading margin	σ	4-8	dB
	Exponent path loss	α	3.7	#
	Standard deviation of the shadow	σ_φ	4	dB

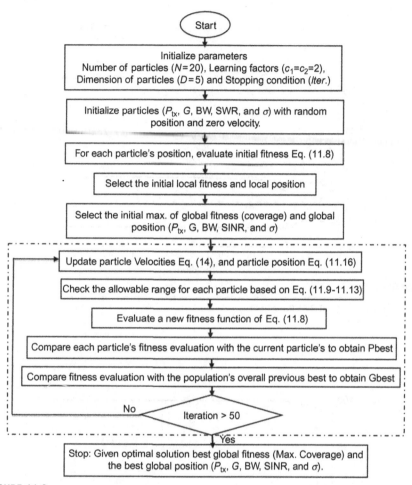

FIGURE 11.6

Flow chart of the operational process of the PSO that has been used to solve the problem.

11.6.3 RESULTS AND DISCUSSION

The PSO has been used to maximize coverage under five different constraints: (i) the transmission power of the eNB (P_{tx}), (ii) the total antenna gain (G), (iii) BW, (iv) the SINR, and (v) shadow fading (σ). The impact of these parameters on the coverage is shown in Figure 11.7. It is clear from Figure 11.7 that the coverage increases when some of these parameters are increased and others are decreased to balance and maintain the maximum coverage.

The most important parameters concerning maintaining coverage at the edge of the cell, where the SINR is low and where the shadowing is high, are the P_{tx} and G. When these parameters increase, the coverage also increases, as shown in Figure 11.8.

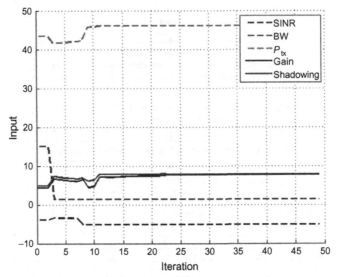

FIGURE 11.7

The behavior of constraint parameters.

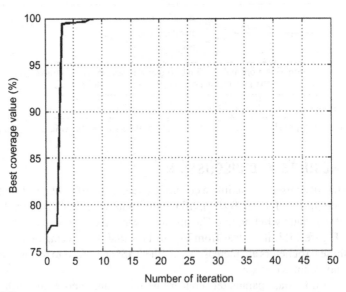

FIGURE 11.8

The behavior of fitness function—coverage.

The optimal transmission power P_{tx} and the antenna gain G achieved a maximum coverage at a maximum radius of 1.5 km, where the SINR was the lowest, at -5.1 dB, and the shadowing, at 7.9 dB, is 46 dBm and 7.7 dB. Because the BW is proportional to the noise power, the optimum BW at the edge is 1.4 MHz.

For the downlink-data transmissions in an LTE network, the eNB typically selects the modulation and coding scheme (MCS) based on the channel quality indicator feedback characteristics of the UE's receiver, that is, the SINR via a link-adaptation procedure. The SINR requirement translates into a minimum received power P_{min} throughout the cell. Figure 11.9 shows the relation, based on assumptions made in a reference (Sesia et al., 2011), between the radius of the cell P_{min} and the MCS.

It is clear that when the P_{min} decreases, the MCS decreases because the demodulation error rate increases as a result of the increase in both the noise and the interference that often occurs at the edge of a cell. Low-order modulation, such as quadrature phase shift keying (QPSK), is more robust and can tolerate higher levels of interference but provides a lower transmission bit rate, whereas the high-order modulation 64-quadrature amplitude modulation (QAM) offers a higher bit rate but is more susceptible to errors because of its higher sensitivity to interference, noise, and channel estimation errors. Therefore, the high-order modulation 64-QAM is only useful for a sufficiently high SINR. However, the bit rate and data rate depend on the MCS, BW, and the number of antennas. For any system, the data rate is

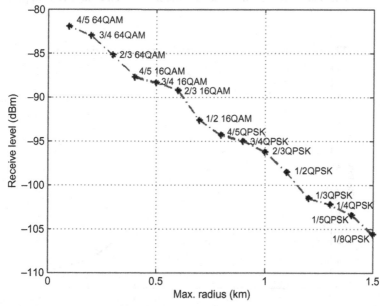

FIGURE 11.9

Cell radii versus receiver sensitivity power for different MCSs, with $P_{tx}=46$ dBm and BW$=1.4$ MHz.

calculated in symbols per second. Furthermore, it is converted into bits per second based on how many bits a symbol can carry, which is dependent on the MCS. It is known that increasing the radius can cause the data rate to decrease because of the low SINR, high path loss, and low MCS level.

Figure 11.10 shows the data rate versus the cell radii, which suggests that when the bandwidth is fixed at 1.4 MHz, increasing the cell radius decreases the number of transferred bits. This study focuses on a cell radius of 1.5 km, which corresponds to that of a cell in a low-traffic case; the lowest modulation rate (QPSK) supports a 1.5-km cell radius. For LTE with a 1.4 MHz BW, this means that there are six resource blocks (RBs), each RB has 12 subcarriers, each subcarrier has seven symbols for a normal cyclic prefix (CP), and the time of the slot is 0.5 ms. Hence, the total number of symbols per RB is $12 \times 7 \times 2 = 168$ symbols per ms; therefore, there are 1008 symbols per ms. When 1/8 QPSK is used (2 bits per symbol), the data rate will be 252 kbps for a single chain, and with 2×2 multiple-input and multiple-output (MIMO) (2T, 2R), the data rate will be two times that of a single chain, that is, 504 kbps.

The number of bits transmitted per joule of energy is the EE. Figure 11.11 shows the EE performance versus the cell radii, where the optimal transmission power of an eNB is 46 dBm, or 40 W. Hence, EE at a cell radius of 1.5 km is 504 kbps $\times$ 0.025 = 12.6 kb/J.

Overall, a high bandwidth can result in greater EE than can a low bandwidth within the same size coverage area because the higher bandwidth can support more

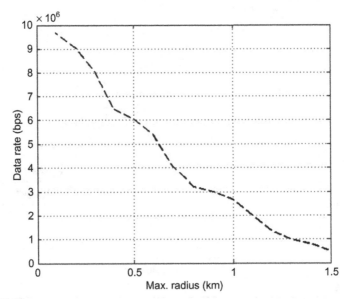

FIGURE 11.10

Data rate versus macrocell radii, with $P_{tx} = 46$ dBm and BW$=1.4$ MHz.

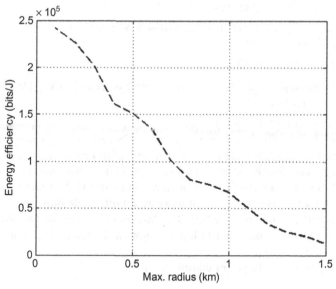

FIGURE 11.11

Energy efficiency versus macrocell radii, with $P_{tx}=46$ dBm and BW $=1.4$ MHz.

RBs; thus, a higher data rate can be achieved. However, designing efficient power management is challenging because of the compromises that must be made between power savings and network performance, that is, high data rates.

11.6.4 ENERGY AND OPEX SAVINGS

Finally, we will briefly discuss both aspects of the operational expenditure (OPEX) and energy savings that are achieved by our proposed scheme. According to Alsharif et al. (2014), the total power consumption of an eNB, $P_{tot}^{macro\_BS}$, is 965 W, and the retail price for energy in Malaysia is RM 0.380/kWh (Malaysian Energy Corporation, 2014). In addition, the emission factor for Malaysia is 0.67552 tCO_2/kWh (Ramli et al., 2013). Suppose that an active cell operates for hr hours per day. The network-wide power consumption can be expressed as

$$P_{Cons}^{day} (\text{W}) = hr_{\text{high\_traffic}}(a) + hr_{\text{low\_traffic}}(a) \qquad (11.17)$$

$$a = N_{BS}^{\text{active}} \times P_{tot}^{macro\_BS} \qquad (11.18)$$

- *Without the switch-off scheme*: All cells (29 cells) are active for 24 h
  a. Compute the daily power consumption
  $$P_{Cons}^{day} = \left(N_{BS}^{\text{active}} \times P_{tot}^{macro\_BS}\right) \times \text{day time} = (29 \times 965) \times 24 = 671.64 \, \text{kW/day}.$$

b. Compute the daily energy cost

$$\text{Cost} = P_{\text{Cons}}^{\text{day}}(\text{kW}) \times \text{energy price } (\text{RM/kW}) = 671.64 \times 0.380$$
$$= \text{RM}255.22 = \$79.^{[1]}$$

c. Compute the daily CO_2 emission

$$CO_2 \text{ emission} = P_{\text{Cons}}^{\text{day}}(\text{kW}) \times \text{Malaysia } CO_2 \text{ emission } (tCO_2/\text{kW})$$
$$= 671.64 \times 0.67552 = 453.71 tCO_2.$$

- *With the proposed switch-off scheme*:

  For a high-traffic load scenario, all cells (29 cells, from Figure 11.3) in the network are active, and the BSs will work with full functionality to provide full coverage as needed to guarantee radio service. This level of traffic load continues for only 13 h (10 a.m. to 11 p.m.), as shown in Figure 11.2. However, in the case of a low-traffic load, 22 cells will be switched off, while only seven cells, the "master cells," will remain active. The "master cells" operate for 24 h (13 h during high-traffic loads and 11 h during low-traffic loads) to provide coverage for the entire region.

  a. Compute the daily power consumption

$$P_{\text{Cons}}^{\text{day}} = \underbrace{13 \ (29 \times 965)}_{\substack{0.4 < \lambda \leq 1 \\ \text{High Traffic}}} + \underbrace{11 \ (7 \times 965)}_{\substack{0 < \lambda \leq 0.4 \\ \text{Low Traffic}}} = 438.11 \text{ kW/day.}$$

  b. Compute the daily energy cost

$$\text{Cost} = P_{\text{Cons}}^{\text{day}}(\text{kW}) \times \text{energy price } (\text{RM/kW}) = 438.11 \times 0.380$$
$$= \text{RM}166.48 = \$51.54.$$

  c. Compute the daily CO_2 emission

$$CO_2 \text{ emission} = 438.11 \times 0.67552 = 295.95 tCO_2.$$

  By applying the proposed switch-off scheme to the cellular network, the energy savings that can be achieved reached 233.53 kW per day, which translates into a 34.77% power reduction, which is in addition to the daily OPEX savings.

11.7 CONCLUSION

In this chapter, we investigated the possibility of reducing the energy consumption of an LTE network based on a switching-on/-off approach, in which cells are switched off during low-traffic periods, while taking into account the cell-coverage area. A PSO technique has been adopted to maximize coverage by increasing the size of some cells to provide coverage for cells that are switched off during low-traffic periods in order to achieve energy savings at the level of the network under the

---

[1]Where the foreign exchange rate is 1 USD$=3.23$ MYR (June 3, 2014).

constraints of parameters that affect the cell-coverage area: the transmission power of an eNB, the total antenna gain, the bandwidth, the SINR, and shadow fading. The simulation results show that when the cell radius increases, the shadowing increases and the SINR decreases, translating into minimum received power, which may impact the detected and decoded signals. However, the transmitted power and antenna gain maintain high coverage at the edge of the cell. This study demonstrated that an energy savings of up to 34.77% could be achieved in addition to the lower OPEXs under a maximum coverage condition.

ACKNOWLEDGMENT

The authors would like to thank the Universiti Kebangsaan Malaysia for the financial support of this work under the Economic Transformation Programme, ETP-2013-072.

REFERENCES

Alsharif, H.M., Nordin, R., Ismail, M., 2013. Survey of green radio communications networks: techniques and recent advances. J. Comput. Netw. Commun. 2013, 1–13.

Alsharif, H.M., Nordin, R., Ismail, M., 2014. Classification, recent advances and research challenges in energy efficient cellular networks. Wirel. Pers. Commun. 77 (2), 1249–1269.

Bhaumik, S., Narlikar, G., Chattopadhyay, S., Kanugovi, S., 2010. Breathe to stay cool: adjusting cell sizes to reduce energy consumption. In: Proceedings of the First ACM SIGCOMM Workshop on Green Networking 2013, New Delhi, pp. 41–46.

Bousia, A., Antonopoulos, A., Alonso, L., Verikoukis, C., 2012. "Green" distance-aware base station sleeping algorithm in LTE-advanced. In: Proceedings of IEEE International Conference on Communications (ICC), Ottawa, ON, pp. 1347–1351.

Chen, T., Yang, Y., Zhang, H., Kim, H., Horneman, K., 2011. Network energy saving technologies for green wireless access networks. IEEE Wirel. Commun. 18 (5), 30–38.

Chiaraviglio, L., Ciullo, D., Meo, M., Marsan, M.A., Torino, I., 2008. Energy-aware UMTS access networks. In: Proceedings of IEEE International Conference on W-GREEN, pp. 1–8.

Chiaraviglio, L., Ciullo, D., Meo, M., Marsan, M.A., 2009. Energy-efficient management of UMTS access network. In: Proceedings of IEEE International Conference on Teletraffic Congress, Paris, France, pp. 1–8.

Debus, W., 2006. RF Path Loss & Transmission Distance Calculations. Axonn, LLC, New York.

Ericsson Corporation, 2013. Ericsson mobility report on the pulse of the networked society. Ericsson Corporation report. Available: http://www.ericsson.com/res/docs/2013/ericsson-mobility-report-june-2013.pdf.

Goldsmith, A., 2005. Wireless Communications. Cambridge University Press, Cambridge.

Gong, J., Zhou, S., Niu, Z., Yang, P., 2010. Traffic-aware base station sleeping in dense cellular networks. In: Proceedings of International Workshop on Quality of Service (IWQoS), Beijing, pp. 1–2.

Johansson, K., Furuskar, A., Karlsson, P., Zander, J., 2004. Relation between base station characteristics and cost structure in cellular systems. In: Proceedings of IEEE International Symposium on Personal, Indoor and Mobile Radio Communications (PIMRC), pp. 2627–2631.

Kennedy, J., Eberhart, R.C., 1995. Particle swarm optimization. In: Proceedings of IEEE International Conference on Neural Networks, Piscataway, NJ, pp. 1942–1948.

Kokkinogenis, S., Koutitas, G., 2012. Dynamic and static base station management schemes for cellular networks. In: Proceedings of IEEE Global Communications Conference (GLOBECOM), Anaheim, pp. 3443–3448.

Lorincz, J., Capone, A., Begusic, D., 2012. Impact of service rates and base station switching granularity on energy consumption of cellular networks. EURASIP J. Wirel. Commun. Netw. 2012 (1), 1–24.

Malaysian Energy Corporation. Available: http://www.tnb.com.my/business/for-industrial/pricing-tariff.html (accessed 11.06.14.).

Marsan, M.A., Chiaraviglio, L., Ciullo, D., Meo, M., 2009. Optimal energy savings in cellular access networks. In: Proceedings of IEEE International Conference on Communications Workshops, Dresden, pp. 1–5.

Niu, Z., Wu, Y., Gong, J., Yang, Z., 2010. Cell zooming for cost-efficient green cellular networks. IEEE Commun. Mag. 48 (11), 74–79.

Ramli, S.S., Hanapei, M.H., Yahya, M.R., Omar, N.A., Almsafir, M.K., 2013. The impact of Gpon technology on power consumption and carbon footprint in Malaysia. J. Purity Util. React. Environ. 2 (2), 1–8.

Sesia, S., Toufik, I., Baker, M., 2011. LTE: The UMTS Long Term Evolution. John Wiley & Sons, New York.

Suarez, L., Nuaymi, L., Bonnin, J.M., 2012. An overview and classification of research approaches in green wireless networks. EURASIP J. Wirel. Commun. Netw. 2012 (1), 1–18.

Xiang, L., Pantisano, F., Verdone, R., Ge, X., Chen, M., 2011. Adaptive traffic load-balancing for green cellular networks. In: Proceedings of IEEE International Symposium on Personal Indoor and Mobile Radio Communications (PIMRC), Toronto, pp. 41–45.

Zhou, S., Gong, J., Yang, Z., Niu, Z., Yang, P., 2009. Green mobile access network with dynamic base station energy saving. In: Proceedings of ACM MobiCom, Beijing, pp. 10–12.

Bio-Inspired Computation for Solving the Optimal Coverage Problem in Wireless Sensor Networks: A Binary Particle Swarm Optimization Approach

12

Zhi-Hui Zhan[1,2,3,4] and Jun Zhang[1,2,3,4]

[1]*Department of Computer Science, Sun Yat-sen University, Guangzhou, China*
[2]*Key Laboratory of Machine Intelligence and Advanced Computing (Sun Yat-sen University), Ministry of Education, Guangzhou, China*
[3]*Engineering Research Center of Supercomputing Engineering Software (Sun Yat-sen University), Ministry of Education, Guangzhou, China*
[4]*Key Laboratory of Software Technology, Education Department of Guangdong Province, Guangzhou, China*

CHAPTER CONTENTS

12.1 INTRODUCTION

Wireless sensor network (WSN) is a very new technology that has become one of the hottest and most challenging research areas recently (Akyildiz et al., 2002; Kulkarni and Venayagamoorthy, 2011; Morsly et al., 2012; Kim and Lee, 2014). The WSN consists of lots of sensor nodes that monitor the area for specialized applications such as battlefield surveillance (He et al., 2004), habitat monitoring (Cerpa et al., 2001), environmental observation (Martinez et al., 2004), health applications (Heinzelman et al., 2004), and many others (Arampatzis et al., 2005; Kuorilehto et al., 2005). The environments of these applications are usually not friendly, and it is difficult to deploy the sensors determinately. Therefore, a lot of nodes are randomly deployed in the area, resulting in more sensors than required. The high density of sensors on the one hand compensates for the lack of exact positioning and improves the fault tolerance, while on the other hand may cause larger energy consumption due to conflicts in accessing the communication channels, maintaining information about neighboring nodes, and some other factors (Shih et al., 2001). Moreover, the sensor node is always equipped with a battery cell that has limited energy. Because the application environments are always not friendly or are difficult to approach, it is not a realistic way to recharge the battery when the energy exhausts. Therefore, research into optimally scheduling the sensor nodes and making the redundant nodes turned off to sleep, in order to save the energy to prolong the network lifetime, has become one of the most significant and promising topics in the WSN (Huang and Tseng, 2005; Cardei and Wu, 2006; Zhan et al., 2012).

A considerable amount of research has been devoted to addressing the energy-efficient problem in the WSN in order to prolong the network lifetime. Most of the research transforms this issue to the optimal coverage problem (OCP) (Huang and Tseng, 2005; Zhan et al., 2010). The OCP is based on the fact that the WSN contains a large number of sensor nodes. As a result, many nodes would share the same monitored regions; some of the nodes are redundant and can be turned off to preserve energy, while the others still work to offer full coverage. Activating only the necessary sensor nodes at any particular moment can save energy. Therefore, the OCP is a fundamental problem in WSN, with the objective of finding a minimal set of nodes to monitor the area, and turning off the redundant nodes to save energy, while at the same time meeting the coverage requirement. This way, not only can the nodes reduce the energy consumption caused by the nodes' confliction or the neighborhood communication, but the network lifetime can also be significantly prolonged because the nodes can be scheduled to work in turn (Cardei and Wu, 2006).

In the literature, different models and assumptions have been introduced to this problem, and various approaches have been proposed for solving the problem. In the works of Zhang and Hou (2005) and Xing et al. (2005), they proved that the full sensing coverage of the network can guarantee the connectivity of the network when the communication range is not shorter than twice the sensing range. As many kinds of wireless sensors can meet this condition (Zhang and Hou, 2005), most of the existing research only concentrates on the coverage problem of the network. Also, only the coverage problem is considered in this chapter. In the literature, approaches such as coverage-based off-duty eligibility rule (Tian and Georganas, 2003), time axis dividing node working schedule (Yan et al., 2003), and probing environment and adaptive sleeping (PEAS) protocol (Ye et al., 2003) have been proposed to address the OCP in finding out a minimal set of nodes to be active. Moreover, some approaches focused on dividing the original deployed sensor nodes into as many disjoint sets as possible and scheduling the sets to work in turn (Slijepcevic and Potkonjak, 2001; Cardie et al., 2002; Liu et al., 2006). Even though minimizing the work nodes can prolong the network lifetime, it is also interesting to investigate the division of the nodes, because the latter is more intuitive for prolonging the network lifetime (Slijepcevic and Potkonjak, 2001). In this chapter, our OCP model and proposed approaches can not only solve the minimizing-active-nodes problem, but also the maximizing-disjoint-sets problem (Zhan et al., 2012). Among the above state-of-the-art approaches, it should be noted that the approaches in Tian and Georganas (2003), Yan et al. (2003), and Slijepcevic and Potkonjak (2001) can guarantee full coverage, whereas the ones in Ye et al. (2003), Cardie et al. (2002), and Liu et al. (2006) failed to do so. Our model and approach are designed to guarantee full coverage.

The motivations for and contributions of our work include the following four aspects:

1. The WSN consists of lots of sensor nodes with very limited energy. It has been a promising and significant research area to solve the OCP in order to save sensor energy and prolong network lifetime.
2. The existing models and their approaches to the OCP are always not easy to understand or implement. Therefore, it is significant and promising to design a simple OCP model to describe the problem, and at the same time propose a simple but effective and efficient approach to solve the problem.
3. The OCP is nondeterministic polynomial-time complete (NP-complete) (Slijepcevic and Potkonjak, 2001; Zou and Chakrabarty, 2005), and some of the traditional approaches have the natural disadvantage of being trapped into local optimal. Therefore, we propose to use a swarm intelligence (SI) algorithm (Kennedy et al., 2001) to solve the problem. Because the SI algorithms are bio-inspired approaches that have strong global search ability, good adaptation, and robustness, it is expected that the SI algorithms can solve the OCP efficiently.
4. Some existing approaches cannot guarantee full coverage when dealing with the OCP (Ye et al., 2003; Cardie et al., 2002; Liu et al., 2006), resulting in disadvantages when used in practical applications. In our work, the model and approach are hence designed to guarantee full coverage.

In this chapter, the OCP is modeled as a 0/1 programming problem. Given a WSN topology with a set of randomly deployed sensor nodes, our model marks the nodes with values of 0 or 1, where 0 means that the node is turned off to sleep and 1 means that the node is active to work. In this way, the OCP is transformed to optimizing a 0/1 string, that is, minimizing the number of 1 and at the same time making sure that the nodes with value 1 can provide full coverage of the area. In order to solve this problem effectively and efficiently, a variant of the SI algorithm named particle swarm optimization (PSO) (Kennedy and Eberhart, 1995; Eberhart and Kennedy, 1995) is adopted. As the standard PSO was designed for the problems in a continuous domain, it is not suitable to directly use the standard PSO to solve the OCP. In our work, we use the discrete binary PSO (BPSO) (Kennedy and Eberhart, 1997) as an approach to solve the OCP. The BPSO approach is promising in solving discrete binary problems (Kennedy and Eberhart, 1997), and therefore it is also suitable for the OCP. Simulations are conducted to evaluate the performance of the proposed BPSO approach. Moreover, the genetic algorithm (GA) designed in our previous study (Zhan et al., 2010) for solving OCP is adopted to compare with the BPSO approach. Experimental results show that our BPSO approach wins both the GA and some state-of-the-art approaches in minimizing the active-sensor number in the WSN. Experiments are also conducted to evaluate the performance of the proposed approach in different scale and different density networks. The experimental results show that the BPSO approach is more robust and more efficient than the GA approach at minimizing the active-sensor number over different sensing ranges and different deployed-node numbers. More important, by extending the previous work in Zhan et al. (2010), another contribution of this chapter is to extend the BPSO approach to maximize the disjoint-set number of a WSN (Zhan et al., 2012). Therefore, this chapter has contributed in evaluating BPSO in solving both OCP and the disjoint-set maximization problem. Experimental results show that BPSO has very good performance in solving OCP and in maximizing the disjoint-set number when compared with the traditional heuristic and GA approach.

The rest of this chapter is organized as follows. Section 12.2 gives the problem formulations of the OCP in the WSN and reviews the related work on this problem. Moreover, the PSO framework is presented. Section 12.3 proposes our methodology that uses the BPSO approach to solve the OCP. Section 12.4 gives the experimental results and comparisons. Moreover, the search behaviors and scalabilities of the BPSO approach are investigated. At last, conclusions are summarized in Section 12.5.

12.2 OPTIMAL COVERAGE PROBLEM IN WSN
12.2.1 PROBLEM FORMULATION

Given an $L \times W$ (length $\times$ width) rectangle area A for monitoring and a great amount of N sensors randomly deployed in the area, using only a subset of M sensors from the N sensors to fully cover the monitored area, the OCP is to determine that the area can be fully covered by the original N sensors. (Because the N sensors are randomly deployed, the area may be not fully covered in the original network topology, so we do not consider this situation in our chapter.) The objective of the OCP is to minimize the number of M.

In order to know whether the area A is fully covered by the sensor network, we assume that the location of the sensor is prior known. This assumption is always used in the sensor network coverage research (Zhang and Hou, 2005; Xing et al., 2005; Tian and Georganas, 2003; Slijepcevic and Potkonjak, 2001; Cardie et al., 2002; Huang and Lu, 2012); for example, this information can be obtained through a low-power global positioning system (GPS) receiver (Meguerdichian et al., 2001a,b; Li et al., 2003), or we can obtain the position information by using the location algorithms (Meguerdichian et al., 2001a,b; Savarese et al., 2001; Patwari et al., 2003). Moreover, the area is divided into grids, and the coverage issue can be transformed to check whether each of the grids is covered by at least one active sensor (Tian and Georganas, 2003).

All the N sensors form the sensor set $S = \{s_1, s_2, \ldots, s_N\}$, where each sensor node s_i has the location (x_i, y_i) and the sensor radius R. For any grid point $g = (x, y) \in A$ in the monitored area, the relationship between the s_i and the g is defined as

$$P(s_i, g) = \begin{cases} 1 & \text{if } (x - x_i)^2 + (y - y_i)^2 \leq R^2 \\ 0 & \text{otherwise} \end{cases} \tag{12.1}$$

where 1 means that the grid g is covered by the sensor s_i, and 0 means the sensor s_i does not cover the grid g. Therefore, for any grid point g, if there exists at least one sensor $s_i (1 \leq i \leq N)$ that makes $P(s_i, g) = 1$, we say that g is covered by the sensor network. In this sense, the monitored area A is fully covered if any grid point g in the area is covered by the sensor network.

In the OCP, the area is monitored by an optimal selection subset $S^* = \{s_i, |s_i$ is selected, $1 \leq i \leq N\}$ with M sensors from the N sensors, with the constraint that the area A is still fully covered by the M sensors, and with the objective of minimizing the value of M, as

$$\begin{aligned} f = \min M \qquad &\text{where } M = |S^*|, \ S^* \subseteq S \\ \text{subject to } (\oplus_{s_i \in S^*} P(s_i, g)) = 1 \ &\forall g \in A \end{aligned} \tag{12.2}$$

Here, the operator $\oplus$ results in a value of 0 if all the elements are 0. Otherwise, the result is 1 if at least one of the elements is 1.

Therefore, the OCP can be modeled as a 0/1 programming problem to determine whether a sensor is selected (with the value 1) or is not selected (with the value 0). The model has the following features and advantages: (1) The 0/1 programming model is easy and intuitive to understand: 1 means the sensor is selected to active, while 0 means the sensor is scheduled to sleep. (2) Unlike some other scheduling algorithms and protocols, our OCP model does not need the neighborhood information of the sensor nodes. This makes the model robust to adapt to different network topologies. (3) The OCP model is naturally suitable to be solved by the evolutionary computation (EC) algorithms such as the GAs and the BPSO. Therefore, good performance can be expected by using the EC algorithms to solve the OCP.

12.2.2 RELATED WORK

The OCP in the WSN has received a lot of attention recently (Huang and Tseng, 2005). In the literature, the OCP may be formulated in various ways for different concerns or particular applications. Generally speaking, there are two types of concerns: one is to answer how well a given deployed sensor network monitors the area (Meguerdichian et al., 2001a,b; Li et al., 2003), and the other is to schedule the sensor nodes in order to prolong the network lifetime (Cardei and Wu, 2006). In this chapter, we focus on the latter concern. This concern deals with the energy-efficient problem (Cardei and Wu, 2006), where the main objective is to minimize the number of active sensor nodes and to maximize the network lifetime. In these applications, the central idea is to turn off some of the sensor nodes in order to conserve energy, while the other active sensor nodes can still meet the application requirements, that is, the full coverage for the monitored area.

Different algorithms and protocols have been proposed to solve the OCP to optimally schedule the sensor nodes. Tian and Georganas (2003) proposed a node-scheduling scheme for energy conservation by using the neighborhood information of nodes to determine whether the node is redundant. However, this approach may lead to excess energy consumption because some of the nodes may not be turned off, even though they are redundant. Yan et al. (2003) used another scheme to schedule the sensor nodes by dividing the time axis into rounds with equal duration, and activating only one of the nodes that share the common monitor area in a time round. Ye et al. (2003) proposed a PEAS protocol to schedule the sensor nodes. In PEAS, some of the nodes are working and some are sleeping. The working nodes will keep active until the energy is exhausted, while the sleeping nodes will wake up in a period of time and send a probing message. If no responses are received, it means that there are no other working nodes in the neighborhood, so the nodes turn to work; otherwise, the nodes continue to sleep. This protocol has very good extensibility and can be applied to a larger scale network. However, it cannot guarantee full coverage of the monitored area.

The above approaches only focus on minimizing the number of working nodes. Even though this objective has significant influence on the network lifetime, it is still interesting to research dividing the sensor nodes into disjoint sets and letting the sets work successively. Slijepcevic and Potkonjak (2001) may have been the first to address this problem. In their approach, the area is first divided into different fields, with each field containing the grids that are covered by the same set of sensor(s). Then, the approach uses a heuristic approach that gives more attention to the critical element (i.e., the field covered by the minimal number of sensor nodes) to determine a full coverage set. These nodes are then removed, and another pass is performed on the rest nodes to find another full coverage set. This process goes on until the rest nodes cannot provide full coverage for the area. Later, Cardie et al. (2002) proposed to use a graph-coloring mechanism to compute the disjoint dominate sets. However, their approach cannot guarantee full coverage. Also, Liu et al. (2006) used a random-allocation strategy to allocate the nodes to different disjoint sets. Nevertheless, their approach cannot always provide full coverage.

12.2.3 BIO-INSPIRED PSO

SI algorithms are biologically inspired optimization approaches that have fast developed in recent years and have been successfully applied to various real-world applications (Kennedy et al., 2001; Li et al., 2015). The most significant advantages of using SI optimization approaches are not only that they are flexible and adaptive to the problems, but also that they have strong global search ability and robust performance. Therefore, it is significant and promising to investigate the performance of the SI approaches in dealing with the OCP. In this chapter, we try to use the representative SI algorithms—particle swarm optimization—to solve the OCP.

The PSO algorithm was proposed by Eberhart and Kennedy in 1995. The particles cooperate to search for the global optimum in an N-dimension hyperspace. The current status of each particle i is denoted by a velocity vector $V_i = [v_{i1}, v_{i2}, \ldots, v_{iN}]$ and a position vector $X_i = [x_{i1}, x_{i2}, \ldots, x_{iN}]$. Moreover, the particle i will keep a vector $Pbest_i = [p_{i1}, p_{i2}, \ldots, p_{iN}]$ to store its personal historically best position found so far, that is, the position with the best fitness value that the particle has found. The best one of all the $Pbest_i$ in the whole population is regarded as $Gbest = [g_1, g_2, \ldots, g_N]$. In the process of the PSO algorithm, the V_i and X_i are initialized randomly and are updated as Equations (12.3) and (12.4) generation by the guidance of $Pbest_i$ and $Gbest$.

$$v_{ij} = v_{ij} + c_1 \times r_{1j} \times \left(p_{ij} - x_{ij}\right) + c_2 \times r_{2j} \times \left(g_j - x_{ij}\right) \tag{12.3}$$

$$x_{ij} = x_{ij} + v_{ij} \tag{12.4}$$

The c_1 and c_2 are acceleration coefficients, which are commonly set as 2.0 (Eberhart and Kennedy, 1995), or can be adaptively controlled according to different evolutionary states (Zhan et al., 2009). The r_{1j} and r_{2j} are two randomly generated values in range [0, 1] for the jth dimension. There is a parameter $V_{\max}$ with a positive value. This parameter is used to clamp the velocity in a certain range. That is, if the updated velocity v_{ij} is out of the range $[-V_{\max}, V_{\max}]$, it is set to the corresponding bound. The flowchart of PSO is shown as Figure 12.1.

12.3 BPSO FOR OCP

12.3.1 SOLUTION REPRESENTATION AND FITNESS FUNCTION

The OCP is modeled as a 0/1 programming problem, and therefore, the individual (i.e., a particle in the BPSO) is coded as a binary string of 0 and 1. The length of the binary string is the same as the sensor-node number N, and the representation is

$$X = [x_1, x_2, \ldots, x_N] \text{ where } x_j = 0 \text{ or } 1 \tag{12.5}$$

In Equation (12.5), $x_j = 1$ means that the jth sensor is selected to active, and $x_j = 0$ means that the jth sensor is selected to sleep.

The objective of the problem is to minimize the number of active sensor nodes. Therefore, the fitness function can be simply defined as Equation (12.2), as in Section 12.1.

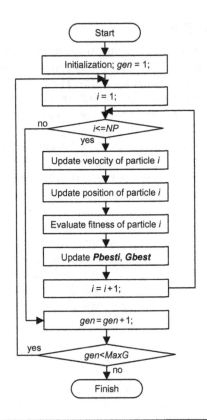

FIGURE 12.1

Flowchart of the PSO algorithm.

12.3.2 INITIALIZATION

In this phase, a check is first carried out to make sure that the area can be fully covered by the original sensor network with all the nodes active. Otherwise, the optimization process reports a fail result and terminates.

In the initialization, NP individuals are randomly generated. For each dimension j, the value of x_j is set as 0 or 1 randomly. Moreover, the velocity in BPSO is randomly initialized as a real value within the velocity range $[-V_{max}, V_{max}]$.

One thing to note is that the initialized individual X may be infeasible because it cannot provide full coverage of the area. In this case, a repair procedure would be performed on the individual to make it feasible. The pseudocode of the repair procedure is given as Figure 12.2 and is described as follows.

Given an infeasible individual X to be repaired, it is supposed that the X is feasible if the values of all the dimensions of X are 1. First, a random integer value k in range $[1, N]$ is generated as the starting dimension, where N is the total sensor-node number. Then, the procedure checks the dimensions of X from k to find the first dimension k^* with the value of 0. This dimension is forced to set as 1. Then the

```
01. //Input: A infeasible solution X
02. //Output: A feasible solution X
03. Procedure Repair
04. {
05.      int k=rand()%N;
06.      while (X is infeasible){
07.          while (xₖ==1){
08.              k=k+1;
09.                  if (k>=N) k=k−N;
10.          }//end of while (xₖ==1)
11.          xₖ=1;
12.      }//end of while (X is infeasible)
13. }
```

FIGURE 12.2

The pseudocode of the repair procedure.

new X is checked to see whether it is feasible. If the X is still infeasible, the procedure goes on finding another dimension k^{**} with the value of 0 and forces it to be 1. The procedure terminates until the X is feasible.

12.3.3 BPSO OPERATIONS

The BPSO-related operations include the velocity update and position update for each particle. These operations are described as follows, according to the implementation of the BPSO proposed in Kennedy and Eberhart (1997).

The BPSO performs the velocity update similar to Equation (12.3), which is used in the standard PSO. The only difference is that the value for x_{ij}, p_{ij}, and g_j is 0 or 1. Also, the velocity is to be clamped within the range of $[-V_{max}, V_{max}]$. The jth dimension of velocity, v_{ij}, is regarded as the probability of the jth position x_{ij} being the value of 1. Therefore, the value of v_{ij} should be mapped into the interval of [0.0, 1.0]. In the proposal of Kennedy and Eberhart (1997), the sigmoid function is used to obtain this transformation:

$$p_{ij} = \text{sigmoid}(v_{ij}) = \frac{1}{1+e^{-v_{ij}}} \tag{12.6}$$

With the value of p_{ij} obtained, the BPSO performs the position update as

$$x_{ij} = \begin{cases} 1 & \text{if rand}() < p_{ij} \\ 0 & \text{otherwise} \end{cases} \tag{12.7}$$

where rand() is a random value generated from a uniform distribution in the range [0.0, 1.0].

After the velocity- and position-update operations, the particle will be evaluated. The evaluation operation is to count the sensor nodes that the particle uses to fully cover the area, that is, the number of nodes with value 1. It should be noted that some particles may be infeasible after the evolutionary operations; that is, the solution represented by the particle cannot fully cover the area. In this case, we do not repair the particle if its current position X is infeasible. Instead, we do not evaluate the infeasible particle. Because we have made sure that all the particles are initialized feasibly, all the particles will store a feasible solution in their **Pbest** vectors. Therefore, all the particles will eventually return to the search range because they are attracted by the feasible guidance **Pbest** and **Gbest**. Using such a strategy can save a lot of computational time by avoiding the repair procedure.

12.3.4 MAXIMIZING THE DISJOINT SETS

In the practical applications, it is also interesting to find out the maximal disjoint sets of a given sensor network (Zhan et al., 2012). When the maximal disjoint sets are known, the sets can be scheduled to work successively to prolong the network lifetime.

Our OCP model is very suitable to extend to the maximizing-disjoint-set problems in the WSN. This is because minimizing the number of active sensor nodes has a direct impact on the number of disjoint sets. In order to maximize the disjoint sets, we use our BPSO approach to minimize the number of active sensor nodes. These active nodes form the first set. Then these nodes are marked as unavailable, and the rest sensor nodes form another network topology. The approach is performed on the new topology to minimize the number of active sensor nodes, forming a second set. A similar process goes on until the last network topology cannot provide full coverage for the area. In this sense, the number of maximal disjoint sets can be determined.

12.4 EXPERIMENTS AND COMPARISONS
12.4.1 ALGORITHM CONFIGURATIONS

The BPSO approach for solving the OCP is implemented and evaluated. The parameter configurations are described as follows. The population size is 40 and the maximal generation number is 200. The algorithm parameter V_{max} is 6, and acceleration coefficients c_1 and c_2 are both 2.0. These parameter configurations are determined by considering the commonly recommended values in the literature and our empirical study, as shown below.

The investigations on the parameters V_{max} and c_1, c_2 in BPSO are based on the network topology used in Tian and Georganas (2003). That is, the monitored area is a square field (50 m × 50 m), and the sensing range of the sensor node is 10 m. We randomly deploy 100 nodes in the network to make up the topology. Notice that the 100 nodes can provide full coverage. Otherwise, the nodes are deployed again randomly until the network is fully covered. In order to obtain the statistical sense results, we run our approach on 100 different random topologies to minimize the number of active nodes and evaluate the average performance.

The parameters V_{max} and c_1, c_2 in BPSO are investigated, and the results are plotted as Figure 12.3. Figure 12.3a is for the V_{max}. The curve shows that BPSO

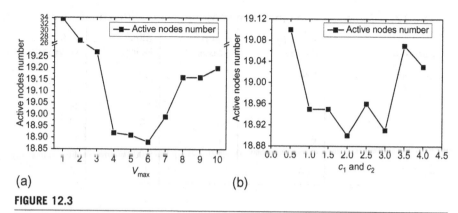

FIGURE 12.3

The BPSO performance based on different parameter configurations. (a) Results on different V_{max} and (b) results on different c_1 and c_2.

performance would be very poor when the V_{max} is too small, for example, set as 1, 2, or 3. Also, the value should not be set too large, for example, 8, 9, or 10. Here, the value 6 for the V_{max} is best for BPSO, which is the same as the one used in Kennedy and Eberhart (1997), and is therefore adopted in our experiments. On the other hand, the BPSO performance is not affected significantly by the values of c_1 and c_2. Nevertheless, Figure 12.3b shows that the value 2.0 is relatively good, and therefore is used in our experiments.

12.4.2 COMPARISONS WITH STATE-OF-THE-ART APPROACHES

In the literature, many approaches concentrate on finding a minimal set of active sensor nodes to monitor the area. Here, we take the representative state-of-the-art work in Tian and Georganas (2003) and Ye et al. (2003) for comparison. These two approaches are both aimed at minimizing the number of active sensor nodes. It should be noticed that the approach in Tian and Georganas (2003) ensures full coverage, while the one in Ye et al. (2003) cannot.

In order to make a fair comparison, we follow the network topology used in Tian and Georganas (2003), which was mentioned in the last subsection; that is, with the monitored area 50 m × 50 m, the sensing range 10 m, and the original-deployed-node number 100. We carried out the simulations on 100 different random topologies, and the mean results are compared.

We show the average number of sleep nodes obtained by the BPSO approach and the approaches of Tian and Georganas (2003) and Ye et al. (2003) in Table 12.1, where the data of approaches Tian and Georganas (2003) and Ye et al. (2003) are derived directly from Tian and Georganas (2003). It can be observed from the table that our BPSO approach has a strong search ability to identify the redundant nodes and let more nodes sleep to save energy. The performance of the PEAS in Ye et al. (2003) relies significantly on the probing range. The sleep-node number

Table 12.1 Comparisons on the sleep-node number of different approaches

Approach	Probing Range	Sleep Nodes	Blind Points	Full Coverage
Tian and Georganas (2003)	N/A	53	0	Y
	3	38	13	N
	4	54	26	N
PEAS Ye et al. (2003)	5	66	68	N
	6	71	91	N
	7	81	100	N
BPSO	N/A	81.12	0	Y

increases as the probing range increases. However, the larger probing range results in more blind points. The approach in Tian and Georganas (2003) can avoid the blind point, but it is not efficient enough to schedule the redundant nodes to sleep. Our proposed BPSO approach not only avoids the blind point, but is also efficient in turning off the redundant nodes to save energy.

12.4.3 COMPARISONS WITH THE GA APPROACH

Because of the stochastic nature of the SI-optimization approaches, it is interesting to investigate the search behaviors of the BPSO approach in solving different network topologies with different node densities and sensing ranges. Therefore, more experiments are conducted in this subsection. In order to demonstrate the robust advantages of BPSO in solving the OCP, the GA approach that was designed for solving OCP in our previous study (Zhan et al., 2010) is adopted herein for comparison. The parameters of the GA are set according to those in Zhan et al. (2010).

In the investigations, the sensing range is set as 8 m, 10 m, and 12 m, while the original-deployed-node number is set as 100, 150, 200, 250, and 300. We test the approaches in 100 random topologies for each configuration, and the average results are presented in Table 12.2. The observations from Table 12.2 show that increasing the sensing range and increasing the number of original deployed nodes can both result in more nodes being scheduled to sleep. This result is consistent with our expectation.

Another interesting investigation is to see whether the number of active nodes remains constant over different numbers of original deployed nodes when the sensing range is fixed. The data in Table 12.3 and the curves in Figure 12.4a show that the behavior of the GA approach does not have a good ability to keep the number of active nodes in a dense network as small as the one in a sparse network. Instead, the number of active nodes increases as the original deployed nodes increase. Such a phenomenon was also observed in the coverage-based off-duty eligibility rule, which was proposed in Tian and Georganas (2003). The authors of Tian and

Table 12.2 The sleep-node number over different deployed-node numbers on different sensing ranges

Nodes	GA					BPSO				
Sensing Range	100	150	200	250	300	100	150	200	250	300
8 m	72.03	118.71	162.41	204.57	245.09	71.10	121.23	171.59	221.36	270.87
10 m	81.74	128.41	172.25	214.37	254.68	81.12	130.96	180.83	230.56	280.38
12 m	86.74	133.73	177.52	218.86	257.75	86.52	136.30	185.95	236.15	285.78

Table 12.3 The active-node number over different deployed-node numbers on different sensing ranges

Nodes	GA					BPSO				
Sensing Range	100	150	200	250	300	100	150	200	250	300
8 m	27.97	31.29	37.59	45.43	54.91	28.90	28.77	28.41	28.64	29.13
10 m	18.26	21.59	27.75	35.63	45.32	18.88	19.04	19.17	19.44	19.62
12 m	13.26	16.27	22.48	31.14	42.25	13.48	13.70	14.03	13.85	14.22

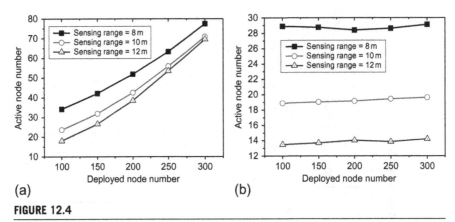

FIGURE 12.4

The active-node number over different sensing ranges and deployed-node numbers. (a) GA and (b) BPSO.

Georganas (2003) claimed that this was caused by the increasing of the edge nodes, which cannot be turned off in their approach. However, the GA and BPSO approaches seem to be less affected by the edge nodes. This can be demonstrated by the performance of BPSO, which has very good ability in keeping a small set of active nodes over different networks. As shown in Figure 12.4b, within the same sensing range, the number of active nodes almost does not change when the original-deployed-node number increases.

12.4.4 EXTENSIVE EXPERIMENTS ON DIFFERENT SCALE NETWORKS

In order to compare the scalability of the GA and BPSO approaches, we evaluate them over different scale networks with different areas, different sensing ranges, and different deployed sensor nodes. In the experiments, five different scales, with the area 100 m × 100 m, 200 m × 200 m, 300 m × 300 m, 400 m × 400 m, and 500 m × 500 m, are considered, where the sensing ranges are 20 m, 40 m, 60 m, 80 m, and 100 m, respectively. In each area, different nodes are deployed to form networks with different densities. The original-deployed-sensor-node numbers are 100, 150, 200, 250, and 300. Therefore, in all, there are $5 \times 5 = 25$ different network configurations in the experiments. In order to obtain results in the statistical sense, we randomly generate 100 networks for each network configuration and optimally schedule the redundant nodes by using both GA and BPSO. The parameters of BPSO are the same as the ones specified in Section 12.4.1. The mean results of the 100 networks of each network configuration are presented, plotted, and compared.

The experimental results of GA and BPSO in solving different scale networks are compared in Table 12.4. The data show that when the sensor nodes are scheduled by using GA, the active-node number increases as the deployed-node number increases.

Table 12.4 Experimental results of GA and BPSO in solving different scale networks

Area	Range	Nodes	GA		BPSO	
			Active	**Sleep**	**Active**	**Sleep**
100 m × 100 m		100	18.48	81.52	19.09	80.91
		150	21.96	128.04	19.09	130.91
	20 m	200	27.77	172.23	19.18	180.82
		250	35.85	214.15	19.11	230.09
		300	45.98	254.02	19.68	280.32
200 m × 200 m		100	18.67	81.33	19.46	80.54
		150	22.32	127.68	19.37	130.63
	40 m	200	28.28	171.72	19.54	180.46
		250	35.98	214.02	19.90	230.10
		300	45.95	254.05	19.95	280.05
300 m × 300 m		100	18.90	81.10	19.58	80.42
		150	22.45	127.55	19.68	130.32
	60 m	200	28.40	171.60	19.63	180.37
		250	36.06	213.94	19.92	230.08
		300	45.44	254.56	20.14	279.86
400 m × 400 m		100	19.07	80.93	19.59	80.41
		150	22.12	127.88	19.86	130.14
	80 m	200	28.59	171.41	19.77	180.23
		250	36.02	213.98	19.89	230.11
		300	45.80	254.20	20.02	279.98
500 m × 500 m		100	18.99	81.01	19.50	80.50
		150	22.29	127.71	19.69	130.31
	100 m	200	28.43	171.57	19.93	180.07
		250	36.31	213.69	19.97	230.03
		300	46.20	253.80	20.10	279.90

However, when the sensor nodes are scheduled by using BPSO, the active-node numbers over different deployed-node numbers are almost the same. For example, when dealing with the 300 m × 300 m network, the active-node numbers are 19.58, 19.68, 19.63, 19.92, and 20.14 over the deployed-node numbers 100, 150, 200, 250, and 300, respectively. This means that BPSO is more robust than GA in reducing the active nodes in the dense network with a large number of deployed sensor nodes. We also plot the comparisons in Figure 12.5. Because the results are similar, we only plot the results of the 300 m × 300 m network. Figure 12.5a compares the active-node number versus the deployed-node number, while Figure 12.5b compares the percentage of sleep nodes versus the deployed-node number. The figures show that in the GA approach, the percentage of sleep nodes increases first and then decreases when node density increases,

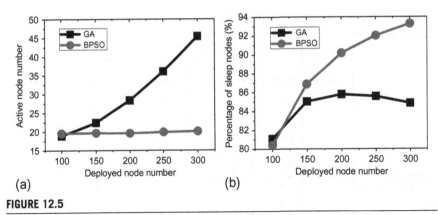

FIGURE 12.5

The active node number over different sensing ranges and deployed-node numbers.
(a) Active number and (b) sleep percentage.

while in the BPSO approach, the sleep node percentage keeps increasing as the node density increases, indicating the good performance of BPSO in scheduling as many sensor nodes as possible to sleep in order to save energy.

12.4.5 RESULTS ON MAXIMIZING THE DISJOINT SETS

Finding out the maximal disjoint sets of the deployed nodes and then scheduling the sets to work in turn is significant for the lifetime of the network. Even though the work in Tian and Georganas (2003) and Ye et al. (2003) has addressed the problem of reducing the active nodes, it does not give a solution to the problem of maximizing the disjoint sets. The work in Slijepcevic and Potkonjak (2001), Cardie et al. (2002), and Liu et al. (2006) addresses the disjoint-set-maximization problem, but only the approach used in Slijepcevic and Potkonjak (2001) can guarantee full coverage. Therefore, we compare our approaches and the one of Slijepcevic and Potkonjak (2001) in dealing with the disjoint-set-maximization problem.

We adopt the same simulation environment as in Slijepcevic and Potkonjak (2001), where the monitored area is a 500 m × 500 m square. Different numbers of sensor nodes and different sensing ranges are tested, as shown in Table 12.5. For each combination configuration of the sensor nodes N and the sensing range R in Table 12.5, we randomly generate three network topologies and run the approaches three independent times for each topology. The mean results are calculated and compared with the upper bound of the disjoint-set number for each topology, as shown in Tables 12.6 and 12.7 for the GA approach and the BPSO approach, respectively.

The upper bound of the disjoint-set number for each topology can be determined as follows. Because the area has been divided into grids, for each grid, we can find out the number of sensors that cover it. Compare all these numbers, and the minimal

Table 12.5 The simulation environments

Environment	Sensor Nodes N	Sensing Range R	Area
E1	100	200	500 m × 500 m
E2	120	200	500 m × 500 m
E3	150	200	500 m × 500 m
E4	180	150	500 m × 500 m
E5	300	80	500 m × 500 m
E6	400	80	500 m × 500 m

one is the upper bound of the disjoint-set number for this network topology. This method is also always used to study the k-cover of the WSN (Huang and Tseng, 2005; Wan and Yi, 2006). The k-cover of the WSN tells that every point of the monitored area is covered by at least k different sensor nodes. Therefore, the nodes can be at most divided into k disjoint sets, with the constraint that each set can fully cover the whole area. In this sense, the value k can be the upper-bound number of the disjoint sets for this network topology. However, k is a rough number, and the true upper-bound number may be smaller than k.

The data presented in Tables 12.6 and 12.7 show that BPSO generally outperforms GA in solving the disjoint-set-maximization problem. In most of the test cases, the BPSO approach can obtain the upper-bound number of disjoint sets. In the tables, "Mean" means the average number of disjoint sets of the three runs in each case, and "UP" means the maximal number of disjoint sets of that case. Moreover, the "Run–Mean" is the average value of the three "Mean" values of the three cases in each environment, while the "UP–Mean" is the average value of the three "UP" values of the three cases in each environment. Therefore, the "Accuracy" is the result of (Run–Mean)/(UP–Mean), to indicate how close the approach can obtain the results to the global optima. The **boldface** in the tables indicates that the obtained results equal the global optima. The results in Tables 12.6 and 12.7 show that BPSO can obtain higher accuracy values than GA does in all the test environments. Moreover, the accuracy values obtained by BPSO are almost higher than 96%, indicating the promising performance of BPSO.

We also compare the results obtained by the simulated-annealing (SA) approach and the heuristic approach (Slijepcevic and Potkonjak, 2001) in Table 12.8. The results of GA and BPSO are the "Run–Mean" values in Tables 12.6 and 12.7, respectively. The results show that both the GA and BPSO approaches outperform the SA approach. The **boldface** in Table 12.8 indicates the best solutions of the four approaches. When comparing the performance of the heuristic approach in Slijepcevic and Potkonjak (2001) with GA and BPSO, the GA outperforms the heuristic on environment 1 and environment 4, while the BPSO outperforms the heuristic on environments 1, 2, and 4. This shows that BPSO is more promising in maximizing the disjoint-set number of WSN.

Table 12.6 Experimental results of GA in solving the disjoint-set maximization problem

Environment	Cases	Run 1	Run 2	Run 3	Mean	UP	Run–Mean	UP–Mean	Accuracy (%)
E1: 100-200	Case 1	10	10	10	10	11	10.22	11.67	87.57
	Case 2	11	10	10	10.33	12			
	Case 3	10	11	10	10.33	12			
E2: 120-200	Case 1	13	13	10	12	15	9.33	12.00	77.75
	Case 2	10	7	8	8.33	11			
	Case 3	7	8	8	7.67	10			
E3: 150-200	Case 1	13	10	13	12	16	10.44	14.33	72.85
	Case 2	12	10	10	10.67	15			
	Case 3	9	10	7	8.67	12			
E4: 180-150	Case 1	8	5	6	6.33	9	7.33	9.67	75.80
	Case 2	7	7	**8**	7.33	8			
	Case 3	8	8	9	8.33	12			
E5: 300-80	Case 1	**3**	2	2	2.33	3	1.89	3.00	63.00
	Case 2	1	1	1	1	3			
	Case 3	2	**3**	2	2.33	3			
E6: 400-80	Case 1	3	**4**	3	3.33	4	3.22	4.33	74.36
	Case 2	3	3	**4**	3.33	4			
	Case 3	2	4	3	3	5			

Table 12.7 Experimental results of BPSO in solving the disjoint-Set maximization problem

Environment	Cases	Run 1	Run 2	Run 3	Mean	UP	Run–Mean	UP–Mean	Accuracy (%)
E1: 100-200	Case 1	10	11	11	10.67	11	11.22	11.67	96.14
	Case 2	11	12	11	11.33	12			
	Case 3	12	11	12	11.67	12			
E2: 120-200	Case 1	15	15	14	14.67	15	11.89	12.00	99.08
	Case 2	11	11	11	11.00	11			
	Case 3	10	10	10	10.00	10			
E3: 150-200	Case 1	15	15	15	15.00	16	13.78	14.33	96.16
	Case 2	15	15	14	14.67	15			
	Case 3	12	11	12	11.67	12			
E4: 180-150	Case 1	9	9	8	8.67	9	9.44	9.67	97.62
	Case 2	8	8	8	8.00	8			
	Case 3	11	12	12	11.67	12			
E5: 300-80	Case 1	3	2	3	2.67	3	2.78	3.00	92.66
	Case 2	3	3	3	3.00	3			
	Case 3	2	3	3	2.67	3			
E6: 400-80	Case 1	4	4	4	4.00	4	4.33	4.33	100.00
	Case 2	4	4	4	4.00	4			
	Case 3	5	5	5	5.00	5			

Table 12.8 Comparisons of different approaches in solving the disjoint-set maximization problem

Environment	SA Slijepcevic and Potkonjak (2001)	Heuristic Slijepcevic and Potkonjak (2001)	GA	BPSO
E1: 100-200	6.8	9.7	10.22	**11.22**
E2: 120-200	7	11.5	9.33	**11.89**
E3: 150-200	10.5	**18.5**	10.44	13.78
E4: 180-150	2.9	6.6	7.33	**9.44**
E5: 300-80	2.6	**4.3**	1.89	2.78
E6: 400-80	3.2	**4.5**	3.22	4.33

12.5 CONCLUSION

The OCP in WSN has been formulated as a 0/1 programming problem, and the BPSO approach has been designed to solve the problem. In the OCP model, 0 stands for sleep and 1 stands for active. Therefore, the model is not only simple and easy to understand, but it also has a natural map from the representation to the real network topology. The approaches to solve the OCP are based on SI algorithms, and therefore they are adaptive to the problem, with strong global search ability and good robustness. We have described the implementation details of using the BPSO approach to solve the problem. The performance is evaluated and compared with state-of-the-art approaches and the GA approach. The experimental results have shown the effectiveness and efficiency of the proposed BPSO approach. The performance BPSO is also evaluated over a set of networks with different scales and different node densities. The experimental results show that BPSO is more robust and more efficient than GA in reducing the active-node number to save the network energy, especially when the node density increases.

In future work, we will try to use the new techniques recently proposed for PSO, such as the orthogonal-learning PSO (Zhan et al., 2011), machine learning techniques (Zhang et al., 2011), aging mechanism (Chen et al., 2013), cooperative strategy (Li et al., 2015), or modified BPSO (Yang et al., 2014), to enhance the BPSO approach for designing a more powerful SI approach to solve the OCP problem in WSN. Moreover, the bi-velocity discrete PSO that is efficient for solving the 0/1 programming problem can be adopted for OCP optimization (Shen et al., 2014). The OCP can also be extended to multiobjective problems, and multiobjective PSO can be applied in future work (Zhan et al., 2013).

ACKNOWLEDGMENTS

This work was supported in part by the National Natural Science Foundation of China (NSFC) with No. 61402545, the NSFC Key Program with No. 61332002, the NSFC for Distinguished

Young Scholars with No. 61125205, and the National High-Technology Research and Development Program (863 Program) of China No. 2013AA01A212. For additional information regarding this chapter, please contact corresponding author Zhi-Hui Zhan (zhanzhh@mail.sysu.edu.cn).

REFERENCES

Akyildiz, I.F., Su, W., Sankarasubramaniam, Y., Cayirci, E., 2002. Wireless sensor networks: a survey. Comput. Netw. 38, 393–422.

Arampatzis, T., Lygeros, J., Manesls, S., 2005. A survey of applications of wireless sensors and wireless sensor networks. In: Proceedings of the 13th Mediterranean Conference on Control and Automation, pp. 719–724.

Cardei, M., Wu, J., 2006. Energy-efficient coverage problems in wireless ad-hoc sensor networks. Comput. Commun. 29, 413–420.

Cardie, M., MacCallum, D., Cheng, X., Min, M., Jia, X., Li, D., Du, D.-Z., 2002. Wireless sensor networks with energy efficient organization. J. Interconnect. Netw. 3, 213–229.

Cerpa, A., Elson, J., Hamilton, M., Zhao, J., 2001. Habitat monitoring: application driver for wireless communications technology. ACM SIGCOMM Comput. Commun. Rev. 31, 20–41.

Chen, W.N., Zhang, J., Lin, Y., Chen, N., Zhan, Z.H., Chang, H., Li, Y., Shi, Y.H., 2013. Particle swarm optimization with an aging leader and challengers. IEEE Trans. Evol. Comput. 17, 241–258.

Eberhart, R.C., Kennedy, J., 1995. A new optimizer using particle swarm theory. In: Proceedings of the Sixth International Symposium Micromachine Human Science, pp. 39–43.

He, T., Krishnamurthy, S., Stankovic, J.A., Abdelzaher, T., Luo, L., Stoleru, R., Yan, T., Gu, L., Hui, J., Krogh, B., 2004. Energy-efficient surveillance system using wireless sensor networks. In: Proceedings of the Second International Conference on Mobile Systems, Applications, and Services, pp. 270–283.

Heinzelman, W.B., Murphy, A.L., Carvalho, H.S., Perillo, M.A., 2004. Middleware to support sensor network applications. IEEE Netw. 18, 6–14.

Huang, Z., Lu, T., 2012. A particle swarm optimization algorithm for hybrid wireless sensor networks coverage. In: Proceedings of the IEEE Symposium on Electrical & Electronics Engineering, pp. 630–632.

Huang, C.F., Tseng, Y.C., 2005. A survey of solutions to the coverage problems in wireless sensor networks. J. Internet Technol. 6, 1–8.

Kennedy, J., Eberhart, R.C., 1995. Particle swarm optimization. In: Proceedings of the IEEE International Conference Neural Networks, pp. 1942–1948.

Kennedy, J., Eberhart, R.C., 1997. A discrete binary version of the particle swarm algorithm. In: Proceedings of the IEEE International Conference on System, Man, and Cybernetics, pp. 4104–4109.

Kennedy, J., Eberhart, R.C., Shi, Y.H., 2001. Swarm Intelligence. Morgan Kaufmann, San Mateo, CA.

Kim, Y.G., Lee, M.J., 2014. Scheduling multi-channel and multi-timeslot in time constrained wireless sensor networks via simulated annealing and particle swarm optimization. IEEE Commun. Mag. 52, 122–129.

Kulkarni, R.V., Venayagamoorthy, G.K., 2011. Particle swarm optimization in wireless-sensor networks: a brief survey. IEEE Trans. Syst. Man Cybern. C Appl. Rev. 41, 262–267.

Kuorilehto, M., Hannikainen, M., Hamalainen, T.D., 2005. A survey of application distribution in wireless sensor networks. EURASIP J. Wirel. Commun. Netw. 5, 774–788.

Li, X.Y., Wan, P.J., Frieder, O., 2003. Coverage in wireless and ad-hoc sensor networks. IEEE Trans. Comput. 52, 753–762.

Li, Y.H., Zhan, Z.H., Lin, S., Zhang, J., Luo, X.N., 2015. Competitive and cooperative particle swarm optimization with information sharing mechanism for global optimization problems. Inf. Sci. 293, 370–382. http://dx.doi.org/10.1016/j.ins.2014.09.030.

Liu, C., Wu, K., Xiao, Y., Sun, B., 2006. Random coverage with guaranteed connectivity: joint scheduling for wireless sensor networks. IEEE Trans. Parallel Distrib. Syst. 17, 562–575.

Martinez, K., Hart, J.K., Ong, R., 2004. Environmental sensor networks. IEEE Comput. Soc. 37, 50–56.

Meguerdichian, S., Koushanfar, F., Potkonjak, M., Srivastava, M., 2001a. Coverage problems in wireless ad-hoc sensor networks. In: IEEE Infocom, pp. 1380–1387.

Meguerdichian, S., Slijepcevic, S., Karayan, V., Potkonjak, M., 2001b. Localized algorithms in wireless ad-hoc networks: location discovery and sensor exposure. In: Proceedings of the Second ACM International Symposium on Mobile Ad Hoc Networking & Computing, pp. 106–116.

Morsly, Y., Aouf, N., Djouadi, M.S., Richardson, M., 2012. Particle swarm optimization inspired probability algorithm for optimal camera network placement. IEEE Sens. J. 12, 1402–1412.

Patwari, N., Hero III, A.O., Perkins, M., Correal, N.S., O'Dea, R.J., 2003. Relative location estimation in wireless sensor networks. IEEE Trans. Signal Process. 51, 2137–2148.

Savarese, C., Rabaey, J.M., Beutel, J., 2001. Location in distributed ad-hoc wireless sensor networks. In: Proceedings IEEE International Conference on Acoustics, Speech, and, Signal Processing, pp. 2037–2040.

Shen, M., Zhan, Z.H., Chen, W.N., Gong, Y.J., Zhang, J., Li, Y., 2014. Bi-velocity discrete particle swarm optimization and its application to multicast routing problem in communication networks. IEEE Trans. Ind. Electron. 61, 7141–7151.

Shih, E., Cho, S.H., Ickes, N., Min, R., Sinha, A., Wang, A., Chandrakasan, A., 2001. Physical layer driven protocol and algorithm design for energy-efficient wireless sensor networks. In: Proceedings of the Seventh Annual International Conference on Mobile Computing and Networking, pp. 272–287.

Slijepcevic, S., Potkonjak, M., 2001. Power efficient organization of wireless sensor networks. In: IEEE International Conference on Communications (ICC), pp. 472–476.

Tian, D., Georganas, N.D., 2003. A node scheduling scheme for energy conservation in large wireless sensor networks. Wirel. Commun. Mob. Comput. 3, 271–290.

Wan, P.J., Yi, C.W., 2006. Coverage by randomly deployed wireless sensor networks. IEEE/ACM Trans. Netw. 52, 2658–2669.

Xing, G., Wang, X., Zhang, Y., Lu, C., Pless, R., Gill, C., 2005. Integrated coverage and connectivity configuration for energy conservation in sensor networks. ACM Trans. Sens. Netw. 1, 36–72.

Yan, T., He, T., Stankovic, J.A., 2003. Differentiated surveillance for sensor networks. In: ACM First International Conference on Embedded Networked Sensor Systems (SenSys), pp. 51–62.

Yang, J., Zhang, H., Ling, Y., Pan, C., Sun, W., 2014. Task allocation for wireless sensor network using modified binary particle swarm optimization. IEEE Sens. J. 14, 882–892.

Ye, F., Zhong, G., Lu, S., Zhang, L., 2003. PEAS: a robust energy conserving protocol for long-lived sensor networks. In: Proceedings of the International Conference on Distributed Computing Systems (ICDCS), pp. 28–37.

Zhan, Z.H., Zhang, J., Li, Y., Chung, S.H., 2009. Adaptive particle swarm optimization. IEEE Trans. Syst. Man Cybern. B Cybern. 39, 1362–1381.

Zhan, Z.H., Zhang, J., Fan, Z., 2010. Solving the optimal coverage problem in wireless sensor networks using evolutionary computation algorithms. In: Proceedings of the Eighth International Conference on Simulated Evolution and Learning, Kanpur, India, pp. 166–176.

Zhan, Z.H., Zhang, J., Li, Y., Shi, Y.H., 2011. Orthogonal learning particle swarm optimization. IEEE Trans. Evol. Comput. 15, 832–847.

Zhan, Z.H., Zhang, J., Du, K.J., Xiao, J., 2012. Extended binary particle swarm optimization approach for disjoint set covers problem in wireless sensor networks. In: Proceedings of the Conference on Technologies and Applications of Artificial Intelligence, pp. 327–331.

Zhan, Z.H., Li, J., Cao, J., Zhang, J., Chung, H., Shi, Y.H., 2013. Multiple populations for multiple objectives: a coevolutionary technique for solving multiobjective optimization problems. IEEE Trans. Cybern. 43, 445–463.

Zhang, H., Hou, J.C., 2005. Maintaining sensing coverage and connectivity in large sensor networks. Ad Hoc Sens. Wirel. Netw. 1, 89–124.

Zhang, J., Zhan, Z.H., Lin, Y., Chen, N., Gong, Y.J., Zhong, J.H., Chung, H., Li, Y., Shi, Y.H., 2011. Evolutionary computation meets machine learning: a survey. IEEE Comput. Intell. Mag. 6, 68–75.

Zou, Y., Chakrabarty, K., 2005. A distributed coverage- and connectivity-centric technique for selecting active nodes in wireless sensor networks. IEEE Trans. Comput. 54, 978–991.

Clonal-Selection-Based Minimum-Interference Channel Assignment Algorithms for Multiradio Wireless Mesh Networks

13

Su Wei Tan[1], Sheng Chyan Lee[2], and Cheong Loong Chan[2]

[1]*Faculty of Engineering, Multimedia University, Cyberjaya, Selangor, Malaysia*
[2]*Faculty of Engineering and Green Technology, Universiti Tunku Abdul Rahman, Kampar, Perak, Malaysia*

CHAPTER CONTENTS

13.1 INTRODUCTION

A wireless mesh network is a multihop wireless network formed by a number of stationary wireless mesh routers. These routers are connected wirelessly using a mesh-like backbone structure. Some of the routers function as a wireless access point for clients (e.g., laptops and smart devices with wireless access) to attach themselves to the network. The clients transmit and receive data via the backbone mesh network. To connect to external networks such as the Internet, one or more routers are connected to the wired network and serve as gateways. Figure 13.1 illustrates a sample wireless mesh network consisting of six mesh routers, two of which also function as gateways.

By leveraging the commodity of IEEE 802.11 (more commonly known as Wi-Fi) hardware, wireless mesh networking reduces the dependency on wired infrastructure, and hence is being used for providing low-cost Internet access to low-income neighborhoods and scarcely populated areas. The interested reader is referred to Akyildiz et al. (2005) for other application areas of wireless mesh networks.

One key challenge in adopting wireless mesh networking is the capacity of effective throughput that can be offered to the clients. Due to the broadcast nature of the wireless medium, signals transmitted from different devices over the same channel (frequency band) will result in collision, which in turn causes data loss. Hence, multiple access techniques such as time division multiple access, frequency multiple access, or random access are required to coordinate the transmissions over the channel. It is well known that the effectiveness of random access techniques used in IEEE 802.11 networks degrades as the number of devices increases. To reduce the interference, the devices may transmit over different nonoverlapping channels provisioned in the IEEE 802.11 standards. In other words, the capacity of a wireless mesh network can be increased by equipping the routers with multiple radio interfaces, each of which is tuned to a different channel.

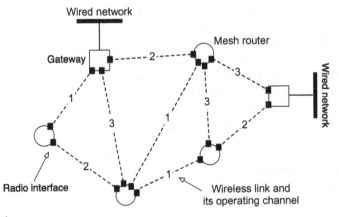

FIGURE 13.1

A multiradio wireless mesh network with channel assignment.

The attainable network capacity of a multiradio wireless mesh network is dependent on how various channels are assigned to each radio interface to form a mesh network with minimum interference. This is referred to as the channel assignment problem. The channel assignment must fulfill the constraints that the number of channels assigned to a router is at most the number of interfaces on the router, and the resultant mesh network remains connected. This problem is known to be nondeterministic polynomial-time hard (NP-hard) (Subramanian et al., 2008). A sample channel assignment fulfilling the constraints is also given in Figure 13.1.

Channel assignment techniques developed for wireless mesh networks can broadly be classified into two categories: (1) *Dynamic* and (2) *Quasistatic* (Subramanian et al., 2008). In the dynamic approach, every router is equipped with a single radio interface. The interface is dynamically switched from one channel to another between successive data transmissions. While the technique allows routers with a single interface to exploit the additional capacity offered by the available channels, it cannot be achieved using commodity hardware that does not provide fast channel-switching capability. For cost and practical considerations, we focus on wireless mesh networks that use off-the-shelf wireless cards. Hence, we adopt the quasistatic approach in which channels are assigned to router interfaces statically. However, the channel assignment can be updated if significant changes to traffic load or network topology are detected.

The channel assignment problem can be solved centrally or in a distributed fashion. This chapter focuses on centralized algorithms. Various approaches have been proposed for the problem, such as the greedy graph theoretic-based algorithm (Marina and Das, 2005), genetic algorithm (Chen et al., 2009), and greedy and Tabu-based algorithms (Subramanian et al., 2008). Subramanian et al. (2008) compared their centralized algorithms with lower bounds obtained from semi-definite programming (SDP) and linear programming formulations. While the results show that their algorithms outperform the algorithm proposed in Marina and Das (2005), a large performance gap with the lower bounds is observable. This suggests room for further improvement.

In this chapter, we investigate the use of artificial immune algorithms as an optimization tool for the problem. In de Castro and Timmis (2003), immune algorithms are classified as population-based and network-based according to the adaptation procedures used. Our study focuses on algorithms developed based on the clonal selection principle, a population-based approach. Basically, clonal-selection-based algorithms evolve a population of individuals, typically called B-cells, to cope successfully with antigens representing locations of unknown optima of a given function. At each generation, each B-cell in the population is subject to a series of procedures consisting of cloning, affinity maturation, metadynamics, and possibly aging (collectively known as clonal selection and expansion). Details of these procedures will be explained in Section 13.3.

As will be discussed later, the channel assignment problem can be viewed as a variant of the graph-coloring problem. In Cutello et al. (2003), an immune algorithm was applied to the graph-coloring problem, with competitive results to those obtained by the best evolutionary algorithms. Furthermore, the immune algorithm achieves this without the need for specialized crossover operators. Motivated by this, in Tan (2010), we proposed an immune algorithm as the strategy to evolve and improve solutions obtained using a simple greedy channel-assignment procedure. The evolution strategy is based on CLONALG (de Castro and Von Zuben, 2002), a popular clonal-selection-based algorithm.

This chapter extends the work on several fronts. First, two widely used clonal selection algorithms are investigated in addition to CLONALG. Specifically, we consider the B-cell algorithm (BCA) developed by Kelsey et al. (2003a,b), and a class of immune algorithms grouped under the title of "Cloning, Information Gain, Aging" (CLIGA), by Cutello et al. (2003, 2005b), (Cutello and Nicosia, 2005). The chosen algorithms have been successfully applied to various optimization problems. Second, a total of 18 variants are implemented for the chosen algorithms. The variants exhibit differences in the ways the populations are maintained and evolved. Systematic comparison among the variants provides insights on the strategy that works best for our problem. Third, a simple Tabu-based local search operator is developed to further improve our channel assignment algorithm.

Through extensive simulations, we show that our algorithms perform better than the genetic algorithm (Chen et al., 2009), graph-theoretic algorithm (Marina and Das, 2005), and Tabu-based algorithm (Subramanian et al., 2008) proposed for the channel assignment problem. Our evaluations also show the behavior of our algorithms in terms of convergence speed, sensitivity to parameter setting, and performance difference among the various variants developed.

The rest of this chapter is structured as follows. In next section, we present the system model and channel assignment problem formulation, and discuss some related proposals. In Section 13.3, we describe our proposed algorithm and its variants in detail. Section 13.4 presents the simulation experiments and results. Section 13.5 concludes the chapter.

13.2 PROBLEM FORMULATION
13.2.1 SYSTEM MODEL
We study the problem of assigning channels to backbone mesh routers in a wireless mesh network. The channel assignment problem is solved centrally at a channel assignment server (CAS) using information (i.e., connectivity and interference between the routers) provided by the routers on a periodic basis, or upon significant changes in the topology. These routers are assumed to run some link-aware routing protocols (e.g., Draves et al., 2004) to overcome short-term fluctuations in wireless link quality. The computed channel assignment will then be disseminated to the routers for making the necessary changes. This chapter focuses on solving the channel assignment problem, i.e., computing a low-interference channel assignment based on the collected information. The interested reader is referred to

Ramachandran et al. (2006) for further discussion on practical considerations in information collection and dissemination in a centralized wireless mesh network.

The location of the CAS has no direct impact on the quality of the channel assignment solution. It does, however, impact the amount of control traffic to be introduced into the network. A wireless mesh network may have multiple gateways. It is assumed that one of the gateways is assigned as the CAS. In this work, we assume that the CAS is colocated with the gateway router. This is because the gateway is typically strategically placed in the network, hence it is suitable for control traffic distribution. For large networks (e.g., with size in the order of hundreds), the network can be clustered such that each cluster is served by one CAS. Hence, this work focuses on channel assignment for nodes within a cluster.

Following typical deployment scenarios, the backbone structure and client access are assumed to use two different Wi-Fi standards, which operate on two distant central frequencies. In particular, the backbone mesh routers use 5.0 GHz (i.e., IEEE 802.11-a/n), and the clients use 2.4 GHz (i.e., IEEE 802.11-b/g). This ensures that the transmissions from the backbone routers and the clients are not interfering with each other. Hence, our study focuses on the channel assignment for the mesh routers only.

We consider a wireless mesh network with stationary mesh routers that are arbitrarily distributed on a plane. Each router is equipped with one or multiple IEEE 802.11 radio interfaces. Following Marina and Das (2005), all radio interfaces use omnidirectional antennas and have identical transmission ranges (denoted by R). The connectivity between the routers (nodes) is modeled using an undirected graph, $G(V,E)$, where V denotes the set of nodes and E denotes the set of connectivity links in the network. A connectivity link (i, j) indicates that both nodes i and j are within each other's *communication range*. Henceforth, we refer to G as *connectivity graph*. For two nodes of a link $(i, j) \in E$ to be able to communicate, both nodes must have one of their respective radio interfaces tuned to a common channel.

Due to the broadcast nature of the wireless medium, transmissions by nodes that are within each other's *interference range* may interfere, and this results in data loss. The interference between the transmissions is typically represented by using an interference model. The interference model defines the degree of interference due to a transmission over a link over other links in the network. Many interference models have been considered and proposed in the literature, most notably the physical and the protocol interference models (Gupta and Kumar, 2000). We consider the conflict graph model (Jain et al., 2003) in this work. The discussion in this chapter is independent of any specific interference model, as long as the interference model can be defined on pairs of links.

For ease of exposition, a binary interference model (Subramanian et al., 2008) in which two links either interfere or do not interfere is considered. Given an interference model, a conflict graph can be used to represent the interference between the links. A conflict graph is defined by first creating a set of vertices V_c corresponding to the links in the connectivity graph, i.e., $V_c = \{l_{ij}|(i, j) \in E\}$. Next, an edge is placed between two vertices (say, l_{ij} and l_{pq}) in the conflict graph if the corresponding links

$((i, j)$ and $(p, q))$ interfere with each other. The weight of the edge indicates the degree of interference between those links, which is in turn due to the amount of traffic transmission over the links. For simplicity, we assume all nodes have equal traffic load, hence all links have unity weight. The conflict graph is represented as $G_c(V_c, E_c)$, where E_c denotes the set of edges as defined above. The conflict graph can be used to represent any interference model. As in Marina and Das (2005), we associate the terms "node" and "link" with the connectivity graph, and use the terms "vertex" and "edge" for the conflict graph.

We show the concept of connectivity graph and interference graph between two nodes in Figure 13.2(a). Each node has a communication range and interference range of R and R', respectively. Figures 13.2(b) and (c) further illustrate the concept for a five-node line topology. In Figure 13.2(c), the conflict graph has four vertices, each representing a link in the network. In this chapter, we assume that the nodes use IEEE 802.11 reliable unicast, where nodes' access to links is controlled by the request-to-send/clear-to-send (RTS/CTS) control frames. Because of this, a transmission on link (a, b) interferes with links (b, c) and (c, d) but not with (d, e), thus giving the conflict graph in Figure 13.2(c).

13.2.2 CHANNEL ASSIGNMENT PROBLEM

The channel assignment problem can be investigated by using a connectivity graph. The number of interfaces on the nodes in the connectivity graph limits the number of unique channels that can be assigned to each link. The total number of channels that can be assigned to each link must not exceed the number of interfaces on the node.

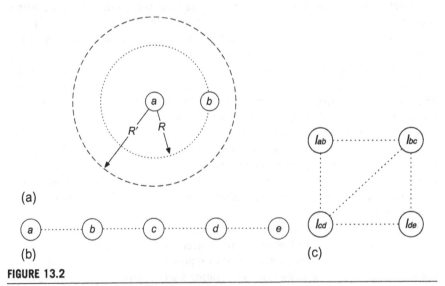

(a)

(b)

(c)

FIGURE 13.2

(a) Illustration of communication range and interference range. (b) Connectivity graph.
(c) The corresponding conflict graph.

The number of pairs of links that are assigned to the same channel is considered to contribute interference to the network. The links can be represented by the edges of a conflict graph. Interference is implied if the edges connecting two vertices are assigned to the same channel. The total interference can thus be determined by considering those vertices assigned to a common channel. For example, links (a, b), (b, c), and (c, d) in Figure 13.2(b) are assigned to channel 1, while link (d, e) operates on channel 2. Network interference is contributed only by the first three links, resulting in total network interference of 3. Minimizing the total network interference is thus an essential optimizing objective.

Consider a wireless mesh network depicted by connectivity graph $G(V,E)$ and conflict graph $G_c(V_c, E_c)$. A set of K channels in the system is denoted by $K = \{1, 2, \ldots, K\}$, while R_i represents the number of radio interfaces on node $i \in V$. Formally, the channel assignment problem is to determine a function $f: V_c \rightarrow K$ to minimize the overall network interference $I(f)$:

$$\text{Minimize} \quad I(f) = |\{(u, v) \in E_c | f(u) = f(v)\}| \tag{13.1}$$

subject to the below interface constraint,

$$\forall i \in V, |\{k | f(e) = k \text{ for some } e \in E(i)\}| \leq R_i \tag{13.2}$$

If we regard the assignment of channels to vertices as coloring of vertices, and that vertices (V_c) depict links (E) in the connectivity graph G, the problem can be regarded as coloring of links in G. Unfortunately, standard edge-coloring formulation is unable to capture the interface constraint. Because the number of edge incidents on a node may exceed the number of interfaces at the node, existing solutions for the problem are not directly applicable here. We adopt the example in Marina and Das (2005) to demonstrate how the interface constraint may limit the coloring choices. Figure 13.3 exhibits a four-node connectivity graph, with one interface per node. Suppose that there are two colors in the system, 1 and 2, which are assigned to link (a, b) and link (c, d), respectively. This occupies the only interface available at the nodes, and results in nodes a and b being disconnected from nodes c and d. In order to preserve network connectivity, only one color (either 1 or 2) should be used in this example.

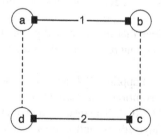

FIGURE 13.3

Example showing how arbitrarily chosen colors can break the connectivity.

Adapted from Marina and Das (2005).

In Subramanian et al. (2008), the problem is shown to be NP-hard, as it is reducible to an NP-hard max K-cut problem. In the following, we sometimes use the term "color" in place of "channel" for ease of exposition.

13.2.3 RELATED CHANNEL ASSIGNMENT ALGORITHMS

We review three closely linked works in this section. It is important to note that other formulations of the channel assignment problem are available for wireless mesh networks, such as the joint channel assignment and routing problems in Mohsenian-Rad and Wong (2007) and Raniwala et al. (2004). The interested reader is referred to Crichigno et al. (2008) for a more comprehensive survey of the related work.

Marina and Das (2005) proposed a polynomial-time heuristic known as connected low-interference channel assignment (CLICA). The main idea of CLICA is based on a notion of degree of flexibility, which marks the flexibility for a node to make coloring decisions. A priority value is assigned to each of the nodes as their respective degree of flexibility, and coloring decisions are made on a node-by-node basis in the order of priority. Each coloring decision is concocted in a greedy fashion: a node picks a color that is locally optimal (i.e., with minimum interference) for each of the links' incident from itself. The interface constraint is imposed on each coloring decision.

Subramanian et al. (2008) proposed a Tabu-based algorithm, a centralized algorithm, and a distributed greedy algorithm for the problem. Unlike CLICA, their algorithms operate on the conflict graph. The Tabu algorithm consists of two phases. In phase one, the Tabu search-based technique (Hertz and de Werra, 1987) developed for the graph-coloring problem is applied to find a good solution not subject to the interface constraints. Consequently, the solution obtained may violate the interface constraints. This is remedied in phase two, in which a "merge" operation is applied on nodes wherein the interface constraints are violated. Our algorithm has a similar structure to this algorithm: the first phase in which the immune algorithm is applied to find a good solution that may contain interface constraint violations, and a second phase that repairs the solution. We briefly discuss the contrasts that differentiate our algorithm from theirs. First, we devise a greedy decision in coloring the vertices. Then, our first phase attempts to impose the interface constraints, and each violation is given a penalty. We believe that in the second phase, this reduces the number of repair operations that generally increases the interference of the solution. Our repair mechanism is similar to theirs.

Subramanian et al. (2008) applied SDP and linear programming (LP) formulations to find lower-bound solutions for comparison with their greedy heuristics, Tabu-based algorithm, and CLICA. The results show that the Tabu-based algorithm mostly outperforms other algorithms, and is near to the lower bounds, especially when the number of interfaces and channels is high. However, a performance gap can be observed between the best performing heuristic and the lower bounds when the number of interfaces and channels is low.

In practice, the number of radio interfaces per node is likely to be small (e.g., from 2 to 4), owing to interference created by broad crosstalk and radio leakage between the radios (Robinson et al., 2005). In terms of number of channels, IEEE 802.11-b/g and 802.11-a networks allocate 3 and 12 orthogonal channels, respectively. Consequent to the interference problem mentioned above and potential interference by other networks, the actual number of usable orthogonal channels may be lessened (Subramanian et al., 2008). The newly standardized 802.11-n, which uses twice the bandwidth of 802.11-a/b/g, further reduces the number of usable orthogonal channels in the system.

Chen et al. (2009) used a different method to formulate the channel assignment problem. Their approach involved a two-objective optimization problem: minimizing the network interference and maximizing the network connectivity. The first objective is similar to our objective (Equation (13.1)). The second objective is to maximize the number of valid channel assignments for links in the connectivity graph such that the network stays connected.

The authors developed a genetic algorithm based on the well-known nondominated sorting genetic algorithm II (NSGA-II) (Deb et al., 2002). Following their notations, an individual is represented as a matrix A of dimension $N \times C$, where

$$a_{ij} = \begin{cases} 1 & \text{if channel } j \text{ is used in node } i \\ 0 & \text{otherwise} \end{cases} \tag{13.3}$$

In the above, N and C represent the number of nodes and the number of available channels, respectively. In the proposed algorithm, a simple two-point crossover operator is used to generate some new individuals, which are subsequently subject to mutation. The mutation is done by randomly selecting part of an individual and performing inversion of bits to the chosen locations. The next generation of individuals is then selected using nondominated sorting with elitism. With a two-objective problem formulation, the best individuals are those that are nondominated by other individuals in the objective functions. In our problem formulation, we assign channels to conflict vertices (i.e., connectivity links); thus, connectivity is ensured. The genetic algorithm's solution for our problem can be obtained by first choosing those individuals with maximum connectivity, and following by selecting the individual with minimum interference out of these individuals.

In Section 13.4, CLICA, the genetic algorithm proposed in Chen et al. (2009), and Tabu in Subramanian et al. (2008) are compared with our proposal.

13.3 CLONAL-SELECTION-BASED ALGORITHMS FOR THE CHANNEL ASSIGNMENT PROBLEM

In Tan (2010), we proposed and studied a channel assignment algorithm based on CLONALG (de Castro and Von Zuben, 2002), a well-known clonal-selection-based algorithm. This work further considers two other popular clonal selection algorithms, namely BCA (Kelsey et al.,, 2003a,b) and a variant of the "Cloning, Information Gain, Aging" family of algorithms (Cutello et al., 2003, 2005b; Cutello and Nicosia, 2005). In addition, an optional local search procedure is developed.

The above algorithms are chosen because they follow a similar computation framework in solving a problem, i.e., by repeated application of cloning, mutation, and selection cycle to a population of candidate solutions (B-cells). Yet they are different in a number of aspects, for example, in how the candidates for cloning are selected, in type of mutation operators they use, and whether or not they include aging as part of the evolution process. By considering the different implementation options for the algorithms (to be discussed in Section 13.3.3), a number of variants for these algorithms can be created, and are studied in this chapter.

Our channel assignment algorithm consists of two phases. In the first phase, one of the immune algorithms is applied to find a good solution, which may contain interface constraint violations. The solution is then repaired in the second phase.

13.3.1 PHASE ONE

In general, phase one consist of the following basic steps:

1. *Initialization*: create an initial random population of individuals (B-cells), P
2. *Main loop*: $\forall x \in P$, do:
   a. Affinity evaluation
   b. Clonal selection and expansion:
      - Cloning
      - Affinity maturation
      - Aging
      - Metadynamics
3. Local search
4. *Cycle*: repeat Step 2 until a stopping criterion is met.

In the above, Steps 1, 2(a), 3, and 4 are common to all the clonal selection algorithms considered. The difference between the algorithms lies in the clonal selection and expansion procedure, i.e., Step 2(b). In particular, the algorithms may perform all or some of the operations (i.e., cloning, affinity maturation, aging, and metadynamics). In addition, the order in which the procedures are being executed may vary. Phase one is stopped after a predefined number of generations. The best solution obtained is presented in phase two. We next discuss the details of Steps 1, 2, and 3.

13.3.1.1 Initialization

As in a typical clonal selection algorithm formulated for optimization (e.g., CLONALG), we consider only the existence of antibodies (individuals or B-cells) to represent the set of candidate solutions. Each B-cell is defined as a vector of integers of finite length $L = |V_c|$, representing the permutations of vertices in the conflict graph G_c. The initial population is randomly created. Let P represent the B-cells in the current population and N represent the size of the population.

13.3.1.2 Affinity evaluation

The affinity for a given solution is defined as the fractional network interference, I_{frac},

$$I_{\text{frac}} = \frac{I(f)}{|V_c|} \tag{13.4}$$

where $I(f)$ is the network interference defined in Equation (13.1) and $|E_c|$ is the total number of edges in the conflict graph. This represents the number of conflicts that remain after channel assignment, as compared to the number of conflicts in the single-channel network.

The affinity for a given B-cell is calculated by first translating the vertex ordering into channel assignment, using a simple greedy assignment strategy. The algorithm visits and assigns channels to vertices one by one in the order presented in the B-cell receptor. The channel assignment is done such that a given vertex v is assigned a channel c that is found to introduce the least increase in the interference to v and the neighbors of v. In addition, the chosen channel must obey the interface constraint on v.

The pseudocode of the greedy channel selection procedure is given in Figure 13.4. The inputs to the procedure include the set of available channels K, the conflict vertex currently being visited v, and the two end nodes (i, j) of the link represented by v. The procedure uses the pickMinInterferenceChannel(L) routine to select a channel from the given list, L. The list L is calculated based on the number of unused interfaces available at nodes i and j as follows: First, if both i and j have unused interfaces, this indicates that we are free to select any channel from K; thus, L is set to K. Otherwise, L should consist of channels that have been assigned to i and/or j. In particular, if i (or j) has no unused interfaces, L is set to the channels that had already been assigned to i (or j). In summary, the conditions enforced the interface constraints at both i and j. Figure 13.5 shows the working of the pickMinInterferenceChannel(L) procedure. Lines 5 and 6 sum up the count that each

```
1:  IF i.unusedInterfaces ≥ 1 AND j.unusedInterfaces ≥ 1
2:      c = v.pickMinInterferenceChannel(K)
3:  ELSE
4:      IF i.unusedInterfaces ≥1 AND j.unusedInterfaces == 0
5:          C = v.getChannelsAssigned(j)
6:          c = v. pickMinInterferenceChannel (C)
7:      ELSIF i.unusedInterfaces == 0 AND j.unusedInterfaces ≥ 1
8:          C = v.getChannelsAssigned(i)
9:          c = v.pickMinInterferenceChannel(C)
10:     ELSE
11:         C = v.getCommonChannels(i, j)
12:         c = v.pickMinInterferenceChannel(C)
13:     END
14: END
```

FIGURE 13.4

Greedy channel selection procedure.

```
 1: min =∞
 2: FOREACH l ∈ L
 3:     sum = 0
 4:     FOREACH e ∈ Adj[v]
 5:         IF l == e.head.channel
 6:             sum = sum + 1
 7:         END
 8:     END
 9:     IF sum < min
10:         min = sum
11:         c = l
12:     END
13: END
```

FIGURE 13.5

The pickMinInterferenceChannel procedure.

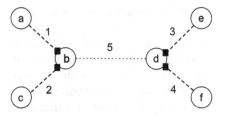

FIGURE 13.6

Sample scenario where a valid channel cannot be found for link (*b*, *d*).

channel $l \in L$ is assigned to each of the neighbors of v. The channel with the least count (i.e., interference) is recorded in line 11. The neighbors of v are represented by adjacency list Adj[v].

It is interesting to point out that the greedy channel selection procedure may not always result in a valid channel, c. Specifically, this occurs at line 11 of Figure 13.4, when nodes i and j do not have any common channel. Figure 13.6 illustrates a sample visiting order that leads to this problem. For ease of exposition, the visiting order of links in the connectivity graph is given in the figure as the numbers over the links. Suppose that $K=4$ and all nodes have two radio interfaces. It can be seen that a different channel can be assigned to each of the links (*a*, *b*), (*b*, *c*), (*d*, *e*), and (*d*, *f*), as the corresponding nodes have unused interfaces during the visit. By the time link (*b*, *d*) is visited, both *b* and *d* have used up all their respective interfaces. Because no common channel exists between the nodes, no channel can be assigned to the link. This problem affects the calculation of affinity (see Equations (13.1) and (13.4)) for a given solution. In order to account for this, a penalty value is introduced as a measure of interference. The penalty value is set as the number of neighbors of v, i.e., the worst case scenario where the assigned channels interfere with all existing neighbors.

13.3.1.3 Clonal selection and expansion
In this step, all B-cells in the current population P are subject to all or some of the following operations:

- *Cloning*. Each B-cell in P is cloned multiple times to produce a clone population.
- *Affinity maturation*. Each of the clones undergoes a hypermutation process.
- *Aging*. In this operation, all B-cells exceeding a maximum age will be deleted from the population. The age of a B-cell is increased at each generation.
- *Metadynamics*. This models the ability of the immune system in continuous production and recruitment of novel structures. Typically, some randomly generated B-cells are added to the population.

CLONALG, BCA, and CLICA each adopts slightly different operations, as described below

13.3.1.3.1 CLONALG
In CLONALG, the following operations are executed in order:

i. Cloningii
ii. Affinity maturationiii
iii. Metadynamics

Cloning. Following the optimization version of CLONALG (de Castro and Von Zuben, 2002), each B-cell will be cloned $\text{round}(\beta \cdot N)$ times. This results in a total of N_c clones for population P,

$$N_c = \sum_{i=1}^{N} \text{round}(\beta \cdot N) \tag{13.5}$$

where β is a configurable parameter, N is the total number of B-cells, and round() is the operator that rounds its argument toward the closest integer.

Affinity maturation. In this step, each of the clones undergoes a hypermutation process. The number of mutations M is determined by a mutation potential. In Tan (2010), we adopted the mutation potential used in CLONALG, which calculates the mutation rate α as

$$\alpha = e^{(-\rho \cdot f)} \tag{13.6}$$

where α is a configurable parameter and f is the fitness function value normalized in [0,1]. Cutello et al. (2005a) compared the original CLONALG (uses Equation (13.6)) with a variant that uses the following mutation potential:

$$\alpha = \frac{1}{\rho} e^{(-f)} \tag{13.7}$$

Originally proposed in de Castro and Timmis (2002). They tested both versions with four different classes of optimization and pattern recognition problems, and showed that the setting of the mutation rate and the parameter ρ is crucial for the algorithm performance. Both Equations (13.6) and (13.7) are considered in this chapter, and are referred to as hyper1 and hyper2, respectively. It can be seen that the mutation rate α is inversely proportional to the affinity; i.e., the higher the affinity, the smaller the value of α, and vice versa. This is a distinct characteristic of hypermutation in both equations.

The mutation is done by a simple random mutation operating on the vertex visiting order. In particular, the mutation operator randomly chooses M times two positions of a clone, and their respective elements are swapped. Given L as the length of the clone receptor and the affinity f of a clone, M is given by

$$M = \lceil L \cdot \alpha \rceil \qquad (13.8)$$

The above mutation operator is referred to as m-point random-swap, as multiple swaps are performed on each B-cell. In this chapter, we also investigate a simpler mutation operator, which allows only one swap for each B-cell. This is referred to as 1-point random-swap. This operator loops through each position of a B-cell to find the first swap position with probability α. Once a position is found, the loop is ended and swapping is done with a randomly chosen target.

Following the mutation operation, the mutated clones are added to population P. After that, B-cells in P undergo an elitist selection in which the best N B-cells are retained.

Metadynamics. Following CLONALG, the d lowest affinity B-cells are replaced by some newly created random solutions.

13.3.1.3.2 BCA
In BCA, the following operations are executed in order:

  i. Cloning
 ii. Metadynamics
iii. Affinity maturation

Cloning. In BCA (Kelsey et al.,, 2003b), each B-cell x is cloned to produce a clonal pool, C_x, and all adaptation takes place within C_x. Similar to CLONALG, the size of C_x can be given by round($\beta \cdot N$). In the literature, $\beta = 1$ is typically considered.

Metadynamics. In order to maintain diversity within the search, a clone is selected at random from C_x, and elements in the chosen clone will be randomized.

Affinity maturation. BCA introduced a novel mutation operator, which operates on contiguous regions of a vector. Applied to function optimization problems where each B-cell is represented as a vector of bit string of 64 bits, the operator randomly chooses a site (called hotspot) within the vector, along with a random length. Each element within the contiguous region is subject to individual mutation. The motivation behind this operator is to offer a more focused search.

We adapt the contiguous mutation operator to our problem as follows. First, a random site and a random length are chosen. Then, elements within the contiguous region are randomized. Note that the operator does not consider the possible influence of the affinity. To take this into consideration, we propose to calculate the length of the contiguous region by Equation (13.8). This results in two simple variants that use Equations (13.6) and (13.7) to calculate the mutation rate α, respectively. The above three operators are referred to as contiguous-random, contiguous-hyper1, and contiguous-hyper2, respectively.

Following the mutation operation, the mutated clones are compared to their respective parents. A parent B-cell will be replaced if there exists a clone with a better affinity.

13.3.1.3.3 CLIGA

In CLIGA, the following operations are executed in order:

  I. Cloning
 ii. Affinity maturation
iii. Aging

Cloning and affinity maturation. Similar to CLONALG, CLIGA generates for each B-cell round($\beta \cdot N$) number of clones. Unlike CLONALG, which performs cloning for all B-cells in P, cloning is done stochastically in CLIGA for each B-cell based on a cloning potential. In Cutello et al. (2003), a truncated exponential $V(f(x)) = e^{-k(l-f(x))}$ is used, where $f(x)$ is the affinity value of B-cell x, k is a configurable parameter, and l is the length of the vector representing a B-cell. In the affinity maturation step, a straight-line mutation potential $M(f(x)) = 1 - (l/f(x))$ is considered.

In this chapter, CLIGA is considered mainly to study the impacts of aging on the algorithm performance. Hence, we adopt the approach taken by CLONALG for both cloning and affinity maturation in our implementation of CLIGA. Specifically, static cloning is used instead of stochastic cloning, and mutation potential Equation (13.6) is used.

Aging. In this step, old B-cells will be eliminated. When the aging operation is in place, each B-cell is assigned an age, which is increased on each new generation. In addition, a newly created clone inherits the age of its parent, and its age is reset if it is successfully mutated. In previous work (Cutello et al., 2003, 2005a,b; Cutello and Nicosia, 2005), two classes of aging operators have been considered: (1) static and (2) stochastic. In static aging, B-cells with ages exceeding a maximum age τ will be eliminated; in stochastic aging, each B-cell is subjected to elimination with a probability $P_{die}(\tau) = (1 - e^{(-ln(2)/\tau)})$. Within each class, two approaches are possible: (1) pure aging and (2) elitist aging. In elitist aging, when a new population for the next generation is generated, we do not allow the elimination of B-cells with the best affinity. On the other hand, pure aging does not impose such restrictions. Consequently, there are four variants of aging operators in total:

1. Static-pure aging
2. Static-elitist aging
3. Stochastic-pure aging
4. Stochastic-elitist aging

We experiment with each of the variants in our evaluations.

At the end of the aging operation, if only $(N' < N)$ B-cells survived, the $(\mu + \lambda)$-selection operator is used to create $(N - N')$ B-cells to maintain a constant population size. In the original CLIGA, a stopping criterion based on information gain is used. This is not included in our implementation of CLIGA, so as to have the same stopping criterion (i.e., number of generations) used for our CLONALG and BCA implementations.

13.3.1.4 Local search

The purpose of the local search operation is to improve a given B-cell x produced by the clonal selection and expansion procedure for a maximum of I_{tabu} iterations. In general, any local search method may be used. In this chapter, we employ a simple local search based on the widely used Tabu search (Glover and Laguna, 1997).

Like any local search method, Tabu search relies on a definition of neighborhood. In our case, a neighbor for a B-cell x can be generated by swapping any two elements of x. Starting with the given B-cell, our Tabu search procedure proceeds iteratively to visit a series of locally best B-cells following the neighborhood. At each iteration, N_{tabu} neighbors are generated randomly for the current B-cell (say x) and the best neighbor is chosen to replace x, even if the former does not improve the current one. In order to prevent the search from getting trapped in cycling or local optima, a memory structure called Tabu list is used. The list keeps all visited B-cells, and these B-cells are forbidden from being revisited during the next T_{tabu} iterations. T_{tabu} is called the Tabu tenure. In Section 13.4, we investigate the impacts of the local search operation and its parameter setting to algorithm performance.

As the local search operation increases the computational needs of the algorithm, the operation is applied only to the population's best B-cell at each generation. The B-cell obtained from the operation will replace the original B-cell if it has a better affinity value.

13.3.2 PHASE TWO

We recall that the solution obtained after the first phase (immune algorithm) may violate interface constraints. The interface constraint violation is eliminated in phase two by applying a "merge" procedure (Subramanian et al., 2008) on each unassigned conflict vertex.

The merge procedure works as follows. Let i and j denote the underlying nodes for the link represented by a conflict vertex in question. To start, the procedure randomly picks a node between i and j. Suppose that node i is chosen. The goal is to reduce the number of channels assigned to i by one so that a new channel can be assigned to link (i, j). First, two channels c_1 and c_2 incident on i are picked. Next, all links that are assigned with c_1 are switched to c_2. The switching of c_1 to c_2 is done recursively on all links that are "connected" to the links whose channel has been switched from c_1 to c_2. This is done in order to preserve the interface constraints. We said that two links are connected if they are both incident on a common node. The propagation of the merging process essentially ensures that for any node j, either

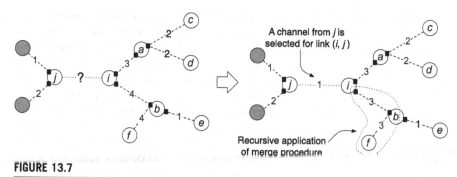

FIGURE 13.7

Sample topology before and after a merge operation.

all or none of the links incident on j with c_1 are switched to c_2. At the completion of one merge procedure, the number of distinct channels incident on i and other nodes is reduced by at most one without a new channel being introduced. Thus, repeated application of the procedure is guaranteed to resolve all interface constraints. At the end of the procedure, a channel c_3 from j is selected for use by link (i, j). In the worst case, a complete merge operation may visit all nodes and their incident links, thus having a complexity of $O(|V||E|)$.

Because a merge operation may result in increase in network interference, channels c_1, c_2, and c_3 are selected for the merge procedure such that they give the least increase in network interference. Figure 13.7 shows the channel assignment of a sample topology before and after the merge operation. In the example, $c_1 = 4$, $c_2 = 3$, and $c_3 = 1$.

13.3.3 VARIANTS OF THE CHANNEL ASSIGNMENT ALGORITHM

Based on the discussion on the clonal selection and expansion procedure, we can see that several implementation options are possible for CLONALG, BCA, and CLIGA, respectively. Table 13.1 lists all the variants considered in this chapter.

Table 13.1 Variants considered for CLONALG, BCA, and CLIGA

No.	Variants Considered for		
	CLONALG	**BCA**	**CLIGA**
1	rs-1p-h1		Static-pure
2	rs-1p-h2	rs-1p-h1	Static-elitist
3	rs-mp-h1	rs-mp-h1	Stochastic-pure
4	rs-mp-h2	rs-mp-h2	Stochastic-elitist
5	Contiguous-r	Contiguous-r	
6	Contiguous-h1	Contiguous-h1	
7	Contiguous-h2	Contiguous-h2	

For CLONALG and BCA, we focus on different combinations of mutation operators and mutation potentials, which result in the following variants: rs-1p-h1, rs-1p-h2, rsmp-h1, rs-mp-h2, contiguous-random, contiguous-h1, and contiguous-h2. In the above, *rs* and *contiguous* refer to the random-swap and contiguous region-swapping operators, respectively. For random-swap, *1p* and *mp* refer to the 1-point and m-point mutations described in Section 13.3.1.3, and *h1* and *h2* refer to the mutation potentials used. For contiguous region swapping, *r*, *h1*, and *h2* denote how the length of the contiguous region is calculated (i.e., randomly), and calculated using Equation (13.8) with mutation potentials hyper1 and hyper2, respectively. In Tan (2010), we studied only the rs-mp-h1 operator for CLONALG. In the rest of this chapter, this version is referred to as our *base* algorithm.

With the above variants for CLONALG and BCA, we are able to access the impacts of the different mutation operators on each algorithm, as well as impacts due to the ways the clones are maintained and evolved between the algorithms.

For CLIGA, in order to focus our study on the performance impacts of different aging operators, the rs-mp-h1 mutation operator as in CLONALG is employed for the clonal selection and expansion procedure. The four aging operators (static-pure, static-elitist, stochastic-pure, and stochastic-elitist) discussed previously are used. In other words, we are assessing the impacts of the aging operators on CLONALG.

13.4 PERFORMANCE EVALUATION

In this section, we present the results obtained for our base algorithm and its variants. First, we compare our base algorithm (rs-mp-h1 for CLONALG as used in Tan (2010)) against CLICA (Marina and Das, 2005) and the genetic-based channel assignment algorithm (GA) described in Chen et al. (2009). For ease of exposition, the base algorithm is referred to as IA. The performance of the algorithms is judged in terms of fractional network interference (4). We also investigate the behaviors of IA in terms of convergence, sensitivity to parameter setting, and the performance of different variants due to various clonal selection and expansion procedures.

In the simulations, we subject the algorithms to varying numbersof available channels (3 to 12) and radio interfaces per node (from 2 to 4). For each pair of radio interfaces and channel settings, we run each algorithm over three sets of topologies of 30, 40, and 50 nodes. Each set consists of 100 random topologies (representing the connectivity graphs), with average degrees ranging from 3.68 to 4.44, and minimum and maximum degrees of 1 and 7, respectively. The interference-to-communication-range ratio of the nodes is set to 2. For the results to be presented, each data point represents average calculated results for each topology. For IA and GA, the simulation with a given topology is repeated five times using different random seeds. The results are plotted with a 95% confidence interval.

Unless specified otherwise, the algorithms use the following configurations. IA: maximum generation, $maxgen=20$; population size $N=5$ or $N=10$; $\beta=1.0$; $\rho=1.0$; and $r=2$. In addition, the local search operator introduced in

Section 13.3.1.4 is disabled. For GA: *maxgen* = 100 (refer to as GA-100) or *maxgen* = 500 (GA-500); N = 40; mutation rate is 0.01; and crossover rate is 0.9, following Chen et al. (2009). There are two configurable parameters for Tabu: number of neighbors = 100 and Tabu size = 10. The number of iterations for Tabu is set to $|V_c|$, the number of conflict vertices as suggested in Subramanian et al. (2008). We have conducted independent experiments to find the best parameter settings for both GA and Tabu. The results shown in Figures 13.8 and 13.9 indicate that

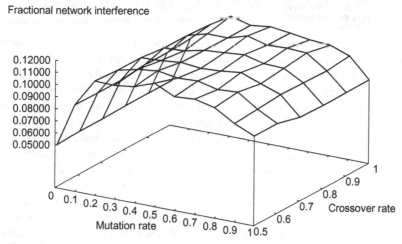

FIGURE 13.8

Impact of parameter settings on GA-500. Number of channels = 6.

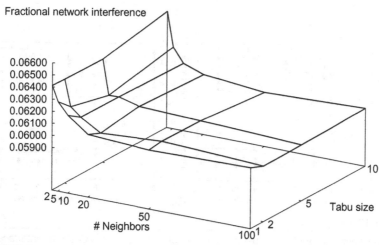

FIGURE 13.9

Impact of parameter settings on Tabu. Number of channels = 6.

the chosen settings indeed provide good average performance for the algorithms. For CLICA, the inputs to the algorithms are the connectivity graph, conflict graph, and the set of available channels. It has no configurable parameters.

In the following, we present results obtained with 30-node random topologies. Similar performance trends were observed in topologies of other sizes.

13.4.1 COMPARISON WITH OTHER CHANNEL ASSIGNMENT ALGORITHMS

Figure 13.10 depicts the performance of the algorithms for cases in which the number of radio interfaces per node is (a) equal to 3; and (b) uniformly assigned from 2 to 4, inclusively. First, it is clear that both versions of IA (N=5 and N=10) give the best performance when the number of channels is less than 8. With 8 channels, Tabu is on par with IA (N=5), and it outperforms all algorithms for a larger number of channels. As explained in Section 13.2.3, a small number of channels is expected in an actual deployment environment. It is worth pointing out that while a large N (i.e., 10) results in better performance as expected, IA with N=5 gives very competitive results. For example, the average performance advantage for the version with N=10 over N=5 is merely 3.78% for case (a). The same measure over GA-100, GA-500, and CLICA is 28.27%, 21.74%, and 16.81%, respectively. Comparing Figures 13.10(a) and (b), we can see that the performance advantage of IA is more prominent in the case where the nodes have different numbers of radio interfaces. In practice, the nodes are expected to have different capabilities and have different numbers of radio interfaces.

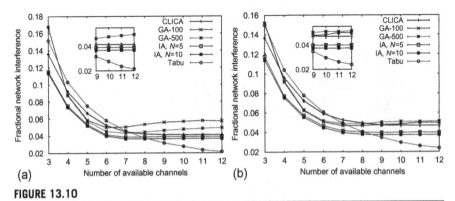

(a)

(b)

FIGURE 13.10

Fractional network interference performance: graph size = 30; number of radio interfaces per node: (a) fixed to 3; (b) uniformly assigned from 2 to 4.

From both plots, we can see that as the number of channels increases from 3 to 6, the network interference for all algorithms decreases rapidly. However, subsequent increase in channels results in little improvement for CLICA and IA. Increasing the channels basically increases the flexibility for the algorithms to assign new channels for improvement of the network interference. However, new channels cannot be blindly added, because they may break the connectivity of the topology, as explained in Section 13.2.2. This explains the performance trend observed. It is interesting to point out that GA actually performs worse for larger numbers of channels. We believe that this is because the GA is formulated without much domain-specific knowledge; hence it requires more iterations to improve the solutions. This can be seen by comparing the curves for both versions of GA (GA-100 and GA-500).

To understand the behavior of Tabu, the graph-coloring program by Cuberson (2010) is used to obtain the channels (colors) needed to color the topologies without creating interference. For the set of 30-node topologies, the number of colors ranges from 12 to 28, with an average of 18.5. Thus, as more and more channels are available, the solutions obtained by Tabu's first phase approximate better the graph-coloring solutions.

We next investigate the performance of the algorithms under a heavier interference level. This is done by increasing the interference-to-communication range from 2 to 3. The results in Figure 13.11 show that both versions of IA consistently outperform other algorithms. It is interesting to point out that the fractional interference obtained by IA, CLICA, and GA for 7 and more channels is lower than the corresponding values in the case with lower interference levels (Figure 13.10(b)). Similar performance trends are observed when the interference-to-communication range is further increased to 4.

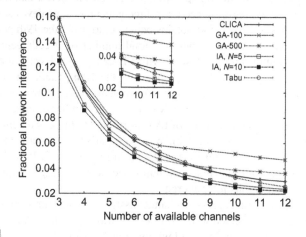

FIGURE 13.11

Fractional network interference performance: network under heavy interference.

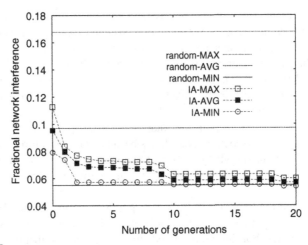

FIGURE 13.12

Convergence performance of IA.

13.4.2 CONVERGENCE OF IA

In Figure 13.12, we show the convergence property of IA with $N=5$ and $d=0$. The curves IA-MAX, IA-AVG, and IA-MIN represent the maximum, average, and minimum achieved affinity in each population, respectively. It is clear that IA can rapidly improve the solutions within the first few generations. In the same figure, we plot also the maximum, average, and minimum affinity for 5000 randomly generated solutions (denoted by random-MAX, random-AVG, and random-MIN in the figure). Comparing this with IA, which has an initial population of 5 B-cells, it is clear that IA manages to obtain reasonably good performance after three generations, achieves similar performance at the 10th generation, and obtains better solutions at the 19th generation. To give a better picture of this, we count the total number of B-cells generated by IA throughout the evolution process. It starts with 5 B-cells. At each generation, 25 clones are created (5 for each original B-cell). Thus, at the 10th generation, a total of 255 B-cells have been created. In other words, IA is able to evolve from a small set of random solutions to reasonably good ones within a small number of iterations.

We believe the fast convergence of IA is attributed to the greedy channel assignment scheme used (Section 13.3.1.2). As a comparison, the convergence performance for GA for two different runs is shown in Figure 13.13. It is clear that GA, which is formulated without much domain-specific knowledge, requires a much larger number of iterations to converge.

In addition to fast convergence, a channel assignment algorithm should have low running time, as it is expected to run on the gateway. The algorithms (IA, CLICA, and GA) are implemented in Java and without optimization. For Tabu, we modified the C-based Tabu graph-coloring routine in Cuberson (2010) to color the given topology with a fixed number of colors not subject to interface constraints. The result is

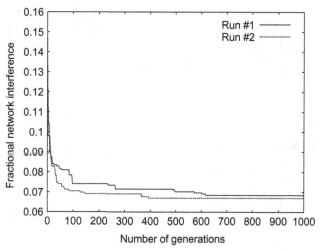

FIGURE 13.13

Convergence performance of GA.

then fed to our Java program to repair the interface violations, using the same routine used by IA. When tested on a laptop with 1.66 GHz CPU and 1.5 GB of memory, for the largest topologies (50 nodes) tested, both Tabu and CLICA managed to complete all problems within 0.1 s. This is followed by IA with 5 and 10 B-cells, which solve the problems within 0.6 and 2 s, respectively. GA requires the longest time: within 30 s for GA-100 and 200 s for GA-500.

Figure 13.14 shows the running time of IA against the number of channels. We report the results obtained from two sets of 50-node random topologies: sparse (S) topologies with average degrees ranging from 3.68 to 4.44, and dense (D) topologies with average degrees ranging from 9.28 to 10.24. It is clear that increases in density, which increase the problem size, increase the running time of the algorithm. However, even with 10 B-cells, IA managed to solve the problems within 30 s. The results also show that the number of channels has little impact on IA running time.

In a centralized channel assignment system, the channel assignment algorithm is invoked on a periodic basis (e.g., in the order of minutes or more) or upon significant changes in the topology. In addition, the running time can be significantly reduced with proper optimization and by rewriting the program using languages like C or C++. We thus believe IA is a viable solution for centralized channel assignment.

13.4.3 IMPACT OF PARAMETER SETTING

Next, we investigate the sensitivity of IA to the setting of β, ρ, and r. Recall that β is the cloning factor that determines the number of clones to be generated Equation (13.5), ρ controls the shape of the mutation rate as in Equations (13.6) and

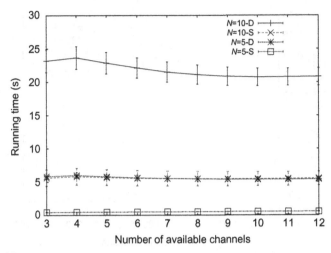

FIGURE 13.14

Running time of IA with topologies of different density against number of channels.

(13.7), and r is the number of low-affinity B-cells to be replaced. We discuss the results obtained from topologies with 30 nodes, where the number of available channels is 6 and the number of radio interfaces is uniformly distributed from 2 to 4.

In Tan (2010), we have observed marginal performance difference for small values of β and ρ (i.e., from 1 to 3). From Equation (13.5), it is clear that a larger value of β increases the number of clones generated. Because this results in higher computational cost and memory usage, $\beta = 1$ is used throughout this chapter.

Figure 13.15 depicts the impacts of ρ when mutation potentials, hyper1 (Equation (13.6)) and hyper2 (Equation (13.7)), are used. It is clear that the algorithm behaves quite differently for hyper1 and hyper2. With hyper1, the algorithm performance improves when ρ increases from 1 to 5 and degrades gradually for larger values of ρ; with hyper2, the performance initially improves as with hyper1 but begins to degrade sharply after $\rho = 4$. This can be explained by comparing the plots of both equations, as shown in Figure 13.16. Both equations use the inverse of an exponential function to model the inverse relationship between mutation rate and affinity of a B-cell, and use ρ to control the smoothness of the inverse exponential (de Castro and Timmis, 2003). For the given range of normalized affinity f ([0, 1]), the mutation rate always begins from 1 for hyper1, and the rate of reduction in mutation rate increases sharply when ρ increases. For hyper2, the starting point of the mutation rate is scaled by ρ. For a given ρ, the mutation rate decreases gradually with the affinity; for large values of ρ (e.g., 100), the decrease in mutation rate is almost flat for the given affinity range. Thus, we believe the poor performance of hyper2 for larger values of ρ observed in Figure 13.15 is because the inverse relationship between the mutation rate and affinity is diminishing. On the other hand, the smaller performance difference observed for hyper1 is because low-affinity

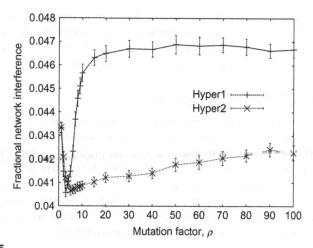

FIGURE 13.15

Impact of mutation potentials and parameter ρ on algorithm performance.

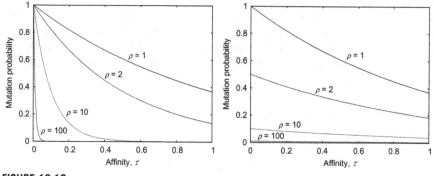

FIGURE 13.16

Mutation potentials: (a) hyper1: $e^{-\rho f}$; (b) hyper2: $\frac{1}{\rho}e(-f)$.

B-cells are always given a chance for further improvement. We note that the observation may be problem specific. Cutello et al. (2005a) found that when solving the classic one-counting problem with hyper2, large ρ is necessary to obtain optimum solutions.

We also investigated the impact of number of random replacements, r. As suggested in de Castro and Von Zuben (2002), r should range from 5% to 20% of N, as very high values of r may result in a random search through the affinity landscape. The results show that for IA with $N=10$, r ranges from 0 to 4 yield very similar results, in which the best and the worst fractional network interference observed differ only by a value of 0.0006. Hence, the results are not shown.

In all the experiments conducted, we found that results obtained with $\beta = 1$, $\rho = 3$, and $r = 2$ give reasonably good performance as compared to other settings that increase the computational needs of IA. Thus, only results obtained with these parameters are discussed in the rest of the following sections.

13.4.4 IMPACT OF LOCAL SEARCH

In this section, we investigate the impact of the Tabu-based local search operator on IA. We first explore the impact of its configurable parameters: number of neighbors for current B-cell (N_{tabu}), Tabu tenure (T_{tabu}), and maximum number of iterations (I_{tabu}). For our problem, we have found that Tabu tenure has little influence on the algorithm performance; hence, the results are omitted. Both N_{tabu} and T_{tabu} increase the number of search points. In general, increasing their values gives better algorithm performance, as shown in Figure 13.17 with results obtained for 6 available channels. Because this increases computational and memory needs, small values are recommended.

Table 13.2 compares the performance of IA with ($N_{tabu} = 5$ and $I_{tabu} = 5$) and without the local search operator. In the table, the best, average, and worst fractional network interference observed for different numbers of channels are given. We can see that both versions can mostly find the same best performance, but local search always provide better average and worst-case results. The channel assignment problem is expected to be solved in a resource-limited environment; thus, we do not expect to have the luxury of finding multiple solutions and picking the best. Hence, a better average performance is preferable. This consideration will be adopted in comparing the variants of IA.

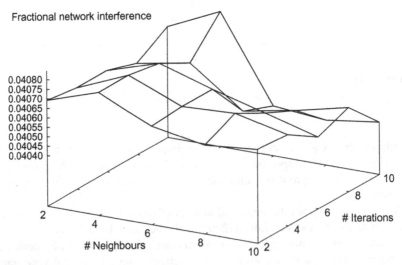

FIGURE 13.17

Impact of number of neighbors N_{tabu} and number of iterations I_{tabu} for local search.

Table 13.2 Impact of local search operator

# Channels	Variants Considered for CLONALG	
	Without Local Search	With Local Search
3	0.0980392157	0.0964052288
	0.1145034700	0.1134886383
	0.1278337531	0.1263418662
4	0.0588235294	0.0588235294
	0.0760746261	0.0747878301
	0.0912469034	0.0879438481
5	0.0375816993	0.0375816993
	0.0554897488	0.0545898644
	0.0685383980	0.0668868704
6	0.0261437908	0.0261437908
	0.0445705716	0.0438344833
	0.0571428571	0.0562947799
7	0.0212418301	0.0212418301
	0.0392756268	0.0387557529
	0.0555555556	0.0518558952
8	0.0212418301	0.0212418301
	0.0372260489	0.0368650675
	0.0521428571	0.0545851528
9	0.0212418301	0.0212418301
	0.0366187763	0.0362849702
	0.0521428571	0.0500000000
10	0.0212418301	0.0212418301
	0.0365203192	0.0360815604
	0.0521428571	0.0500000000
11	0.0212418301	0.0212418301
	0.0365178166	0.0360197939
	0.0542476970	0.0500000000
12	0.0212418301	0.0212418301
	0.0365188503	0.0360071777
	0.0542476970	0.0500000000

13.4.5 VARIANTS OF CHANNEL ASSIGNMENT ALGORITHM

This section presents the results obtained with different variants of IA, as discussed in Section 13.3.3. In the experiments, the local search operator is disabled. We first look at the results (see Tables 13.3 and 13.4) for CLONALG variants. For variants developed based on the random-swap operator, they mostly are able to locate similar best results. In addition, the marginal performance difference is observed

Table 13.3 Comparing variants for CLONALG (Part I)

# Channels	Variants Considered For CLONALG			
	rs-1p-h1	**rs-1p-h2**	**rs-mp-h1**	**rs-mp-h2**
3	0.0980392157	0.0947712418	0.0964052288	0.0964052288
	0.1145034700	0.1147327655	0.1139123002	0.1148973891
	0.1278337531	0.1312964492	0.1271676301	0.1275805120
4	0.0588235294	0.0588235294	0.0604575163	0.0604575163
	0.0760746261	0.0761663474	0.0752367413	0.0763045746
	0.0912469034	0.0924855491	0.0879438481	0.0895953757
5	0.0375816993	0.0375816993	0.0375816993	0.0392156863
	0.0554897488	0.0556955306	0.0551647208	0.0560926781
	0.0685383980	0.0697770438	0.0681255161	0.0689512799
6	0.0261437908	0.0277777778	0.0277777778	0.0261437908
	0.0445705716	0.0447019976	0.0444294995	0.0452818643
	0.0571428571	0.0578034682	0.0573905863	0.0565648225
7	0.0212418301	0.0212418301	0.0212418301	0.0212418301
	0.0392756268	0.0393330155	0.0394541456	0.0405857320
	0.0555555556	0.0524017467	0.0524360033	0.0578602620
8	0.0212418301	0.0212418301	0.0212418301	0.0212418301
	0.0372260489	0.0373724940	0.0375890117	0.0388727819
	0.0521428571	0.0529475983	0.0530864198	0.0589519651
9	0.0212418301	0.0212418301	0.0212418301	0.0212418301
	0.0366187763	0.0368849606	0.0369718124	0.0386389650
	0.0521428571	0.0528571429	0.0550000000	0.0600436681
10	0.0212418301	0.0212418301	0.0212418301	0.0212418301
	0.0365203192	0.0367632487	0.0369086213	0.0387160470
	0.0521428571	0.0528571429	0.0562947799	0.0620022753
11	0.0212418301	0.0212418301	0.0212418301	0.0212418301
	0.0365178166	0.0367490059	0.0368986998	0.0387387083
	0.0542476970	0.0528571429	0.0634595701	0.0614124872
12	0.0212418301	0.0212418301	0.0212418301	0.0212418301
	0.0365188503	0.0367522900	0.0369108451	0.0387128846
	0.0542476970	0.0528571429	0.0634595701	0.0600436681

among the variants. The best average result is given by rs-mp-h1 for a small number of channels (3 to 6), and rs-p1-h1 for a larger number of channels (7 to 12). For variants based on a contiguous region-swapping operator, a greater performance difference among the variants can be observed. Overall, the best average result is given by contiguous-h2. By comparing the best performer with random-swap and contiguous region swapping, we can see that the random-swap operator gives a better performance.

Table 13.4 Comparing variants for CLONALG (Part II)

# Channels	Variants Considered for CLONALG		
	Contiguous-h1	Contiguous-h2	Contiguous-r
3	0.0996732026	0.0964052288	0.0980392157
	0.1177143031	0.1151739709	0.1150299961
	0.1325350950	0.1279933939	0.1288191577
4	0.0604575163	0.0588235294	0.0604575163
	0.0788410144	0.0704290520	0.0765159248
	0.0913098237	0.0913098237	0.0912469034
5	0.0408496732	0.0375816993	0.0408496732
	0.0583600759	0.0559077545	0.0561875551
	0.0714285714	0.0689512799	0.0685404339
6	0.0277777778	0.0261437908	0.0261437908
	0.0472770489	0.0447509449	0.0453696396
	0.0597269625	0.0578034682	0.0573905863
7	0.0212418301	0.0212418301	0.0212418301
	0.0418940712	0.0393875651	0.0402905252
	0.0587248322	0.0506329114	0.0535499398
8	0.0212418301	0.0212418301	0.0212418301
	0.0399871015	0.0373679861	0.0383552269
	0.0549382716	0.0511278195	0.0529010239
9	0.0212418301	0.0212418301	0.0212418301
	0.0394737878	0.0368172450	0.0378086616
	0.0546075085	0.0502183406	0.0526315789
10	0.0212418301	0.0212418301	0.0212418301
	0.0395252698	0.0366990687	0.0378599293
	0.0547594678	0.0502183406	0.0557451650
11	0.0212418301	0.0212418301	0.0212418301
	0.0394558007	0.0367228894	0.0378259546
	0.0568065507	0.0502183406	0.0557451650
12	0.0212418301	0.0212418301	0.0212418301
	0.0394661303	0.0367320142	0.0378252084
	0.0568065507	0.0502183406	0.0557451650

Tables 13.5 and 13.6 show the results for BCA variants. For variants based on the random-swap operator, the best average result is given by rs-1p-h1 for most cases. For the contiguous operator, the overall best average result is given by contiguous-h2. By comparing the best performers from the random and contiguous operators, we again see that the random-swap gives better performance. It is interesting to note that single-point mutation outperforms multipoint mutations for both CLONALG and BCA. Because it requires a lower number of computations, it is recommended over the multipoint mutation operator.

Table 13.5 Comparing variants for BCA (Part I)

# Channels	Variants			
	rs-1p-h1	rs-1p-h2	rs-mp-h1	rs-mp-h2
3	0.0964052288	0.0964052288	0.0964052288	0.0964052288
	0.1140382557	0.1142868899	0.1139301317	0.1142868899
	0.1255161024	0.1275805120	0.1263418662	0.1275805120
4	0.0588235294	0.0588235294	0.0571895425	0.0588235294
	0.0752030525	0.0755161354	0.0751558507	0.0755161354
	0.0891824938	0.0875309661	0.0871180842	0.0875309661
5	0.0392156863	0.0375816993	0.0392156863	0.0375816993
	0.0552075923	0.0553225928	0.0550622771	0.0553225928
	0.0672997523	0.0685383980	0.0672997523	0.0685383980
6	0.0261437908	0.0277777778	0.0277777778	0.0277777778
	0.0442814940	0.0444893138	0.0442592731	0.0444893138
	0.0565648225	0.0557390586	0.0549132948	0.0557390586
7	0.0212418301	0.0212418301	0.0212418301	0.0212418301
	0.0390087439	0.0392432313	0.0392023916	0.0392432313
	0.0501535312	0.0517633675	0.0528571429	0.0517633675
8	0.0212418301	0.0212418301	0.0212418301	0.0212418301
	0.0370075857	0.0371570835	0.0374998817	0.0371570835
	0.0511744966	0.0518558952	0.0534698521	0.0518558952
9	0.0212418301	0.0212418301	0.0212418301	0.0212418301
	0.0363891963	0.0365888323	0.0368872599	0.0365888323
	0.0506257110	0.0518558952	0.0614334471	0.0518558952
10	0.0212418301	0.0212418301	0.0212418301	0.0212418301
	0.0363973953	0.0364851003	0.0367122058	0.0364851003
	0.0496724891	0.0552712385	0.0527123849	0.0552712385
11	0.0212418301	0.0212418301	0.0212418301	0.0212418301
	0.0363680943	0.0364592610	0.0368185732	0.0364592610
	0.0496724891	0.0511744966	0.0557830092	0.0511744966
12	0.0212418301	0.0212418301	0.0212418301	0.0212418301
	0.0364309574	0.0364798934	0.0368330900	0.0364798934
	0.0562947799	0.0542476970	0.0557830092	0.0542476970

Table 13.7 shows the results for CLIGA variants. We can see that the performance difference among the variants is rather small, and the overall best average is provided when stochastic-elitist aging is used. From the above, we extract the results of the best of each category, and show them in Table 13.8 for better exposition. We can see that BCA has the best performance, followed by CLONALG, and finally CLIGA. As both CLONALG and BCA use the same mutation operator

Table 13.6 Comparing variants for BCA (Part II)

# Channels	Variants		
	Contiguous-r	**Contiguous-h1**	**Contiguous-h2**
3	0.0964052288	0.0980392157	0.0964052288
	0.1148736897	0.1157792334	0.1144052353
	0.1271676301	0.1284062758	0.1263418662
4	0.0604575163	0.0604575163	0.0588235294
	0.0763651020	0.0771070019	0.0757674706
	0.0891824938	0.0895953757	0.0891824938
5	0.0392156863	0.0408496732	0.0392156863
	0.0561177864	0.0567425292	0.0553304975
	0.0685383980	0.0706028076	0.0681255161
6	0.0277777778	0.0277777778	0.0277777778
	0.0452724539	0.0457858027	0.0443760414
	0.0565648225	0.0569777044	0.0549132948
7	0.0212418301	0.0228758170	0.0212418301
	0.0402542652	0.0405132073	0.0389640041
	0.0553691275	0.0527728086	0.0500000000
8	0.0212418301	0.0212418301	0.0212418301
	0.0386643770	0.0385266188	0.0370731123
	0.0585893060	0.0527728086	0.0503355705
9	0.0212418301	0.0212418301	0.0212418301
	0.0382105364	0.0381804454	0.0364339792
	0.0584061135	0.0545590433	0.0506653019
10	0.0212418301	0.0212418301	0.0212418301
	0.0380668647	0.0380316321	0.0364612741
	0.0617173524	0.0574516496	0.0540386803
11	0.0212418301	0.0212418301	0.0212418301
	0.0381196284	0.0380540131	0.0363960476
	0.0617173524	0.0574516496	0.0502183406
12	0.0212418301	0.0212418301	0.0212418301
	0.0381194523	0.0380871678	0.0363903571
	0.0617173524	0.0574516496	0.0502183406

(rs-1p-h1), this suggests that by performing the affinity maturation and metadynamics within separate clone pools, BCA is better suited for our problem. However, we note that the performance advantage of BCA to CLONALG is rather small (0.57%) when compared with its advantage over CLIGA (20.05%). Because the version of CLIGA investigated mainly differs from CLONALG by the aging operator, the results suggest that aging has a negative impact on our problem.

Table 13.7 Comparing variants for CLIGA

# Channels	Variants Considered for CLIGA			
	Static-pure	Static-elitist	Stochastic-pure	Stochastic-elitist
3	0.1013071895	0.1013071895	0.1013071895	0.1013071895
	0.1263939187	0.1263939187	0.1264007597	0.1263939187
	0.1403798514	0.1403798514	0.1403798514	0.1403798514
4	0.0685131195	0.0685131195	0.0685131195	0.0685131195
	0.0856218255	0.0855988546	0.0856003980	0.0855958639
	0.1070422535	0.1070422535	0.1070422535	0.1070422535
5	0.0473856209	0.0473856209	0.0473856209	0.0473856209
	0.0641939322	0.0641871537	0.0641900623	0.0641758343
	0.0852017937	0.0852017937	0.0852017937	0.0852017937
6	0.0343137255	0.0343137255	0.0343137255	0.0343137255
	0.0536172284	0.0536328238	0.0536245766	0.0536369715
	0.0783009212	0.0783009212	0.0783009212	0.0783009212
7	0.0261437908	0.0261437908	0.0261437908	0.0261437908
	0.0480002035	0.0480152550	0.0480390157	0.0480147024
	0.0726714432	0.0726714432	0.0726714432	0.0726714432
8	0.0212418301	0.0212418301	0.0212418301	0.0212418301
	0.0457299547	0.0457487579	0.0457249207	0.0457105233
	0.0674547983	0.0674547983	0.0674547983	0.0674547983
9	0.0212418301	0.0212418301	0.0212418301	0.0212418301
	0.0454082498	0.0454047378	0.0453691235	0.0453407948
	0.0737898465	0.0737898465	0.0682593857	0.0682593857
10	0.0212418301	0.0212418301	0.0212418301	0.0212418301
	0.0455009427	0.0454862064	0.0454653765	0.0454153240
	0.0737898465	0.0737898465	0.0685772774	0.0685772774
11	0.0212418301	0.0212418301	0.0212418301	0.0212418301
	0.0454941912	0.0454794549	0.0454323005	0.0454085726
	0.0737898465	0.0737898465	0.0685772774	0.0685772774
12	0.0212418301	0.0212418301	0.0212418301	0.0212418301
	0.0454941912	0.0454779895	0.0454586250	0.0454085726
	0.0737898465	0.0737898465	0.0685772774	0.0685772774

13.5 CONCLUDING REMARKS

We investigate artificial immune-based algorithms for channel assignment in multi-radio wireless mesh networks. Three well-known clonal selection algorithms are considered to evolve and improve solutions created using a greedy channel assignment strategy. To limit the computational complexity and memory usage, we

Table 13.8 Comparing the best of variants for CLONALG, BCA, and CLIGA

# Channels	Variants		
	CLONALG	BCA	CLIGA
3	0.0980392157	0.0964052288	0.1013071895
	0.1145034700	0.1140382557	0.1263939187
	0.1278337531	0.1255161024	0.1403798514
4	0.0588235294	0.0588235294	0.0685131195
	0.0760746261	0.0762000525	0.0855958639
	0.0912469034	0.0891824938	0.1070422535
5	0.0375816993	0.0392156863	0.0473856209
	0.0554897488	0.0552075923	0.0641758343
	0.0685383980	0.0672997523	0.0852017937
6	0.0261437908	0.0261437908	0.0343137255
	0.0445705716	0.0442814940	0.0536369715
	0.0571428571	0.0565648225	0.0783009212
7	0.0212418301	0.0212418301	0.0261437908
	0.0392756268	0.0390087439	0.0480147024
	0.0555555556	0.0501535312	0.0726714432
8	0.0212418301	0.0212418301	0.0212418301
	0.0372260489	0.0370075857	0.0457105233
	0.0521428571	0.0511744966	0.0674547983
9	0.0212418301	0.0212418301	0.0212418301
	0.0366187763	0.0363891963	0.0453407948
	0.0521428571	0.0506257110	0.0682593857
10	0.0212418301	0.0212418301	0.0212418301
	0.0365203192	0.0363973953	0.0454153240
	0.0521428571	0.0496724891	0.0685772774
11	0.0212418301	0.0212418301	0.0212418301
	0.0365178166	0.0363680943	0.0454085726
	0.0542476970	0.0496724891	0.0685772774
12	0.0212418301	0.0212418301	0.0212418301
	0.0365188503	0.0364309574	0.0454085726
	0.0542476970	0.0562947799	0.0685772774

consider small population sizes (5 and 10) and a small number of iterations (20). Simulation results show that the algorithms perform better than genetic-based and graph-theoretic-based algorithms. By comparing variants of the clonal selection algorithm, we found that a simple single-point random-swap mutation operator is able to give a good performance. For the problem considered, better performance can be obtained by performing the mutation and metadynamics procedures in separate clone pools, as in the BCA. We also found that aging operators can result in poorer performance. In addition, our algorithm can be further improved by using a Tabu-based local search operator. Because the channel assignment problem considered has many similarities to the graph-coloring problem, our findings may be adopted for the study of the problem using immune algorithms.

REFERENCES

Akyildiz, I., Wang, X., Wang, W., 2005. Wireless mesh networks: a survey. Comput. Netw. 47 (4), 445–487, Elsevier.

Chen, J., Jia, J., Wen, Y., Zhao, D., Liu, J., 2009. A genetic approach to channel assignment for multi-radio multi-channel wireless mesh networks. In: 1st ACM/SIGEVO Summit on Genetic and Evolutionary Computation, 39-46, Shanghai, China.

Crichigno, J., Wu, M.Y., Shu, W., 2008. Protocols and architectures for channel assignment in wireless mesh networks. Ad Hoc Netw. 6, 1051–1077.

Cuberson, J., (2010), Graph coloring programs: Available at: http://webdocs.cs.ualberta.ca/.

Cutello, V.G., Nicosia, G., 2005. The clonal selection principle for in silico and in vivo computing. In: Recent Developments in Biologically Inspired Computing, Idea Graph Publishing, 104-146.

Cutello, V., Nicosia, G., Pavone, M., 2003. A hybrid immune algorithm with information gain for the graph coloring problem. In: GECCO, LNCS 2723, pp. 171–182.

Cutello, V., Narzisi, G., Nicosia, G., Pavone, M., 2005a. Clonal selection algorithms: a comparative case study using effective mutation potentials. In: The 4th International Conference on Artificial Immune Systems (LNCS3627), 12-28.

Cutello, V., Narzisi, G., Nicosia, G., Pavone, M., 2005b. An immunological algorithm for global numerical optimization. In: 7th International Conference on Artificial Evolution (EA), Lille, France, 284-295.

De Castro, L.N., Timmis, J., 2002. An artificial immune network for multimodal function optimization. In: Proceeding of IEEE Congress on, Evolutionary Computation (CEC'02).

de Castro, L.N., Timmis, J., 2003. Artificial immune systems as a novel soft computing paradigm. Soft. Comput. 7, 526–544.

de Castro, L., Von Zuben, F., 2002. Learning and optimization using the clonal selection principle. IEEE Trans. Evol. Comput. 6 (3), 239–251.

Deb, K., Pratap, A., Agarwal, S., Meyarivan, T., 2002. A fast and elitist multiobjective genetic algorithm: NSGA-II. IEEE Trans. Evol. Comput. 6, 182–197.

Draves, R., Padhye, J., Zill, B., 2004. Routing in multi-radio, multi-hop wireless mesh networks. In: Proc. of ACM MobiCom.

Glover, F., Laguna, M., 1997. Tabu search. Kluwer Academic Publishers, Boston, MA.

Gupta, P., Kumar, P.R., 2000. The capacity of wireless networks. IEEE Trans. Inf. Theory 46 (2), 388–404.

Hertz, A., de Werra, D., 1987. Using Tabu search techniques for graph coloring. Computing 39 (4), 345–351.

Jain, K., Padhye, J., Padmanabhan, V.N., Qiu, L., 2003. Impact of interference on multi-hop wireless network performance. In: Proc. of IEEE/ACM MobiCom, California.

Kelsey, J., Timmis, J., Hone, A., 2003a. Chasing chaos. In: Proceeding of IEEE Congress on, Evolutionary Computation (CEC'03), 413–419.

Kelsey, J., Timmis, J., Hone, A., 2003b. Immune inspired somatic contiguous hypermutation for function optimisation. In: Prodeeding of Genetic and Evolutionary Computation Conference (GECCO'03), 207-218, Chicago, IL, USA.

Marina, M.K., Das, S.R., 2005. A topology control approach for utilizing multiple channels in multi-radio wireless mesh networks. In: Proc. of IEEE International Conference on Broadband Networks (BroadNets), Boston.

Mohsenian-Rad, A.H., Wong, V.W.S., 2007. Joint logical topology design, interface assignment, channel allocation, and routing for multi-channel wireless mesh networks. IEEE Trans. Wireless Commun. 6 (12), 4432–4440.

Ramachandran, K.N., Belding, E.M., Almeroth, K.C., Buddhikot, M.M., 2006. Interference-aware channel assignment in multi-radio wireless mesh networks. In: Proc. of IEEE International Conference on Computer Communications (INFOCOM), 1-12, Barcelona, Spain.

Raniwala, A., Gopalan, K., Chiueh, T.C., 2004. Centralized channel assignment and routing algorithms for multi-channel wireless mesh networks. Mobile Comput. Commun. Rev. 8 (2), 50–65.

Robinson, J., Papagiannaki, K., Diot, C., Guo, X., Krishnamurthy, L., 2005. Experimenting with a multi-radio mesh networking testbed. In: Proc. of International Workshop Wireless Network Measurements (WiNMee).

Subramanian, A.P., Gupta, H., Das, S.R., Cao, J., 2008. Minimum interference channel assignment in multiradio wireless mesh networks. IEEE Trans. Mobile Comput. 7 (12), 1459–1473.

Tan, S.W., 2010. An immune algorithm for minimum interference channel assignment in multi-radio wireless mesh networks. In: The 9th International Conference on Artificial Immune Systems (LNCS6209), Edinburgh, UK.

Index

Note: Page numbers followed by *f* indicate figures and *t* indicate tables.

Printed in the United States
By Bookmasters